Annals of the ICRP

Proceedings of the Second International Symposium on the System of Radiological Protection

Editor-in-Chief
C.H. CLEMENT

Associate Editor
N. HAMADA

PUBLISHED FOR

The International Commission on Radiological Protection

by

SAGE

CONTENTS

SAGE

Annals of the ICRP

Editorial

ICRP'S SECOND INTERNATIONAL SYMPOSIUM

The International Commission on Radiological Protection (ICRP) has evolved in important ways in recent years, although it has remained the same in many ways too. Its first priority is still the promotion of radiological protection for the public good. It is still an independent organisation, formally a not-for-profit organisation registered with the UK Charity Commission. ICRP still has nearly 250 members, who are among the top experts in radiological protection science and policy from around the world.

Although ICRP's *raison d'être* has not changed, significant strides have been made to modernise the way in which the organisation operates.

In 2011, the first ICRP strategic plan was released, covering 2011–2017. Notably, this plan included six initiatives, including 'holding regular ICRP symposia'.

2012 saw the first open call for nominations for Committee membership (another initiative from the strategic plan). The result is that half of the Committee members in the current term (2013–2017) are new.

In the same year, ICRP re-organised how formal relations with other organisations are managed, to be more inclusive and more effective. The first annual meeting of liaison organisations was held in 2013, and today ICRP has formal relations with 18 organisations.

ICRP's first code of ethics was released in 2014, to help ensure that our work is conducted with the highest of moral standards.

Since 2011, biennial international symposia have been a regular feature. ICRP 2011: the First ICRP Symposium on the International System of Radiological Protection was held in Bethesda, MD, USA.

The subject of these proceedings, ICRP 2013: the Second International Symposium on the System of Radiological Protection, was held in Abu Dhabi, United Arab Emirates (UAE). It was hosted by the Federal Authority for Nuclear Regulation and kindly supported by: the Emirates Nuclear Energy Corporation; the UAE Armed Forces; the Abu Dhabi Health Authority; the German Federal Ministry for the Environment, Nature Conservation, and Nuclear Safety (specifically to make these proceedings open access); the Canadian Nuclear Safety Commission; Cameco; Landauer Europe; and the Japan Radioisotope Association.

This series of international symposia is a cornerstone in ICRP's efforts to engage more closely with the radiological protection community. Each symposium has attracted hundreds of people from all corners of the world. These proceedings play an important role in broadening this engagement beyond those who could attend in person.

Remarkably, these proceedings include a paper for almost every presentation made during ICRP 2013. In fact, these proceedings are more than complete, as included herein is one additional paper from ICRP 2011 that was accidentally not included in the 2011 Proceedings. We apologise sincerely for this oversight that is only now being corrected.

The papers here are the work of the individual authors. They are not recommendations of ICRP and do not necessarily represent the views of ICRP. However, being a complete set, they offer a good representation of the programme of ICRP 2013.

A debt of gratitude is owed to all presenters, and authors of the papers in these proceedings, for making ICRP 2013 a great success. Thank you as well to those who worked behind the scenes, particularly: ICRP Executive Assistant Lynn Lemaire; former ICRP Assistant Scientific Secretary Michiya Sasaki; ICRP Interns Taylor Whitter, Tudor Dragea, Rob Shakelton, Andrea Bauer, and Robert Martin; and all the FANR staff who worked tirelessly to make everything run as smoothly as possible.

Most importantly, thank you to everyone who attended ICRP 2011 and ICRP 2013. Your positive feedback inspires us to continue: 95% of those who responded to a questionnaire after ICRP 2013 recommended attendance at ICRP 2015: the Third International Symposium on the System of Radiological Protection in Seoul, Korea, October 20–22, 2015. Plans have even commenced for ICRP 2017. Stay tuned for more!

Christopher Clement
ICRP Scientific Secretary
Editor-in-Chief

SAGE

ICRP
Annals of the ICRP

ICRP: past, present, and future

C. Cousins

Department of Radiology, Box 218, Addenbrooke's Hospital NHS Trust, Hills Road, Cambridge CB2 0QQ, UK; e-mail: claire.cousins@addenbrookes.nhs.uk

Abstract–The International Commission on Radiological Protection (ICRP) is a premier international organisation for the protection of workers, patients, and the public against ionising radiation. It was established in 1928 to advance, for the public benefit, the science of radiological protection, with its work in the early years focusing mainly on occupational exposure in medicine. The name 'International Commission on Radiological Protection' was adopted in 1950 to reflect the wider and more diverse areas of work that were being undertaken. ICRP has published 13 sets of general recommendations and these form the basis of radiation safety standards worldwide. ICRP consists of the Main Commission and five standing committees. There are 84 official members of the Main Commission and Committees 1–5, but more than 200 members who work with ICRP through its task groups. ICRP has developed a strategic plan for 2011–2017 and has made progress with some of its initiatives. These include establishing close liaison with other radiological protection organisations, responding to the needs and concerns of the radiological protection community, and identifying areas of work that require scrutiny of science and practice to produce relevant recommendations. This strategy means that ICRP will continue to be a leader in radiological protection for many years to come.

Keywords: ICRP; Radiological protection; Recommendations

1. INTRODUCTION

The International Commission on Radiological Protection (ICRP) has been in existence since 1928, and has continuously provided recommendations and guidance on a wide range of subjects. Some of these subjects include the radiological protection of workers, patients, and the public against ionising radiation. These recommendations are incorporated into legislation, regulations, and standards worldwide.

This paper does not necessarily reflect the views of the International Commission on Radiological Protection.

The diversity of work undertaken has expanded over the years and continues to evolve. There are challenges facing all radiological protection organisations, some of which will not be easy to resolve. ICRP will be included in this arena as it continues to be a leading radiological protection authority in the future.

2. ICRP PAST

ICRP was established in Stockholm in 1928 during the Second International Congress of Radiology (ICR), and the initial name was the 'International X-ray and Radium Protection Committee' (IXRPC). The original committee had very few members, two of whom were medical doctors (Lindell, 1996). In the early years, the Committee (or Commission as it was called from 1934) was not very active and met for only 1 day every 3 years at the ICR. Only two members of the IXRPC survived World War II, including the first Chairman, Rolf Sievert (Clark and Valentin, 2009).

In 1950, the Commission was given its present name, the International Commission on Radiological Protection, and a broader remit of work beyond the deleterious effects of x rays in medicine. The six sub committees established at this time were condensed to five sub committees by 1953. These were developed to work on topics including permissible doses for external and internal radiation; protection against x rays generated at potentials up to 3 million volts; protection against x rays above 3 million volts, as well as beta and gamma rays; and the handling and disposal of radioactive isotopes (ICRP, 1955).

The committee structure was re-organised again in 1962, creating four of the committees that exist today, although with revised names and mandates. Committee 5, for protection of the environment, was added in 2005. Since 1928, there have been 12 chairs of the Main Commission (11 individuals, as Rolf Sievert served at two different time periods), with the current chair being the only woman. Scientific secretaries have also been appointed since 1928 and there have been 10 to date. The first full-time, paid scientific secretary was established in 1962, and the presence of such an employee greatly improved the efficiency of the work at ICRP.

Since its establishment, ICRP has written well over 100 publications on all aspects of radiological protection, and a short history of ICRP and the evolution of its policies are included in *Publication 109* (Clark and Valentin, 2009). There have been 13 fundamental recommendations on which regulations are based, the latest being the 2007 Recommendations (ICRP, 2007). The original publications were printed in scientific journals in the areas of medicine and physics, but ICRP has had its own numbered series of reports since 1959, and these have been produced as a separate journal, *Annals of the ICRP*, since 1977.

3. ICRP PRESENT

ICRP aims to advance, for the public benefit, the science of radiological protection, in particular by providing recommendations and guidance on all aspects of protection against ionising radiation. For many years, the organisation has been a registered charity in the UK. ICRP consists of the Main Commission, with a chair and 12 members, and five standing committees on radiation effects, radiation doses, protection in medicine, application of ICRP's recommendations, and protection of the environment.

The Main Commission and the Scientific Secretariat of ICRP work together to direct, organise, and oversee the work of the committees, task groups, and working parties. All ICRP members are volunteers who serve 4-yearly terms. The membership consists of eminent scientists and policy makers in the field of radiological protection. Although there are 84 members of the Main Commission and Committees 1–5, the wider membership of those involved in task groups and working parties extends to over 200 experts from 30 countries. A new term began on 1 July 2013 with, for the first time, a process of open nominations and elections for membership. At any one time, there are 20–30 task groups and working parties working on many different aspects of radiological protection.

ICRP now has two paid employees: the Scientific Secretary and, since 2009, the Executive Assistant. Additional administrative support is provided by the position of assistant secretary, filled on a multiyear term basis as a cost-free staff loan from the Central Research Institute of the Electric Power Industry of Japan, and by the position of intern filled on a 4-monthly rotation by students from the Canadian Nuclear Safety Commission.

The system of radiological protection that has evolved since 1928 is the basis for radiological protection standards, guidance, programmes, and practice worldwide. The objective of the work of ICRP is to contribute to an appropriate level of protection against the detrimental effects of ionising radiation exposure without unduly limiting the benefits associated with the use of radiation. In preparing its recommendations, ICRP considers advances in scientific knowledge, evolving social values, and practical experience. Formulating standards, regulations, and codes of practice is the responsibility of other national and international organisations. The recommendations of ICRP are largely adopted by the International Atomic Agency into the Basic Safety Standards, and are incorporated into the national legislation of many countries.

4. ICRP FUTURE

In 2011, ICRP developed, for the first time, a strategic plan to extend to the end of the term of membership in 2017. This was devised to outline some historical and operational information regarding ICRP, as well as highlighting challenges, objectives, and initiatives regarding the policy and future direction of ICRP in coming years. The main areas of challenge include keeping abreast of technological developments in medicine, the non-cancer effects of radiation (e.g. cardiovascular), natural sources of radiation, integrating protection of the environment into the system of radiological protection, and ensuring the maintenance of knowledge and expertise in radiological protection amongst professionals.

Initiatives have already begun to address some of these challenges. ICRP aims to raise awareness of radiological protection in medicine by continuing to produce focused publications on specific topics, and by increasing ICRP participation in medical meetings wherever possible.

ICRP will continue to seek some of the best-informed experts in its membership through a process of nomination and election, but realises that maintaining knowledge in radiological protection in younger and future generations is, in itself, a challenge. Interest in the subject of radiological protection has to be promoted and encouraged with research and other educational opportunities.

A system of protection for the environment has been developed by Committee 5, and it is important that this is fully integrated into the existing system of radiological protection. Task Group 82 (joint work between Committees 4 and 5) on 'Integrating the system of protection for human and non-human species' is now complete and awaiting publication.

There are currently many other organisations working in the field of radiological protection, and ICRP recognises that collaboration with these organisations will promote and achieve the best practices in radiological protection for the benefit of the public, workers, and patients. The aim is to promote the one system of radiological protection that should be applicable to all. ICRP has long had a close relationship with these other radiological protection organisations. However, the number of such organisations continues to increase, and ICRP has established working partners as 'special liaison organisations', who are invited to share information and advance radiological protection into the future.

ICRP aims to promote its work and stimulate discussion on different areas of radiological protection, and hence has established a biennial International Symposium on the System of Radiological Protection. The first symposium was held in Bethesda, MD, USA in 2011 and the second symposium was held in Abu Dhabi, United Arab Emirates in 2013. These have included several topical and challenging issues related to radiological protection in the programme. The third symposium is already being planned and will be held in Seoul, Korea in 2015.

ICRP is a small charitable organisation whose publications are produced by volunteer expert members with limited extra time to promote the work at appropriate forums. The publications of ICRP are not currently freely available and are relatively expensive, usually being purchased by organisations and institutions rather than individuals. ICRP is aware of this issue and is developing strategies to increase its financial reserve, and hopefully allow distribution of its publications at low or no cost.

5. CONCLUSIONS

ICRP has a long tradition of providing recommendations and guidance on radiological protection and is a well-known and widely respected organisation. The membership is drawn from experts worldwide, and ICRP prides itself on maintaining its independence from government and other external influences. The challenges faced by many organisations working in radiological protection are numerous, and ICRP is actively involved in many different arenas. ICRP will continue to be a leading authority in radiological protection, and by collaborating closely with other relevant organisations, will impact on the evolving system of radiological protection far into the future.

REFERENCES

Clark, R.H., Valentin, J., 2009. The history of ICRP and the evolution of its policies. ICRP Publication 109. Ann. ICRP 39(1), 75–110.

ICRP, 1955. Recommendations of the ICRP. Br. J. Radiol. (Suppl. 6).

ICRP, 2007. The 2007 Recommendations of the International Commission on Radiological Protection. ICRP Publication 103. Ann. ICRP 37(2–4).

Lindell, B., 1996. The history of radiation protection. Radiat. Prot. Dosim. 68, 83–95.

SAGE

Annals of the ICRP

International Commission on Radiological Protection Committee 1: current status and future directions

W.F. Morgan

Biological Sciences Division, Pacific Northwest National Laboratory, 902 Battelle Blvd, PO Box 999, MSIN J4-02, Richland, WA 99352, USA; e-mail: wfmorgan@pnnl.gov

Abstract–Commission 1 of the International Commission on Radiological Protection considers the risk of induction of cancer and heritable disease (stochastic effects) together with the underlying mechanisms of radiation action. Committee 1 also considers the risks, severity, and mechanisms of induction of tissue/organ damage and developmental defects (deterministic effects). The Committee was significantly revamped in 2013 and last met in Abu Dhabi in October 2013. Committee 1 evaluated progress on two ongoing task groups: Task Group 64 'Cancer Risk from Alpha Emitters' and Task Group 75 'Stem Cell Radiobiology'. Following approval from the Main Commission, Committee 1 established two new task groups: Task Group 91 'Radiation Risk Inference at Low Dose and Low Dose Rate Exposure for Radiological Protection Purposes' and Task Group 92 'Terminology and Definitions'. This article presents a synopsis of the current status of Committee 1 and outlines the tasks that Committee 1 may undertake in the future.

Keywords: ICRP; Stochastic effects; Deterministic effects

1. INTRODUCTION

Committee 1 of the International Commission on Radiological Protection (ICRP) considers the risk of induction of cancer and heritable disease (stochastic effects) together with the underlying mechanisms of radiation action. Committee 1 also considers the risks, severity, and mechanisms of induction of tissue/organ damage and developmental defects (deterministic effects).

The Committee meets on an annual basis, and also meets in conjunction with the Main Commission every second year. For example, in 2011, the Main Commission

This paper does not necessarily reflect the views of the International Commission on Radiological Protection.

met in Washington, DC, USA. This was followed by the 1st International Symposium on the System of Radiological Protection and the five subcommittees, Committees 1–5, met independently after the symposium. In 2013, the Main Commission and all five subcommittees met in Abu Dhabi, United Arab Emirates. At this meeting, there was also the highly successful 2nd International Symposium, where Committee 1 organised a session on 'Tissue Reactions: the Road from Science to Protection'. Details can be found at http://www.icrp.org/page.asp?id=184. Manuscripts from each presenter in this session can be found in these proceedings.

In the intervening years between joint meetings with the Main Commission and the other subcommittees, a member of Committee 1 usually hosts the annual meeting. In 2010, the meeting was hosted by Fiona Stewart in Amsterdam, The Netherlands, and in 2012, Sisko Salomaa hosted the meeting in Helsinki, Finland. In 2014, Quanfu Sun and former member of Committee 1 Ping-Kun Zhou hosted the meeting in Beijing, China.

2. COMMITTEE 1

Following the open nominations process adopted by the Main Commission in 2012, Committee 1 underwent some significant changes in 2013 as a result of the subsequent elections. Prior to 2013, Committee 1 had only nine members; however, the Main Commission recognised the need for additional expertise on Committee 1 and increased the number of members to 16 (Fig. 1).

3. RESPONSIBILITIES AND ACTIVITIES OF COMMITTEE 1

In addition to being aware of recent publications and issues in their own area of expertise, members of Committee 1 are also expected to be aware of publications of various bodies, e.g. the United Nations Scientific Committee on the Effects of Atomic Radiation (UNSCEAR) and the US National Council on Radiation Protection, as well as advances in high-profile research programmes, e.g. the outstanding European Union's research programmes – Multidisciplinary European Low Dose Initiative and the Open Project for European Radiation Research Area, and the host of innovative programmes that have been, or are, supported by these oversite groups: the Radiation Effects Research Foundation that studies radiation effects in the atomic bomb survivors from Hiroshima and Nagasaki, the Southern Urals Biophysics Institute, and the US Department of Energy's Low Dose Radiation Research Program to name a few. A useful example of how this works is the recent report on 'Human Radiosensitivity' published by researchers at Public Health England (PHE) in the UK (AGIR, 2013).

Individual radiosensitivity is a key issue, and as scientific advances in 'omics' technologies (transcript omics, genomics, proteomics, metabolomics, lipidomics, etc.) move forward at a rapid pace, it is conceivable that, in the future, it may be possible to predict an individual's sensitivity, or resistance, to ionising radiation. The fiscal, legal, and ethical implications for medicine, employment etc. are enormous.

Figure 1. Committe 1 participants at the Abu Dhabi meeting. Back row left to right: Jolyon Hendry (former Committee 1 and Co-Chair of Task Group 75); Dan Stram (USA); Simon Bouffler (UK); Bill Morgan (USA, Chair of Committee 1); Sisko Salomaa (Finland); Wolfgang Dörr (Austria; Chair of Task Group 92); Quanfu Sun (China); Richard Wakeford (UK). Front row left to right: Ohtsura Niwa (former Committee 1 and Co-Chair of Task Group 75, Japan); Dominique Laurier (France); Preetha Rajaraman (India); Tamara Azizova (Russia); Nobuhiko Ban (Japan); Margot Tirmarche (France); Ranajit Chakraborty (USA); Werner Rühm (Germany, Secretary of Committee 1 and Chair of Task Group 91); Michael Hauptmann (Netherlands). Missing: Alice Sigurdson (USA, Vice-Chair of Committee 1).

A summary of this document was presented to Committee 1 by one of the authors, and the Committee felt that there was little need at this time for a specific task group to address this issue further beyond the PHE report.

Committee 1 is fortunate to include world-class experts from many of these organisations and/or committees among its members. In addition to monitoring publications from these various organisations and/or committees, Committee 1 considers the use of biologically based dose–response models for assessing the effects of low-dose ionising radiation exposures. In addition, Committee 1 monitors epidemiological data, non-cancer and potential hereditary effects, advances in radiation-induced DNA damage recognition, DNA repair, and the potential impact of epigenetic effects.

Committee 1 makes a judgement call given the facts and data available as to whether a block of research warrants a task group. An overview of the 'Statement of Task' is presented by the Chair of Committee 1 to the Main Commission; pending their approval, a task group will or will not be established. The task group will consider the evidence in detail and make recommendations based on scientific principles, philosophy, and policy. Any document prepared by a task group must be approved by the sponsoring committee and the Main Commission, and is subsequently sent out for external review. Since many task group members have 'day jobs' and task group membership is voluntary, this can sometimes be a complicated and protracted experience.

4. ONGOING PROJECTS UNDER THE AUSPICES OF COMMITTEE 1

Please note: the Chair, membership, corresponding members, and consultants are listed on the Committee 1 website (http://www.icrp.org/icrp_group.asp?id=7) and will not be included here.

Committee 1's most recent report (2012) was *Publication 118*, entitled 'ICRP statement on tissue reactions and early and late effects of radiation in normal tissues and organs – threshold doses for tissue reactions in a radiation protection context' (ICRP, 2012). This was the subject of the Committee 1 session at the 2nd International Symposium on the System of Radiological Protection, and the manuscripts summarising these presentations are included in these proceedings.

4.1. Task Group 64: Cancer risks from alpha emitters

This is a follow-up report complementing *Publication 115* on 'Lung cancer risk from radon and progeny and statement on radon' (ICRP, 2010).

Task Group 64 will complement and extend *Publication 115* and report on potential risks from the alpha emitters plutonium, uranium, thorotrast, and radium. This task group represents an excellent collaboration between ICRP Committees 1 and 2. Committee 2 is providing essential radiation dosimetry information. The project is a little behind schedule due to the closure of the West Lakes facility in the UK, which delayed the joint analysis of the Sellafield and Mayak worker cohorts. Nevertheless, the task group did meet in Paris in 2014, so this task group appears to be back on track.

4.2. Task Group 75: Stem cell radiobiology

This task group was established to review the current state of knowledge of stem cell radiobiology and the potential impacts on cancer risk. While there have been tremendous advances in understanding of stem cell biology in recent years, there is comparatively little information on radiation effects on stem cells. Task Group 75 is placing emphasis on stem cell radiobiology in relation to carcinogenic radiation risk. The document is in the final stages of preparation, and it is anticipated that it will be submitted to the Main Commission and subsequently sent out for external review in late 2014.

4.3. Task Group 91: Radiation risk inference at low dose and low-dose rate

The detriment-adjusted nominal risk coefficients recommended by ICRP have been based, to a large extent, on data obtained from the atomic bomb survivors in Japan. As their exposure was a single acute exposure, it was thought that the most plausible biological model for the dose–response relationship should be linear quadratic (which implies a larger slope at high doses than at low doses). Many internationally and nationally relevant bodies, such as UNSCEAR, Board on the Effects of Ionizing Radiation (BEIR), and ICRP, have used a dose and dose-rate effectiveness factor (DDREF) for the estimates of coefficients at low doses and protracted dose rates. A value of 2 was used by ICRP for low-dose and low-dose-rate exposures, which are typical in radiation protection. With more epidemiological information becoming available and using Bayesian analysis, UNSCEAR has recently re-evaluated all the available information and has estimated risk coefficients that are similar to the ICRP estimates using high doses and a DDREF value of 2. However, in their 2006 report, the BEIR Committee (NRC, 2006), also using a Bayesian approach, recommended a DDREF of 1.5.

The task group will review the current available information on the estimation of risk coefficient and recommend:

- whether it is desirable to continue to estimate risk at low doses by assessing the slope of the dose response at high doses and then applying a DDREF reduction factor. The alternative is to adopt the UNSCEAR approach of inferring the risk coefficients at low doses by using all available information and the Bayesian analysis to estimate the best expert judgement;
- whether the currently inferred values of the detriment-adjusted nominal risk coefficients recommended by ICRP for an adult and a whole population, namely, values of around 5% Sv^{-1} of effective dose, requires revision; and
- whether such coefficients are applicable to acute, protracted, and prolonged exposure or need a particular correction.

Task Group 91 had its inaugural meeting in Munich in December 2013.

4.4. Task Group 92: Terminology and definitions

Over many years, the terminology and definitions of specific terms used in ICRP publications have evolved, and in some instances, have been used inconsistently. Task Group 92 will review the terminology and definitions from *Publication 103* (ICRP, 2007) onwards and revise as required. This will occur in two phases:

- terminology and definitions where there is consistency within ICRP documents and general agreement that they are correct; and
- terminology and definitions where there is inconsistency between and within ICRP reports, and/or where investigators do not agree that the terminology and definition accurately reflect the scientific message being conveyed.

ICRP interns have prepared both documents and the first phase is currently underway. It is envisaged that the updated terminology and definitions will be a web-based resource for future use.

5. FUTURE DIRECTIONS

A major dilemma for radiation protection is the scientific basis for radiation standards to protect the public from exposures to low levels of ionising radiation (<100 mSv), where there are considerable uncertainties in the epidemiological data. Research in this dose range is important and essential (Morgan and Bair, 2013), and a comprehensive analysis of all scientific data might resolve the controversies in the shape of the radiation dose–response profile at radiation doses <100 mSv.

Additionally, progress in the radiation sciences is moving rapidly. With new techniques and technologies available in radiology and radiation oncology departments, the increasing use of computed tomography scans, threats of radiological terrorism, the increased use of ionising radiation in industry and various security strategies, and concerns over nuclear accidents (e.g. Fukushima) demand that Committee 1 remains diligent and aware of scientific advances in the radiation sciences. To this end, Committee 1 has identified a number of further areas where work may be initiated and where a priority watch will be maintained:

- individual radiosensitivity to ionising radiations (with Committee 3);
- radiation detriment (with Committee 2);
- shape of the dose–response for cancer;
- uncertainties in dosimetry (with Committee 2);
- late effects from particle therapy (with Committee 3);
- relative biological effectiveness for low-energy vs high-energy gammas;
- radiation weighting factors for tritium;
- radiation risk in relation to chemical risk;
- tissue banking, archival tissues, and the use and value of these resources; and
- transgenerational effects.

ACKNOWLEDGEMENTS

The author acknowledges interactions and the hard work of all members of Committee 1, and the support of the Main Commission, the Scientific Secretary, and the various members of his support staff. In addition, the author would like to acknowledge the numerous contributions of the previous Chair of Committee 1, R. Julian Preston. The author would also like to thank him for his support, encouragement, and mentorship over many years.

REFERENCES

AGIR, 2013. Human Radiosensitivity. RCE-21. Health Protection Agency, Chilton. Available at: http://www.hpa.org.uk/Publications/Radiation/DocumentsOfTheHPA/RCE21HumanRadiosensitivity/ (last accessed November 2013).

ICRP, 2007. The 2007 Recommendations of the International Commission on Radiological Protection. ICRP Publication 103. Ann. ICRP 37(2–4).

ICRP, 2010. Lung cancer risk from radon and progeny and statement on radon. ICRP Publication 115. Ann. ICRP 40(1).

ICRP, 2012. ICRP statement on tissue reactions and early and late effects of radiation in normal tissues and organs – threshold doses for tissue reactions in a radiation protection context. ICRP Publication 118. Ann. ICRP 41(1/2).

Morgan, W.F., Bair, W.J., 2013. Issues in low dose radiation biology: the controversy continues. A perspective. Radiat. Res. 179, 501–510.

NRC, 2006. Health Risks from Exposure to Low Levels of Ionizing Radiation: BEIR VII Phase 2. Board on Radiation Effects Research. Washington, DC: National Research Council of the National Academies.

SAGE

ICRP
Annals of the ICRP

Overview of ICRP Committee 2 'Doses from Radiation Exposure'

J. Harrison

Public Health England, Centre for Radiation, Chemical and Environmental Hazards, Chilton, Didcot, Oxon OX11 0RQ, UK;
e-mail: john.harrison@phe.gov.uk

Abstract–Over many years, Committee 2 of the International Commission on Radiological Protection (ICRP) has provided sets of dose coefficients to allow users to evaluate equivalent and effective doses for intakes of radionuclides or exposure to external radiation for comparison with dose limits, constraints, and reference levels as recommended by ICRP. Following the 2007 Recommendations, Committee 2 and its task groups are engaged in a substantial programme of work to provide new dose coefficients for various conditions of radiation exposure. The methodology being applied in the calculation of doses can be regarded as state-of-the-art in terms of the biokinetic models used to describe the behaviour of inhaled and ingested radionuclides, and the dosimetric models used to model radiation transport for external and internal exposures. The level of sophistication of these models is greater than required for calculation of the protection quantities with their inherent simplifications and approximations, which were introduced necessarily, for example by the use of radiation and tissue weighting factors. However, ICRP is at the forefront of developments in this area, and its models are used for scientific as well as protection purposes. This overview provides an outline of recent work and future plans, including publications on dose coefficients for adults, children, and in-utero exposures, with new dosimetric phantoms in each case. The Committee has also recently finished a report on radiation exposures of astronauts in space, and is working with members of the other ICRP committees on the development of advice on the use of effective dose.

Keywords: Reference phantoms; Equivalent dose; Effective dose; Dose coefficients

---

This paper does not necessarily reflect the views of the International Commission on Radiological Protection.

1. INTRODUCTION

The remit of Committee 2 of the International Commission on Radiological Protection (ICRP) is: the development of dose coefficients for the assessment of internal and external radiation exposure; development of reference biokinetic and dosimetric models; and reference data for workers and members of the public. The current membership (2013–2017) is John Harrison (Chairman), François Paquet (Vice-Chairman), Wesley Bolch (Secretary), Mike Bailey, Volodymyr Berkovskyy, Doug Chambers, Marina Degteva, Akira Endo, John Hunt, Chan Hyeong Kim, Rich Leggett, Jizeng Ma, Dietmar Nosske, Nina Petoussi-Henss, and Frank Wissmann.

The Committee has four task groups that are responsible for the production of reports:

- Task Group on Dose Calculations, chaired by Wesley Bolch, is concerned with the development of computational models and reference data needed to assess organ and effective doses from both internal and external radiation sources to both occupational workers and members of the general public.
- Task Group on Internal Dosimetry, chaired by François Paquet, is currently engaged in the revision of biokinetic models, for use in the recalculation of dose coefficients for the inhalation and ingestion of radionuclides, first for workers and subsequently for members of the public.
- Task Group on Effective Dose, chaired by John Harrison, will produce a report to provide guidance on the use of the quantity 'effective dose', and its relation to risk, particularly in the context of applications in medicine but also more widely.
- Task Group on Dose Coefficients for External Environmental Exposures, chaired by Nina Petoussi-Henss, will provide conversion coefficients for the external exposure of members of the public to airborne sources and ground deposits.

Committee 2 works closely with the International Commission on Radiation Units and Measurements (ICRU); the Chairman of ICRU, Hans Menzel, is a member of the ICRP Main Commission. Committee members also support the work of the other ICRP committees, and are currently providing members for the task groups of Committees 1, 3, and 5.

Revisions of ICRP recommendations invariably require recalculation of dose coefficients because changes are made to the radiation and tissue weighting factors used in the calculation of equivalent and effective dose. In addition, improvements to the models used to calculate doses lead to revised values. Work is currently in progress to provide replacement dose coefficients based on the 2007 Recommendations (ICRP, 2007), incorporating a number of important methodological improvements. The following sections provide a short explanation of recent and forthcoming Committee 2 reports.

2. RECENT COMMITTEE 2 REPORTS (2009–2013)

2.1. Adult reference computational phantoms *(Publication 110)*

Menzel and Harrison 2012a gave an outline of the development of ICRP reference phantoms, and explained that *Publication 110* (ICRP, 2009), a joint report with ICRU, provides reference phantoms for the adult male and female derived from imaging data for individuals. The individuals were chosen for their similarity to the external dimensions and organ masses of the reference adult male and female (ICRP, 2002), and the models were subsequently adjusted for consistency with these data. The use of male and female phantoms will replace the use of stylised hermaphrodite phantoms developed by the Medical Internal Radiation Dose Committee of the Society of Nuclear Medicine (Cristy, 1980; Cristy and Eckerman, 1987). Thus, in future calculations, equivalent dose will be calculated separately for males and females, and averaged in the calculation of effective dose (ICRP, 2007).

2.2. Conversion coefficients for radiological protection quantities for external radiation exposures *(Publication 116)*

Publication 116 (ICRP, 2010) is a joint report with ICRU, presenting conversion coefficients for organ and effective doses, for external exposures in occupational settings, calculated using the *Publication 110* phantoms and in accordance with the 2007 Recommendations. The radiations considered are external beams of monoenergetic photons, electrons and positrons, neutrons, protons, pions (negative/positive), muons (negative/positive), and helium ions. The organ dose conversion coefficients tabulated in the report represent ICRP/ICRU recommended values. Comparisons of the protection quantities, equivalent and effective dose, with corresponding operational quantities show the latter to provide conservative estimates of dose in the majority of cases. Annexes and a CD provide detailed supporting information, including equivalent dose coefficients for the lens of the eye and skin.

2.3. Compendium of dose coefficients based on *Publication 60 (Publication 119)*

At the specific request of the International Atomic Energy Agency and European Commission, a compilation of dose coefficients has been provided as *Publication 119* (ICRP, 2012), based on the 1990 Recommendations (ICRP, 1991). These values will be used and referred to in International and European Basic Safety Standards, while new values based on the 2007 Recommendations are being calculated until their use is required.

Publication 119 includes committed effective dose coefficients for inhalation and ingestion intakes of radionuclides by workers, as compiled in *Publication 68* (ICRP, 1994b), calculated using the *Publication 66* (ICRP, 1994a) model of the human respiratory tract and the *Publication 30* (ICRP, 1979) model of the gastrointestinal tract. The biokinetic models used to describe the distribution, tissue retention, and excretion of individual elements and their radioisotopes following their absorption to blood were those developed for the *Publication 30* series (ICRP, 1979, 1980, 1981, 1988), except for cases for which updates were provided in *Publications 56, 67, 69* and

71 (ICRP, 1990, 1993, 1995a, 1995b). *Publication 119* also includes dose coefficients for intakes of radionuclides by members of the public, specifically 3-month-old infants; 1-, 5-, 10-, and 15-year-old children; and adults, as compiled in *Publication 72* (ICRP, 1995c). The biokinetic models used to describe the distribution, retention, and excretion of radionuclides were developed to consider age dependence in the case of radioisotopes of the 31 elements for which dose coefficients were given in *Publications 56, 67, 69*, and *71* (ICRP, 1990, 1993, 1995a, 1995b).

For external radiation exposures, *Publication 119* includes conversion coefficients for use in occupational radiological protection abstracted from *Publication 74* (ICRP, 1996), calculated assuming whole-body irradiation by monoenergetic photons, electrons, and neutrons in a number of idealised standard exposure geometries.

2.4. Assessment of radiation exposure of astronauts in space *(Publication 123)*

Astronauts are exposed to extremely complex radiation fields. The primary radiation field contains high-energy particles with unique high-linear energy transfer (LET) components that are very different from the typical low-energy electron, photon, and neutron radiation fields encountered in occupational exposure environments on earth. Radiation fields in space may arise from outside the solar system within supernova explosions, neutron stars, pulsars, or other high-energy phenomena. These galactic cosmic rays are composed of electrons, protons, alpha particles, and heavier nuclei. In addition, the sun emits particles continuously in the form of protons and electrons, as well as gamma rays, hard and soft x rays, and radiowaves. Solar particle events can be a significant source of high radiation exposure, particularly to astronauts performing extravehicular activities in deep space.

Publication 123 (ICRP, 2013) provides an assessment of radiation exposures in space, considers methods of dose estimation and detection for space radiations, and gives reference doses. The report and the accompanying CD provide an extensive set of organ dose conversion coefficients for a range of space radiation particles, calculated using the *Publication 110* reference computational phantoms and complementing the external dose conversion coefficients given in *Publication 116* (ICRP, 2010). In recommending the use of a radiation weighting factor (w_R) of 20 for all heavy ions of all energies for protection purposes, ICRP (2007) recognised that this may be unduly conservative in situations in which heavy ions make a major contribution to exposures, as experienced by astronauts. *Publication 123* discusses alternatives to the use of w_R, providing guidance on the use of quality factors as a function of either unrestricted LET [Q(L) functions from *Publication 60* (ICRP, 1991)] or particle charge and energy [Q(Z, E) functions defined by National Aeronautics and Space Administration (NASA): Cucinotta et al., 2011].

3. FORTHCOMING COMMITTEE 2 REPORTS

3.1. Reference dosimetric phantoms and radiation transport calculations

Reference phantoms are being developed for children aged 3 months and 1, 5, 10 and 15 years, and for the fetus and pregnant female at gestational ages of 10, 15,

25, and 35 weeks. These dosimetric models, and those for adults provided in *Publication 110* (ICRP, 2009), are being used to provide reference radiation transport data in the form of specific absorbed fractions (SAFs) for radiations emitted from radionuclides retained in body organs and tissues. SAFs represent the deposition of energy in all important organs and tissues (target regions) following emissions from radionuclides retained in body organs and tissues (source regions). These data are used in the calculation of dose coefficients for the inhalation and ingestion of radionuclides by workers and members of the public (see below). They are also used in calculations of doses from radiopharmaceuticals, the responsibility of a task group led by Committee 3.

The calculation of SAFs involves Monte Carlo radiation transport of photons, electrons, and neutrons for an extensive set of source/target organ pairs. Additional work has focused on micro-computed-tomography-based models of electron and alpha particle dosimetry of skeletal tissues, and revisions to electron and alpha particle dosimetry in the dosimetric models of the human respiratory tract (ICRP, 1994a) and alimentary tract (ICRP, 2006).

Reports scheduled for the next few years in this series are:

- internal SAF values for the reference adult male and female;
- computational phantoms for the infant and children;
- internal SAF values for the infant and children;
- computational phantoms for the pregnant female, embryo, and fetus; and
- internal SAF values for the fetus and pregnant female.

3.2. Intakes of radionuclides by workers and the public

Work is in progress to replace *Publications 30* and *68* (ICRP, 1979, 1980, 1981, 1988, 1994b), that give biokinetic data and dose coefficients for occupational intakes of radionuclides by inhalation and ingestion, and *Publications 54* and *78* (ICRP, 1989, 1997), that give information for bioassay interpretation, with a single series of publications on occupational intakes of radionuclides (OIR series).

Part 1 of this series of reports is complete and provides a description of biokinetic and dosimetric methodology, including a summary of the *Publication 100* (ICRP, 2006) alimentary tract model, a detailed description of changes to the *Publication 66* (ICRP, 1994a) respiratory tract model, and an outline of approaches taken to the development of systemic models for elements. The use of bioassay data is also discussed. Subsequent parts will consist of element sections describing element-specific biokinetic models and providing dose coefficients and bioassay data. The series will have accompanying electronic annexes giving detailed information on organ doses and additional bioassay data. Planned publications are as follows:

- Part 2: hydrogen (H), carbon (C), phosphorus (P), sulphur (S), calcium (Ca), iron (Fe), cobalt (Co), zinc (Zn), strontium (Sr), yttrium (Y), zirconium (Zr), niobium (Nb), molybdenum (Mo), and technetium (Tc).

- Part 3: ruthenium (Ru), antimony (Sb), tellurium (Te), iodine (I), caesium (Cs), barium (Ba), iridium (Ir), lead (Pb), bismuth (Bi), polonium (Po), radon (Rn), radium (Ra), thorium (Th), and uranium (U).
- Part 4: lanthanides and actinides.
- Part 5: remaining elements to be considered.

The schedule of work for Committee 2 and its task groups includes replacement of all currently available dose coefficients for ingestion and inhalation of radionuclides by members of the public. Dose coefficients will be provided for:

- infants (3 months), children (1, 5, 10, and 15 years) and adults, replacing *Publications 56*, *67*, *69*, *71*, and *72* (ICRP 1990, 1993, 1995a, 1995b, 1995c).
- embryo/fetus and breast-fed infant following radionuclide intakes by the mother, replacing *Publications 88* and *95* (ICRP, 2001, 2004).

3.3. Dose coefficients for external environmental exposures

A new task group has been set up to develop age-dependent dose conversion coefficients for external radiation exposures, not previously provided by ICRP. Assessment of external exposures of members of the public, including infants and children, is important in the context of accidental releases from nuclear facilities, and more generally. Conversion coefficients for environmental exposures will be computed using ICRP reference dosimetric phantoms.

3.4. Use of effective dose

The task group on the use of effective dose led by Committee 2 has membership from Committees 1, 3, and 4 as well as external experts, in recognition of the central importance of this issue (Menzel and Harrison, 2012b). Experience has shown that 'effective dose', which has been defined and introduced by ICRP for risk management purposes (i.e. for risk limitation and optimisation), is widely used in radiological protection and related fields beyond its original purpose – incorrectly in some cases. Useful guidance on restrictions on the use of the quantity is provided by Committee 2 in Annex B to the 2007 Recommendations (ICRP, 2007). This guidance needs to be further expanded, and proposals made for the control of exposures and risk management in situations where 'effective dose' should not be used. An important focus of the report will be medical exposures (Balonov and Shrimpton, 2012; Harrison and Lopez, 2015).

4. DISCUSSION

ICRP Committee 2 has a large programme of work to provide new dose coefficients following the 2007 Recommendations (ICRP, 2007). Revisions of ICRP recommendations invariably require recalculation of dose coefficients because changes are made to the radiation and tissue weighting factors used in the calculation of equivalent and effective dose. However, it is also important that the methodology used to calculate doses is examined and updated as necessary to ensure the appropriate use of developing scientific knowledge. Committee 2 is therefore in the process of developing a set of reference computational phantoms based on medical imaging data, and revising radiation transport calculations using these models. Biokinetic models used to describe the behaviour of inhaled and ingested radionuclides are being updated, also leading to changes in organ doses and effective dose coefficients. It can be argued that the level of sophistication of these models is greater than required for the calculation of the protection quantities, equivalent and effective dose, because the risk-related adjustments made using radiation and tissue weighting factors are necessarily broad judgements and approximations applied across all body organs and tissues of people worldwide. However, ICRP is at the forefront of developments in this area, and its biokinetic and dosimetric models are used for scientific as well as protection purposes. Thus, ICRP models can provide best estimates of organ and tissue absorbed dose for use in epidemiological studies and assessments of risk to individuals. Biokinetic and dosimetric models can be adjusted for application to individuals or specific population groups rather than ICRP reference persons. Furthermore, it is important that methodology is continually refined and improved to counter suggestions that the dose assessments underestimate risks, particularly for internal exposures.

Effective dose is a radiation protection quantity that is used to set limits, constraints, and reference levels that apply to reference workers or reference members of the public. It provides an elegant solution to the requirement for a single quantity that enables the summation of all radiation exposures, from external exposures and radionuclides entering the body by inhalation or ingestion. Effective dose is used as a risk-related quantity for the optimisation of protection below constraints and reference levels. As our scientific understanding of radiation dosimetry and the risks of radiation continue to develop, it will be important to examine the formulation and use of effective dose as the central quantity in the control of radiation exposure and its continued fitness for purpose.

The estimation of doses and risks to astronauts provides a good example of an application that goes beyond the scope of effective dose. As discussed in *Publication 123* (ICRP, 2013), use of a radiation weighting factor of 20 for heavy ions in the calculation of organ equivalent doses may be an unduly conservative risk adjustment; available evidence suggests peak values of around 10 with substantially lower values at higher energies. *Publication 123* discusses the alternative use of quality factors as a function of LET, and calculation of dose equivalent and effective dose equivalent as preferred risk-related quantities for space applications. An alternative energy-related function used by NASA is also discussed. However, the protection of astronauts is generally related to limits on stochastic risks in terms of radiation-exposure-induced death (REID) rather than effective dose. For example, a limit of 3% REID at the upper 95% confidence level is currently imposed on NASA flight crew. To calculate values of REID, age- and sex-specific risk factors are applied to radiation weighted doses. To obtain best estimates of organ doses, reference phantoms of the type being developed by ICRP can be adjusted to the body dimensions of individual astronauts. The ICRP report on effective dose will explain the use of the protection quantities in the optimisation of protection, as distinct from dose and risk estimation.

REFERENCES

Balonov, M.I., Shrimpton, P.C., 2012. Effective dose and risks from medical x-ray procedures. Ann. ICRP 41(3–4), 129–141.

Cristy, M., 1980. Mathematical Phantoms Representing Children of Various Ages for Use in Estimates of Internal Dose. ORNL Report TM-367. Oak Ridge National Laboratory, Oak Ridge, TN.

Cristy, M., Eckerman, K.F., 1987. Specific Absorbed Fractions of Energy at Various Ages from Internal Photon Sources. ORNL TM-8381/VI-7. Oak Ridge National Laboratory, Oak Ridge, TN.

Cucinotta, F.A., Myung-Hee, Y.K., Chappell, L.J., 2011. Space Radiation Cancer Risk Projections and Uncertainties – 2010. NASA/TP-2011 – 216155.

Harrison, J.D., Marsh, J.W., 2012. Effective dose from inhaled radon and its progeny. Ann. ICRP 41(3/4), 378–388.

Harrison, J.D., Lopez, P.O., 2015. The use of effective dose in medicine. Ann. ICRP 44(1S), 280–287.

ICRP, 1979. Limits for intakes of radionuclides by workers. ICRP Publication 30 (Part 1). Ann. ICRP 2(3/4).

ICRP, 1980. Limits for intakes of radionuclides by workers. ICRP Publication 30 (Part 2). Ann. ICRP 4(3/4).

ICRP, 1981. Limits for intakes of radionuclides by workers. ICRP Publication 30 (Part 3). Ann. ICRP 6(2/3).

ICRP, 1988. Limits for intakes of radionuclides by workers: an addendum. ICRP Publication 30 (Part 4). Ann. ICRP 19(4).

ICRP, 1989. Individual monitoring for intakes of radionuclides by workers. ICRP Publication 54. Ann. ICRP 19(1–3).

ICRP, 1990. Age-dependent doses to members of the public from intake of radionuclides – Part 1. ICRP Publication 56. Ann. ICRP 20(2).

ICRP, 1991. 1990 Recommendations of the International Commission on Radiological Protection. ICRP Publication 60. Ann. ICRP 21(1/3).

ICRP, 1993. Age-dependent doses to members of the public from intake of radionuclides – Part 2. Ingestion dose coefficients. ICRP Publication 67. Ann. ICRP 23(3/4).

ICRP, 1994a. Human respiratory tract model for radiological protection. ICRP Publication 66. Ann. ICRP 24(1–3).

ICRP, 1994b. Dose coefficients for intakes of radionuclides by workers. ICRP Publication 68. Ann. ICRP 24(4).

ICRP, 1995a. Age-dependent doses to members of the public from intake of radionuclides – Part 3. Ingestion dose coefficients. ICRP Publication 69. Ann. ICRP 25(1).

ICRP, 1995b. Age-dependent dose to members of the public from intake of radionuclides – Part 4. Inhalation dose coefficients. ICRP Publication 71. Ann. ICRP 25(3/4).

ICRP, 1995c. Age-dependent doses to members of the public from intake of radionuclides – part 5. Compilation of ingestion and inhalation coefficients. ICRP Publication 72. Ann. ICRP 26(1).

ICRP, 1996. Conversion coefficients for use in radiological protection against external radiation. ICRP Publication 74. Ann. ICRP 26(3/4).
ICRP, 1997. Individual monitoring for internal exposure of workers. ICRP Publication 78. Ann. ICRP 27(3/4).
ICRP, 2001. Doses to the embryo and fetus from intakes of radionuclides by the mother. ICRP Publication 88. Ann. ICRP 31(1/3).
ICRP, 2002. Basic anatomical and physiological data for use in radiological protection reference values. ICRP Publication 89. Ann. ICRP 32(3).
ICRP, 2004. Doses to infants from ingestion of radionuclides in mothers' milk. ICRP Publication 95. Ann. ICRP 34(3/4).
ICRP, 2006. Human alimentary tract model for radiological protection. ICRP Publication 100. Ann. ICRP 26(1/2).
ICRP, 2007. The 2007 Recommendations of the International Commission on Radiological Protection. *ICRP Publication* 103. Ann. ICRP 37(2–4).
ICRP, 2009. Adult reference computational phantoms. ICRP Publication 110. Ann. ICRP 39(2).
ICRP, 2010. Conversion coefficients for radiological protection quantities for external radiation exposures. ICRP Publication 116. Ann. ICRP 40(2–5).
ICRP, 2012. Compendium of dose coefficients based on ICRP Publication 60. ICRP Publication 119. Ann. ICRP 41(Suppl.).
ICRP, 2013. Assessment of radiation exposure of astronauts in space. ICRP Publication 123. Ann. ICRP 42(4).
Menzel, H.-G., Harrison, J.D., 2012a. Doses from radiation exposure. Ann. ICRP 41(3/4), 12–23.
Menzel, H.-G., Harrison, J.D., 2012b. Effective dose: a radiation protection quantity. Ann. ICRP 41(3/4), 117–123.

SAGE

ICRP
Annals of the ICRP

Overview of ICRP Committee 3 'Protection in Medicine'

E. Vañó[a], D.L. Miller[b], M.M. Rehani[c]

[a]*Radiology Department, Complutense University, Madrid, 28040 Spain; e-mail: eliseov@med.ucm.es*
[b]*Center for Devices and Radiological Health, Food and Drug Administration, USA*
[c]*Massachusetts General Hospital, Harvard Medical School, USA*

Abstract–According to the 2011–2017 strategic plan, Committee 3 develops recommendations and guidance for protection of patients, staff, and the public against radiation exposure when ionising radiation is used for medical diagnosis, therapy, or biomedical research. This paper presents an overview of the work that Committee 3 has accomplished in recent years and describes its current work. The International Commission on Radiological Protection reports dealing with radiological protection in medicine from 2000 to the present cover topics on education and training in radiological protection; preventing accidental exposures in radiation therapy; doses to patients from radiopharmaceuticals; radiation safety aspects of brachytherapy; release of patients after therapy with unsealed radionuclides; and managing radiation dose in interventional radiology, digital radiology, computed tomography, paediatrics, cardiology, and other medical specialties. Current work deals with radiological protection in ion beam therapy, occupational protection in brachytherapy, justification in imaging, radiological protection in cone-beam computed tomography, occupational protection in interventional procedures, diagnostic reference levels for diagnostic and interventional imaging, and an update of an earlier publication on doses to patients and staff from radiopharmaceuticals. Committee 3 is also involved in preparation of a document on effective dose and its use in medicine.

Keywords: Radiological protection; Diagnostic; Interventional radiology; Nuclear medicine; Radiation therapy

This paper does not necessarily reflect the views of the International Commission on Radiological Protection.

1. INTRODUCTION

Committee 3 of the International Commission on Radiological Protection (ICRP) is concerned with protection of persons and unborn children when ionising radiation is used for medical diagnosis, therapy, or biomedical research. According to the 2011–2017 strategic plan, Committee 3 develops recommendations and guidance for protection of patients, staff, and the public against radiation exposure in medicine. For the 2013–2017 term, Committee 3 (http://www.icrp.org/icrp_group.asp?id=9) is composed of 16 members, seven of whom are new to the Committee. The members have expertise in different areas of radiological protection in medicine (medical physics, nuclear medicine, radiology, and radiation oncology). The primary qualification for membership is demonstrated expertise in radiological protection. The membership of the committee demonstrates a wide geographical distribution: Belgium (P. Scalliet); Canada (S. Demeter); China [P. Khong (Hong Kong) and B. Yue]; France (M. Bourguignon); Germany (R. Loose); India-Austria (M.M. Rehani); Japan (Y. Yonekura); the Republic of Korea (K. Kang); Spain (P. Ortiz and E. Vañó); Sweden (K.Å. Riklund); the UK (C. Martin); and the USA (K. Applegate, L. Dauer and D.L. Miller).

International organisations and stakeholders are encouraged to propose topics of interest for new publications. In addition, a new mechanism, introduced at the 2013 ICRP symposium, provides opportunities for symposium participants to provide input on suggested topics to the Commission's committees. The members of each committee review the proposals relevant to their committee's area of responsibility, and evaluate the need to produce reports on specific topics. Each committee forwards its recommendations to the Main Commission of ICRP. The work suggested by the committees is subject to approval by the Main Commission.

This paper presents an overview of the work that Committee 3 has accomplished in recent years and describes its current work.

2. ICRP PUBLICATIONS ON PROTECTION IN MEDICINE

Since 2000, Committee 3 has produced, either alone or in conjunction with other ICRP committees, 18 reports dealing with radiological protection in medicine (abstracts are available at http://www.icrp.org/publications.asp):

- *Publication 121*: Radiological protection in padiatric diagnostic and interventional radiology (ICRP, 2013b).
- *Publication 120*: Radiological protection in cardiology (ICRP, 2013a).
- *Publication 117:* Radiological protection in fluoroscopically guided procedures performed outside the imaging department (ICRP, 2010).
- *Publication 113:* Education and training in radiological protection for diagnostic and interventional procedures (ICRP, 2009b).
- *Publication 112:* Preventing accidental exposures from new external beam radiation therapy technologies (ICRP, 2009a).

- *Publication 106:* Dose to patients from radiopharmaceuticals: a third amendment to ICRP *Publication 53* (with an annex on radiation exposure of hands in radiopharmacies) (ICRP, 2008).
- *Publication 105:* Radiological protection in medicine (ICRP, 2007b).
- *Publication 102:* Managing patient dose in multi-detector computed tomography (ICRP, 2007a).
- *Publication 98:* Radiation safety aspects of brachytherapy for prostate cancer using permanently implanted sources (ICRP, 2005b).
- *Publication 97:* Prevention of high-dose-rate brachytherapy accidents (ICRP, 2005a).
- *Publication 94:* Release of patients after therapy with unsealed radionuclides (ICRP, 2004b).
- *Publication 93:* Managing patient dose in digital radiology (ICRP, 2004a).
- *Publication 87:* Managing patient dose in computed tomography (ICRP, 2000d).
- *Publication 86:* Prevention of accidents to patients undergoing radiation therapy (ICRP, 2000c).
- *Publication 85:* Avoidance of radiation injuries from medical interventional procedures (ICRP, 2000b).
- *Publication 84:* Pregnancy and medical radiation (ICRP, 2000a).
- *Supporting Guidance 2:* Radiation and your patient: a guide for medical practitioners (ICRP, 2001a).
- *Supporting Guidance 2:* Reference levels in medical imaging: review and additional advice (ICRP, 2001b).

3. WORK IN PROGRESS

In 2012, Committee 3 updated a work plan that included 10 topics and presented it to the Main Commission. The Committee proposed to carry out this work through four new and two existing task groups and four working parties.

Due to the change in membership of Committee 3 in 2013, the 10 topics in the 2012 work plan were refined during the 2013 Committee 3 meeting, and condensed to eight (four task groups and four working parties). One benefit of this revised work plan is the involvement of the new members of Committee 3. The eight topics in the revised work plan are as follows.

- Task Group 36: Dose to patients from radiopharmaceuticals. Chairs: Sören Mattsson and Dietmar Nosske (from Committee 2)

A compilation of the most relevant information from both current work and published reports on this topic. Several annexes are ready for review by the Main Commission before public consultation: Rb-82 chloride and I-123, I-124, I-125, and I-131 iodide. Other work in progress deals with I-123 FP-CIT (ioflupane), C-11 PiB, F-18 flutemetamol, and C-11 choline.

- Task Group 87: Radiological protection in ion beam radiotherapy. Chair: Yoshiharu Yonekura

This document deals with radiotherapy using protons and carbon ions. The objective is to give recommendations and guidance for the safe operation of medical facilities that use high-energy beam accelerators and for protection of patients and staff at these facilities. Potential incidents and measures for prevention, detection, and mitigation are also provided. The target audience is users, facility planners, and radiation protection specialists. A final draft of this document was discussed at the Main Commission meeting in October 2013 in Abu Dhabi, United Arab Emirates and was accepted by the Main Coimmission for public consultation after some minor changes.

- Task Group 88: Radiological protection in cone-beam computed tomography. Chair: Madan Rehani

This document contains recommendations and guidance on radiological protection for the use of cone-beam computed tomography (CT) in medical and interventional applications. Dental applications are also discussed briefly. The target audience is physicians, other medical staff, and medical physicists. Applications covered include interventional radiology, intra-operative surgery, radiotherapy, breast imaging, and head and neck imaging. The report includes data on radiation doses to patients and workers, and provides guidance and recommendations in accordance with the Commission's principles. Links to other ICRP publications are noted. Committee 3 approved the draft of this document during its 2013 meeting in Abu Dhabi, and it was submitted to the Main Commission for review and consideration for public consultation.

- Task Group 89: Occupational radiological protection in brachytherapy. Chair: Lawrence Dauer

This document will compile existing recommendations and provide guidance to staff on radiological protection in brachytherapy. External beam radiation therapy, when administered in an appropriately shielded facility, results in minimal occupational doses. Brachytherapy is different because of the need for active management of occupational dose. In modern centres, brachytherapy-related occupational dose from a variety of activities is unavoidable (e.g. loading of seeds, plaques, caesium implants, associated fluoroscopy). There is substantial variation in the practice of brachytherapy throughout the world. Some facilities still use older techniques with the potential for high staff doses (e.g. radium use, iridium wires). Technological developments and newer techniques present new occupational radiological protection concerns that need to be addressed. The target audience is clinicians, other medical staff, medical physicists, radiological protection officers, and regulatory authorities. The table of contents and an initial draft outline were discussed during the 2012 Committee 3 meeting in Vienna, Austria. A working draft was prepared for the 2013 Committee 3 meeting in Abu Dhabi.

- Working Party: Justification – framework for justification in medical uses of ionising radiation (including imaging of asymptomatic individuals). Chair: Katrine Åhlström Riklund

In the last decade, the use of ionising radiation for imaging of asymptomatic individuals has increased. A substantial amount of data has been published recently on cardiac CT, positron emission tomography (PET)/CT, and CT colonography. In 2009, Committee 3 began a review to analyse the risks and benefits of imaging asymptomatic individuals. Committee 3 reviewed available guidance, including the position statements of professional societies. The evidence for this practice appears to be weak and difficulties were encountered in providing concrete comments on justification. The focus for this working party is the process of and framework for justification. The target audience is referring physicians (referring medical practitioners) and imaging specialists (radiological medical practitioners). The document will provide practical guidance on level 2 and level 3 justifications for medical professionals. It should not overlap with the work of professional bodies, and will emphasise that radiological protection is neither the only nor the main issue that professionals need to consider. The table of contents was revisited and approved at the 2013 Committee 3 meeting in Abu Dhabi.

- Working Party: Radiological protection in therapy with radiopharmaceuticals. Chairs: Sören Mattsson and Yoshiharu Yonekura

The document is intended to identify situations with unique radiological protection aspects and to provide a framework for quantitative imaging and dosimetry for current and novel treatment approaches. This framework may include: individual dosimetry for planning therapy; test activities and pre-treatment tracers; measurement of whole-body/tumour/organ kinetics; analysis of urine or blood samples; dose calculation based on three-dimensional patient images or patient-like phantoms; evaluation of methods to scale up to therapeutic activity levels; and written directives for the therapy. Variations in patient kinetics at therapeutic levels of activity will be discussed. The target audience is oncologists, other clinicians, nuclear medicine physicians, medical physicists, other staff members, radiation protection officers, and regulatory authorities.

- Working Party: Occupational protection issues in interventional procedures (fluoroscopically guided). Chair: Pedro Ortiz-López

The Commission has provided practical advice, in specific ICRP reports aimed at physicians, on the protection of workers involved in interventional cardiology, paediatric radiology, and fluoroscopic procedures outside the imaging department. There is also a need for advice on occupational protection in interventional procedures directed at other audiences, such as hospital radiological protection officers, medical physicists, professionals involved in dosimetry services, and regulatory authorities. This report will integrate information that is scattered among different reports and publications, and provide detailed information on specific dosimetric aspects of radiological protection (e.g. active dosimetry, radiological protection tools, radiological protection methods, tests of protective devices). This report will also place occupational protection in interventional procedures into the context of the framework of occupational protection in general. The preliminary draft includes the following: purpose and scope; issues and problems; statistics of interventional procedures; trends and overview of exposures; reported tissue effects (eyes, hands, legs); existing problems in monitoring; application of radiological protection principles to occupational exposures in interventions; staff protection, dose constraints and investigation levels for occupational protection (body, eyes, hands); protection of pregnant workers; and exposure monitoring (number and placement of dosimeters, assessment of staff exposure, assessment of doses to the eyes).

- Working Party: Diagnostic reference levels for diagnostic and interventional imaging. Chair: Eliseo Vañó

The Commission introduced diagnostic reference levels (DRLs) in 1991, and published additional supporting guidance in 2001. In 2004, *Publication 93* stated that DRLs for non-digital imaging tasks are not necessarily applicable to specific, similar digital imaging procedures. It recommended a re-evaluation of local DRLs when digital techniques replace conventional analogue imaging techniques. In 2007, *Publication 105* recognised the benefit of DRLs in fluoroscopically guided procedures, but noted the need to take into account the complexity of the procedure when setting these levels. In addition to digital techniques and interventional procedures, new combined imaging techniques such as PET-CT may also benefit from the use of DRLs (this may require different patient-dose-related quantities for PET and CT and different image quality and diagnostic information). Appropriate use of DRLs within the medical community is still rather poor. Committee 3 proposed a document updating existing information and addressing these new aspects of DRL use in interventional imaging, digital radiology, and PET-CT. The document will include different existing methodologies for determining DRL values, including use of full patient dose distributions for optimisation (not just the third quartile). The target audience is medical physicists, radiologists, nuclear medicine specialists, radiographers, industry, and regulatory authorities. The table of contents was discussed and refined during the 2013 Committee 3 meeting in Abu Dhabi. Current Commission recommendations on DRLs will be summarised in an introductory chapter. Some of the topics identified for clarification include: the use of phantoms vs patient dose values; the link between DRLs and image quality or diagnostic information (including post-processing) for different clinical tasks; standardisation and consensus on the levels of complexity for some common procedures and their impact on DRLs; deriving trigger (alarm) levels from DRLs for investigation of high dose values; use of full patient dose distributions, in addition to DRLs, to help with optimisation; balancing the relevance of several dose-related quantities used to set DRLs (e.g. kerma-area product, cumulative air kerma, number of cine or DSA images, fluoroscopy time); recommendations for the use of multiple dose-related quantities simultaneously; and the recommended time interval for updating DRLs and factors to be considered when establishing the time interval.

Committee 3 is also involved in the Task Group on Effective Dose launched by Committee 2 (Pedro Ortiz-López is the member of Committee 3 nominated for this task).

4. TOPICS UNDER CONSIDERATION

During the 2013 Committee 3 meeting in Abu Dhabi, several topics were considered as potential areas of interest for future publications and will be kept under consideration. The topics identified were as follows.

- Individual human sensitivity to ionising radiation (Michel Bourguignon)

The areas covered by this report could include: variation in radiosensitivity – evidence from epidemiological studies; clinical radiosensitivity; radiosensitivity in animals; mechanisms contributing to radiosensitivity; radiosensitivity disorders and familial cancers; cellular radiosensitivity in human subpopulations; human radiosensitivity and ethics; implications of variation in radiosensitivity for radiological protection of workers and the public; implications of variation in radiosensitivity for radiological protection in emergency situations; clinical implications in radiotherapy; clinical implications in diagnosis; clinical implications regarding syndromes; and implications in radiological protection with regard to ethical issues. Both Committee 3 and Committee 1 are interested in this topic. Terms of reference for future work will be drafted and presented at the 2014 meeting.

- Framework for optimisation for individual patients (Madan Rehani)

While the principle of 'as low as reasonably achievable applies' to each medical exposure, there is an assumption that this will lead to overall optimisation. There are many situations where individual events are not optimised. How can overall optimisation be achieved for individual patients who have received multiple medical exposures, some of which are optimised and some of which are not? This requires consideration of optimisation extended throughout an individual patient's lifetime, through monitoring and recording of radiation dose from each exposure, and consideration of the impact of previous and predicted future medical exposures on the effect of the next medical exposure. This may lead to a consideration of risk estimates for individual patients.

- Dose quantities and units in imaging equipment (Pedro Ortiz-López)

The International Electrotechnical Commission has a working group on CT and an agreement has been reached on this topic. The manufacturers of medical equipment are not the only stakeholders interested in the topic. Other companies offer software to collect and process radiation doses from patient exposures based on metadata included in the image header, but some relevant information is missing or is included in proprietary (private) fields. It is appropriate that the Commission takes the initiative and prepares criteria on the information that is needed from the perspective of radiological protection.

- Patient eye dose in CT

The topic was considered of potential interest in light of the Commission's new threshold for the lens of the eye. It was pointed out that the Radiation Protection of Patients website of the International Atomic Energy Agency has good material on this issue. There is some guidance available in German from the German Federal Radiation Protection Committee, and some papers on this topic are now appearing in the scientific literature. Committee 3 will follow this topic.

5. UPDATE OF AN EXISTING DOCUMENT

- Radiation and your patient – a guide for medical practitioners (Sandor Demeter)

The Commission published this document as *Supporting Guidance 2* in 2001. It is not a formal publication of the Commission. This document needs to be updated and extended to include risk communication. The target audience is referring medical practitioners. The intention is to help them in the day-to-day practice of justification of medical imaging, and in communicating effectively with their patients regarding the benefits and risks of imaging examinations and the risks of not performing these examinations when they are indicated.

REFERENCES

ICRP, 2000a. Pregnancy and medical radiation. ICRP Publication 84. Ann. ICRP 30(1).

ICRP, 2000b. Avoidance of radiation injuries from medical interventional procedures. ICRP Publication 85. Ann. ICRP 30(2).

ICRP, 2000c. Prevention of accidents to patients undergoing radiation therapy. ICRP Publication 86. Ann. ICRP 30(3).

ICRP, 2000d. Managing patient dose in computed tomography. ICRP Publication 87. Ann. ICRP 30(4).

ICRP, 2001a. Radiation and your patient – a guide for medical practitioners. ICRP Supporting Guidance 2. Ann. ICRP 31(4), 1–31.

ICRP, 2001b. Reference levels in medical imaging: review and additional advice. ICRP Supporting Guidance 2. Ann. ICRP 31(4), 33–52.

ICRP, 2004a. Managing patient dose in digital radiology. ICRP Publication 93. Ann. ICRP 34(1).

ICRP, 2004b. Release of patients after therapy with unsealed radionuclides. ICRP Publication 94. Ann. ICRP 34(2).

ICRP, 2005a. Prevention of high-dose-rate brachytherapy accidents. ICRP Publication 97. Ann. ICRP 35(2).

ICRP, 2005b. Radiation safety aspects of brachytherapy for prostate cancer using permanently implanted sources. ICRP Publication 98. Ann. ICRP 35(3).

ICRP, 2007a. Managing patient dose in multi-detector computed tomography (MDCT). ICRP Publication 102. Ann. ICRP 37(1).

ICRP, 2007b. Radiological protection in medicine. ICRP Publication 105. Ann. ICRP 37(6).
ICRP, 2008. Radiation dose to patients from radiopharmaceuticals: a third amendment to ICRP Publication 53. ICRP Publication 106. Ann. ICRP 38(1/2).
ICRP, 2009a. Preventing accidental exposures from new external beam radiation therapy technologies. ICRP Publication 112. Ann. ICRP 39(4).
ICRP, 2009b. Education and training in radiological protection for diagnostic and interventional procedures. ICRP Publication 113. Ann. ICRP 39(5).
ICRP, 2010. Radiological protection in fluoroscopically guided procedures performed outside the imaging department. ICRP Publication 117. Ann. ICRP 40(6).
ICRP, 2013a. Radiological protection in cardiology. ICRP Publication 120. Ann. ICRP 42(1).
ICRP, 2013b. Radiological protection in paediatric diagnostic and interventional radiology. ICRP Publication 121. Ann. ICRP 42(2).

SAGE

ICRP
Annals of the ICRP

Application of the Commission's recommendations: the 2013–2017 Committee 4 programme of work

J. Lochard

Centre d'étude sur l'Evaluation de la Protection dans le domaine Nucléaire, 28 rue de la Redoute, 92260 Fontenay-aux-Roses, France; e-mail: lochard@cepn.asso.fr

Abstract–Committee 4 of the International Commission on Radiological Protection (ICRP) is responsible for developing principles, recommendations, and guidance on the protection of people against radiation exposure, and to consider their practical application in all exposure situations. Currently, the Committee's efforts are focused on the completion of a series of future ICRP publications on the implementation of its 2007 Recommendations to the various existing exposure situations. A report on protection against radon exposure was published recently (ICRP *Publication 126*), and two documents on protection against cosmic radiation in aviation, and naturally occurring radioactive material are under development. The programme of work for the forthcoming 2013–2017 period comprises the update of ICRP *Publication 109* on protection of people in emergency exposure situations, and the update of ICRP *Publication 111* on protection of people living in long-term contaminated areas after a nuclear accident, as well as the development of a future ICRP publication on the ethics of radiological protection. It also includes the preparation of task groups on the application of the Commission's recommendations for contaminated sites from past activities and for surface and near-surface disposal of radioactive waste. Another important task for Committee 4 will be to develop a reflection on the tolerability of risk from radiation.

Keywords: Radon; Naturally occurring radioactive material; Nuclear accident; Ethics of radiological protection; Tolerability of radiation risk

This paper does not necessarily reflect the views of the International Commission on Radiological Protection.

1. INTRODUCTION

Committee 4 of the International Commission on Radiological Protection (ICRP) was created in 1962, with the name 'Application of the Commission's Recommendations', when the Commission decided to re-organise its committee system in order to adapt to the changing context of radiological protection and to improve its mode of operation (ICRP, 2009a). In 1977, when the Commission reviewed and updated the names and missions of its committees, Committee 4 kept its name, which is still valid today, and its mission was stated as follows: 'Committee 4 will continue its role of providing advice on the Commission's system of dose limitation, and on protection of the worker and the public. The Committee will also serve as a major point of contact with international organisations concerned with radiation protection'. This basic mission has not changed since that time, but the wording was modified slightly when the first strategic plan of the Commission was prepared, and subsequently adopted in 2011 (ICRP, 2011), to be in coherence with the new approach of the system of radiological protection based on exposure situations. It now reads as: 'Committee 4 develops principles and recommendations on radiological protection of people in all exposure situations'.

In 2013, with the beginning of a new term running until 2017, Committee 4 underwent a substantial renewal with 10 of the 17 members and the Chair being new. The membership reflects a good mixture of operational, academic, and regulatory expertise.

This paper presents an overview of the programme of work of Committee 4 for the 2013–2017 term. The emphasis is on the update of *Publication 109* on protection of people in emergency exposure situations (ICRP, 2009a), and the update of *Publication 111* on protection of people living in long-term contaminated areas after a nuclear accident (ICRP, 2009b) to take into account the lessons from the Fukushima Daiichi nuclear power plant accident, as well as the preparation of a future publication on the ethical foundations of the radiological protection system.

2. PROGRAMME OF WORK OF COMMITTEE 4

In recent years, Committee 4, in parallel with the publication of three reports on protection in geological disposal of long-lived solid radioactive waste (ICRP, 2013), protection of the environment under different exposure situations (prepared jointly with Committee 5) (ICRP, 2014a), and security screening (ICRP, 2014b), has focused its efforts on the development of guidance for 'existing exposure situations'. A first report on protection against radon exposure has also been published (ICRP, 2014c). Further work is underway, which concerns naturally occurring radioactive material (NORM) and cosmic radiation in aviation.

2.1. Work in progress

2.1.1. Naturally occurring radioactive material

Since the renewal of Committee 4, Task Group 76 on radiological protection against enhanced exposure from industrial processes using NORM has been placed under the chairmanship of Jean-François Lecomte (France) with a new membership and a more focused mandate. The objective is to develop a report presenting the general principles that apply to exposure situations resulting from activities involving NORM not deliberately used for their radioactive properties. Without entering into the details of the numerous industrial sectors involved, Task Group 76 will focus on the main steps of the activities that could give rise to significant exposure of workers or the public: extraction and handling of raw material; manufacture of products; generation of by-products; management of waste; and rehabilitation of contaminated sites.

The first discussions of Task Group 76 led to the conclusion that exposure situations resulting from activities involving NORM not deliberately used for their radioactive properties should be considered as existing exposure situations. As for any existing exposure situation, characterisation of the exposures is a prerequisite before any control can be envisaged. Exposures generally fall into the category of public exposure, but can be considered as occupational exposure in certain situations. Establishing a control on exposures resulting from NORM must be justified (i.e. do more good than harm), with all relevant factors being carefully considered, and, if justified, this control must be implemented through the principle of optimisation of protection. This involves identifying exposures that deserve to be reduced given their magnitudes using a restriction on individual doses to reduce inequity in the entire dose distribution. In accordance with the general recommendations of the Commission (ICRP, 2007a), the reference level to restrict exposures must be selected at or below the 1–20 mSv year^{-1} band as appropriate given the prevailing circumstances, and the basic requisites of information, assessment of exposures, and involvement of stakeholders must be applied in a graded approach. Beyond these general principles, Task Group 76 recognises that for reasons of convenience and/or equity, competent authorities may want to manage specific activities or parts of them like planned exposure situations.

2.1.2. Cosmic radiation in aviation

The development of a future ICRP publication on radiological protection against cosmic radiation in aviation is the responsibility of Task Group 83 under the chairmanship of Jacques Lochard (France). A draft report is now available and is being reviewed by Committee 4. Task Group 83 considers that exposure of aircraft crew and passengers to cosmic radiation, including solar eruptions, is an existing exposure situation. Doses received in aeroplanes can be well predicted by computer codes, and experience shows that the average annual dose of aircraft crew is higher than the average annual dose received by workers in the nuclear fuel cycle. Considering the constant increase of passengers in the future and the trend for aircrafts to fly for longer durations and at higher altitude, implementing a strategy for cosmic radiation in aviation is justified with the aim of keeping exposure of passengers and aircraft crew as low as reasonably achievable.

For the practical implementation of this strategy, Task Group 83 is proposing to use a graded approach proportionate to the level of exposure that may be received by passengers and aircraft crew. In view of the current options for the control of exposure during flights, Task Group 83 maintain the previous position of the Commission that the main action to control exposures in aviation is to adapt the flight schedules of the most exposed individuals by combining flight time and route selection.

For the vast majority of passengers who fly occasionally, the doses received can be considered as negligible compared with their total annual exposure due to natural background at ground level. However, Task Group 83 suggests that general information about cosmic radiation should be made available to all passengers. For frequent flyers, the recommendation is to use a graded approach to protection based on the levels of dose that they may receive according to the frequency of flight, and whether they fly on their own initiative or at the request of an employer. This can be achieved by providing individual information to frequent flyers, including access to free dose calculators and adjusting the frequency of flight when relevant. For the protection of aircraft crew, Task Group 83 proposes to maintain the previous advice of the Commission, but to introduce the use of a reference level to be selected by airlines or competent authorities to be consistent with *Publication 103* (ICRP, 2007a) concerning implementation of the optimisation principle for existing exposure situations. Task Group 83 also considers that a reference level may also be useful to some frequent flyers for professional duties whose exposure circumstances are similar to aircraft crew.

Task Group 83 will pay special attention to the issue of protection for pregnant frequent flyers and pregnant aircraft crew to limit exposure to their embryo/fetus. It is proposed that female workers should be informed by their employers about the potential risk to the embryo/fetus from exposure to radiation in order to encourage timely declaration of pregnancy, so that provisions may be made to allow for the adjustment of duties during the remainder of the pregnancy.

Finally, Task Group 83 considers that disseminating accurate and adapted information on the existence of cosmic radiation and its impact in terms of exposure is the best way to raise awareness about protection issues, and to foster informed decisions among all stakeholders considering the benefits they receive from air travel.

2.2. Updates of *Publications 109* and *111*

Following the Fukushima Daiichi nuclear plant accident in March 2011, Committee 4, like the other ICRP committees, has been mobilised to help meet the many challenges that face experts, authorities, and affected populations in Japan (ICRP, 2014d). Several members of Committee 4 participated in the reflection of Task Group 84 created just after the accident to develop recommendations to inform the programme of work of ICRP based on the initial lessons learned from the accident (ICRP, 2012), as well as in the International Expert Symposium on

Radiation and Health Risks and the International Academic Conference on Radiation Health Risk Management held in Fukushima in September 2011 (FMU, 2011) and February 2013 (FMU, 2013), respectively. The last event was the opportunity to discuss the Commission's recommendations for the management of post-accident situations in the light of feedback from the affected territories from the first 2 years since the accident.

However, the main action of Committee 4 was to initiate, at the request of various Japanese members of the Commission, a series of dialogue seminars involving local stakeholders with the support of foreign organisations to find ways to respond to the challenges of long-term rehabilitation of living conditions in the areas affected by the Fukushima accident (Lochard, 2012). Building on the recommendations in *Publication 111* (ICRP, 2009b), this exceptional commitment of the Commission to work practically with the affected communities was seen as an opportunity to share experience and promote radiological protection culture among all those engaged in the recovery process. By October 2013, six dialogue seminars had been organised in Fukushima Prefecture involving experts, authorities, professionals, non-government organisations, and representatives of affected communities in Japan as well as Belarus and Norway.

The dialogue seminars revealed that the human consequences among the affected people in Japan were similar to those observed after the Chernobyl accident in the Commonwealth of Independent States: loss of confidence in the authorities and experts; strong concern for health, especially of children; loss of control over everyday life; fear of being discriminated and abandoned; and apprehension about the future, with the permanent dilemma of 'staying or not' or 'returning or not' in the affected areas (Lochard, 2013). The preliminary lessons highlighted the importance of establishing places for dialogue in affected communities for experts and inhabitants to work together to characterise the radiological situation. This 'co-expertise process' allows individuals to find their own benchmarks progressively with regards to contamination, and to make informed decisions concerning their own protection (self-help protection), thus recovering their autonomy.

Building on the conclusions of Task Group 84 and the first lessons of the dialogue seminars in Fukushima, Committee 4 established a task group to update *Publication 109* on the protection of people in emergency exposure situations (ICRP, 2009a), and *Publication 111* on the protection of people living in long-term contaminated areas after a nuclear accident (ICRP, 2009b). Co-chaired by Michiaki Kai and Toshimitsu Homma (Japan), Task Group 84 is developing its content, taking into account the numerous international developments concerning the protection of workers and the public during the emergency and recovery phases of nuclear accidents. In order to consolidate its work and to get the most from the Fukushima experience, Task Group 84 also plans to hold a series of consultations in Japan in due course, with authorities, experts, professionals, and affected people.

Based on the initial discussions within Committee 4, Task Group 84 will focus its attention on the justification for, and optimisation of, emergency decisions, characterisation of the radiological situation, protection of emergency and recovery responders, decontamination and waste management strategies, management of contaminated foodstuffs and commodities, the shift from the emergency to the existing exposure situation, and the co-expertise process to develop radiation protection culture among the affected population.

An important aspect of the work will be clarification of the consequences on the protection of workers and the public of the situation-based approach introduced in *Publication 103* (ICRP, 2007a) in place of the previous approach based on the distinction between practices and interventions. While this change was taken into account in *Publications 109* and *111* (ICRP, 2009a,b), the reactions of the authorities, as well as those of the experts, professionals, and affected people in Japan, have shown how difficult it is for these stakeholders to think about the situation in any way other than as a planned exposure situation.

2.3. Ethics of radiological protection

Recognising that the system of radiological protection is combining scientific knowledge from different disciplines, a set of values rooted in ethics and social behaviour, and the experience accumulated from the day-to-day practice of radiation protection professionals, Committee 4 decided at the general meeting of the Commission in Porto in November 2009 to initiate a discussion in order to clarify the ethical basis supporting the system. As a first step, Committee 4 reviewed the various theories of ethics, and finally concluded that the system of radiological protection is rooted in the three major ethical theories: virtue, deontological, and utilitarian ethics. It also showed that although explicit considerations about the ethical basis of the system of radiological protection are almost absent from ICRP publications, a constant reflection on the ethical foundations of the system can be found in past writings of several prominent members of the Commission (e.g. Taylor, 1957; Silini, 1992; Lindell, 2001).

Based on these findings, Committee 4 established a working party at its Bethesda meeting in Autumn 2011. Given the importance of adopting a 'cross-cultural' approach, as the ICRP recommendations are intended to be universal, it was also proposed to develop the work in close cooperation with ethicists, philosophers, social scientists, and radiation protection professionals from the different regions of the world. As such, a proposal for cooperation was made to the International Radiation Protection Association (IRPA) in late 2012 to organise, together with ICRP, a series of regional workshops with the support of IRPA Associate Societies on the ethical dimensions of the ICRP radiological protection system. An agreement was established between ICRP and IRPA in early 2013. The first Asian workshop was held in Daejeon, Korea, in August 2013 (ICRP, 2015).

Finally, in October 2013 in Abu Dhabi, the Main Commission approved the creation of Task Group 94 on the ethics of radiological protection under the chairmanship of Deborah Oughton from the University of Life Sciences in Norway, with the purpose of consolidating the recommendations, improving understanding of the system, and providing a basis for communication on radiation risk and its perception. It also endorsed the continuation of regional workshops to collect relevant material for the advancement of the work of Task Group 94, discuss the ideas proposed by the latter, and review the successive versions of the report in preparation, as the work progresses.

The Daejeon workshop was a good opportunity to review the progress of the reflection of Committee 4, and also to confront it with the Asian cultural context. The presentations and discussions confirmed that the system of radiological protection combines the duty to act wisely and reasonably (virtue ethics) while respecting both individual rights (deontological ethics) and the pursuit of collective interest (utilitarian ethics). Furthermore, they strengthened that prudence, beneficence/non-maleficence, justice, and dignity can be considered as the cardinal ethical values underlying the basic principles (justification, optimisation, and limitation) and the main requisites (information, training, monitoring, health surveillance, etc.) of the ICRP system of radiological protection.

Discussions during the workshop also highlighted the values of reasonableness and tolerability; although these are not – strictly speaking – ethical values, they are at the heart of the system and enable the linkage of scientific and ethical components. Incidentally, workshop participants also mentioned that the radiological protection system should not only care about the physical dimension of people's health, but should place more emphasis on the wellbeing of individuals, which includes mental and social aspects. Such enlargement would allow consideration of people's perceptions regarding the sense of security.

The value of dignity was discussed in relation to the principles of bioethics, which are very similar to the ethics of radiological protection. Dignity is an attribute of the human condition, meaning that every individual deserves unconditional respect regardless of age, sex, health, social condition, ethnic origin, and religion. However, dignity is also the corollary of autonomy, which implies freedom and the capacity to deliberately decide and act. In this respect, 'the need to account for the views and concerns of stakeholders when optimising protection', introduced by the Commission for the first time in *Publication 103* (ICRP, 2007a), is likely to contribute to respect the dignity of those for whom the radiological protection system is implemented. The emphasis in the recent Commission publications on notions such as the right to know, informed consent (ICRP, 2007b), and self-help protection (ICRP, 2009b) clearly highlight the importance placed by the Commission on properly informing and, if relevant, training stakeholders to take informed decisions. They undoubtedly strengthen the autonomy and dignity of individuals.

Finally, it is interesting to note the presentation made by Committee 4 member Senlin Liu on the application of classic Confucianism to radiological protection. He showed that this ancient philosophy, based on the virtues of righteousness, ritual, wisdom, and faith, may be used to interpret the radiological protection principles in ethical terms. This presentation largely contributed to strengthen the conviction of the task group members that the ethical values underlying the radiological protection system are widely shared by the different cultures in the world. This is an aspect on which the regional workshops will surely be of great help to confirm or refute this conviction. Following the Korean workshop, a series of other regional workshops are planned in Europe and North America over the course of 2014 and 2015.

The next steps for Task Group 94 will be to analyse, in greater depth, the ethical values and procedures that structure the different components of the ICRP system of radiological protection, such as the types of exposure situations, the categories of exposure, the various dose criteria, and the key requisites. It will also elucidate the ethical dimensions involved in the practical implementation of the system in the different domains: occupational health; medicine; protection of the environment; radioactive waste management; emergencies; and post-accident situations.

Task Group 94 will work closely with the Main Commission, as well as the other committees, given their obvious interest in ethical issues that concern all aspects of radiological protection. The objective is to finalise a report in 2015 in order to have a public consultation by the beginning of 2016, with a final discussion at IRPA14 in Cape Town in May 2016, and adoption of the report for publication by the Main Commission in Autumn 2016.

2.4. Tolerability of radiation risk

The recent experience of Fukushima clearly showed the confusion concerning application of the dose criteria recommended by the Commission since 2007. Dose limits for the public and for occupationally exposed workers (1 and 20 mSv year^{-1}, respectively) that apply to planned exposure situations, regarding exposures resulting from the deliberate introduction and operation of sources used for their radioactive properties, are now generally well accepted all over the world. However, the reference levels introduced in *Publication 103* that apply to emergency and existing exposure situations (ICRP, 2007a), and particularly the framework for their selection (the three bands defined by the values of 1, 20, and 100 mSv), are still misunderstood by many people, including radiological protection professionals, and sometimes even disputed. This de-facto situation has led Committee 4 to reconsider the tolerability of radiological risk, which is a central element of the system of radiological protection. For this purpose, a working party under the chairmanship of Anne McGarry from Ireland was established at the annual meeting in Moscow in September 2012.

The reflections of the working party initially focused on a review of past Commission developments. In *Publication 26* (ICRP, 1977), the acceptability of the risk for workers was judged by comparing the risk associated with radiation exposure with the risk associated with other types of activity recognised for their high level of security, and the acceptability of the risk for the public was judged by comparison with risks commonly accepted in everyday life (risk-based approach). In *Publication 60* (ICRP, 1991), the Commission introduced the so-called 'tolerability of risk model', introducing a distinction between unacceptable, tolerable, and acceptable levels of risks, using the reference to the natural background exposure for the public and a multi-criteria approach for the workers, taking into account several parameters characterising their exposure, to justify the dose limits of 20 mSv and 1 mSv year^{-1}, respectively (risk-informed approach). Although *Publication 103* (ICRP, 2007a) introduced a major evolution of the ICRP system of radiological protection with the introduction of three types of exposure situations, and the generalisation of the optimisation principle in connection with individual dose restrictions to all controllable exposure situations, it is interesting to note that the model of the tolerability of risk is not mentioned explicitly.

This led the working party to explore how it might be possible to articulate the past approaches of the Commission on tolerability with the new framework recommended by *Publication 103* (ICRP, 2007a), particularly to adjust the reference levels for emergency and existing exposure situations. Without going into detail, seen in terms of tolerability of risk, *Publication 103* (ICRP, 2007a) provides a number of indications that allow for the development of a renewed approach to the issue. First, there is the statement that, 'At doses higher than 100 mSv, there is an increased likelihood of deterministic effects and a significant risk of cancer. For this reason, the Commission considers that the maximum value for a reference value is 100 mSv incurred either acutely or in a year' (ICRP, 2007a, Para. 236). Although this statement needs some clarification about how to interpret the 'acutely or in a year', particularly in the context of emergency exposure situations, it gives an indication of the level of exposure above which the Commission considers it undesirable to expose people, and thereby the dose range that can be considered as tolerable. Secondly, *Publication 103* also indicates the criteria to be considered when selecting reference levels: 'the relevant exposure situation in terms of the nature of the exposure, the benefits from the exposure situation to individuals and society, as well as other societal criteria, and the practicability of reducing or preventing the exposures' (ICRP, 2007a, Para. 242).

This last quotation makes it clear that fixing the dose criteria for protection management fundamentally depends on the nature and conditions of exposures, but also the societal aspects. From this point of view, the attitude of exposed people with regard to the exposure situation is certainly a key factor in judging the tolerability of risk levels. The preliminary thoughts of Committee 4 on the subject leads to the distinction of three basic attitudes towards risk depending on the need, or not, for exposed people to act based on the level of risk:

- Quietude: in everyday life, people forget the risk if they are confident in the arrangements put in place to control it, and they trust the institutions and people responsible for this control. This is typically the case for public exposures in planned exposure situations that are completely under control.
- Vigilance: if people are suspicious that something may go wrong, they pay attention to the situation, and in the proven presence of a risk, they do what is reasonably achievable to maintain or mitigate it to a tolerable level. This is typical of occupational exposures in planned exposure situations for which workers must exercise constant vigilance. This is also true, to a lesser extent, for public exposure in existing exposure situations for which exposures must be previously characterised to be controlled.
- Reaction: when facing an imminent danger or being involved in an emergency, people act urgently and in a timely manner to protect themselves and their loved

ones, usually demonstrating solidarity with other affected people. This is typically the case in urgent exposure situations resulting from the loss of control of a source, such as a nuclear accident, or from any unexpected situation.

By building on the above distinction, it is possible to adapt the model of the tolerability of risk of *Publication 60* (ICRP, 1991) to the framework of *Publication 103* (ICRP, 2007a), taking into account the criteria regarding the nature of the exposure situations and their controllability, and the basic attitudes of people towards risk (Fig. 1). Thus, exposures below approximately 100 mSv (acute or per year) could be considered as tolerable and furthermore acceptable when optimised in the range corresponding to the relevant exposure situation.

The approach of *Publication 103* (ICRP, 2007a), as far as the tolerability of risk is concerned, could be called a 'risk and exposure situation informed approach'. This is a preliminary reflection that needs to be further developed by the working party, together with the members of Committee 4 and, ultimately, the other ICRP committees.

2.5. Future work

Following the publication on radiological protection in geological disposal of long-lived solid radioactive waste (ICRP, 2013), Committee 4 received a letter from the Radioactive Waste Management Committee of the Nuclear Energy

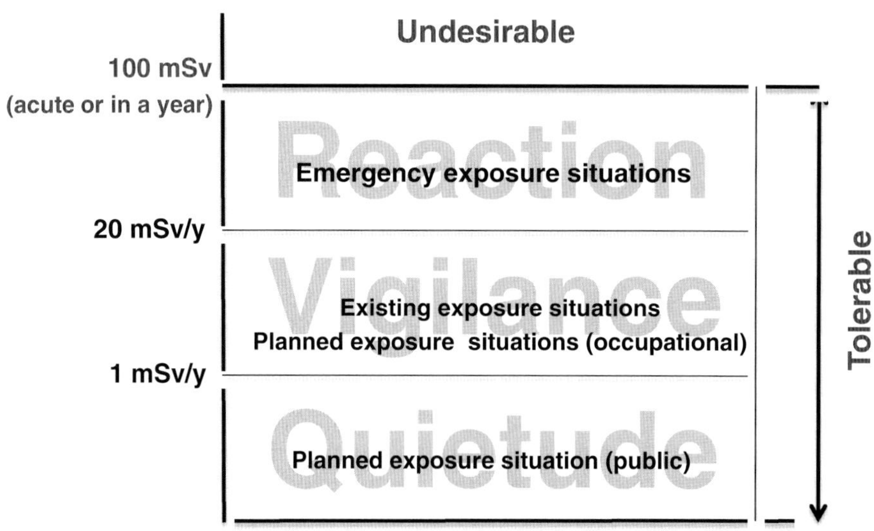

Fig. 1. Adaptation of the tolerability of risk model to the International Commission on Radiological Protection framework for the choice of source-related reference levels according to type of exposure situation.

Agency of the Organisation for Economic Co-operation and Development (OECD) requesting follow-on collaboration to address other forms of land-based disposal of radioactive waste. Committee 4 decided to create a working party to develop the basis for a task group on this subject to be established at their next annual meeting in 2013. The future publication should provide a conceptual framework for the long-term control of exposure over the life cycle of such type of disposal, which could be used as a reference for the waste community when explaining its approach to the public and other stakeholders. Optimisation of protection should be the driving principle, and the publication should also cover the protection of the environment through cooperation with Committee 5. Finally, the work should be developed in close cooperation with the waste community.

Committee 4 has also established a working party to address the issue of contaminated sites resulting from past industrial, military, and nuclear activities. Although presenting some similarities with exposure situations resulting from long-term contaminated areas after a nuclear accident or a radiation emergency, contaminated sites from past activities have specific characteristics about the type of radionuclides, frequent presence of chemical contaminants, conjunction with planned exposure situations, and the fact that, in most cases, there is a historical relationship between the affected population and the sites. Past experience in countries that have addressed the legacy of former nuclear activities has shown that the complex and multi-form dimension of such exposure situations potentially affecting health, environment, economy, and the well-being of people calls for a protection strategy, which, to be successful, must involve local stakeholders. A delicate situation to be addressed by the working party is the one in which communities discover that they live on contaminated sites due to former activities of which they were not previously aware. It is expected that the terms of reference for creation of a new task group of Committee 4 will be ready in time for adoption at the next annual meeting.

Finally, Committee 4 recently discussed the opportunity to initiate a series of 'end-user' oriented publications. In the past, implementation documents have generally focused on broad areas covering multiple exposure situations [practices and interventions in the time prior to *Publication 103* (ICRP, 2007a)]. Committee 4 has already moved towards documents that focus on one particular exposure situation with the series of publications covering key existing exposure situations. Further, comments received during the preparation of the report on security screening (ICRP, 2014b) showed that there is strong interest in a more focused discussion on how the ICRP framework and principles apply in a specific situation that is not covered routinely. Such documents would not attempt to present or endorse specific regulatory approaches, but would elaborate on the principles and framework of protection, which could then be used by organisations in the application of the recommendations. The publications would explain how the principles and recommendations apply in the field or area, and facilitate the connection faced by users between the regulations with which they must comply and the fundamental radiological protection framework and recommendations.

During the Main Commission's meeting with the Special Liaison Organisations (ICRP, 2014e) in Abu Dhabi, the representative of the European ALARA Network expressed the wish for ICRP to develop specific advice on industrial radiography. This is an example of a topic for which Committee 4 might consider preparation of a document. In any case, the reflection on 'end-users' publications continues within Committee 4, and a decision will be taken in consultation with the Main Commission in the near future.

3. CONCLUDING REMARKS

The ICRP system of radiological protection has gradually developed in the 20th Century by incorporating advances in knowledge about the effects of radiation, the evolution of ethical and social values, and feedback experiences from its practical implementation. Until World War II, the Commission only dealt with the protection of medical staff. After the War, the focus was on nuclear energy and radiological protection developed to protect workers inside nuclear installations and the public outside. This resulted in a coherent and effective regime of protection based on solid concepts, principles, and criteria widely shared internationally (ICRP, 1991).

The Chernobyl nuclear accident, followed by the raising concerns about exposure situations inherited from the 1990s, and the threat of 'malevolent events' following the September 11 attacks profoundly questioned the recommendations made in *Publication 60* (ICRP, 1991). Although not explicit, this questioning played an important role in the development of the 2007 Recommendations (ICRP, 2007a). The introduction of three types of exposure situations embracing all controllable exposure situations, the generalisation of the optimisation principle in connection with individual dose restrictions to all these situations, the recommendation to involve stakeholders in the optimisation process, and the widening of the system to protect flora and fauna represent significant evolutions of the new ICRP radiological protection system.

The main task of Committee 4 over recent years has been to assimilate and integrate these evolutions in a series of publications devoted to different exposure situations by focusing, as a priority, on existing and emergency exposure situations for which the regime of protection applied to installations had shown its limits. The Fukushima accident unfortunately occurred before the professional community and all the stakeholders of radiological protection had time to assimilate the new approach to these situations. The confusion that prevailed regarding the use of reference levels to manage the emergency phase and post-accident phases is, from this point of view, emblematic. One of the challenges for the future is undoubtedly to explain the ICRP system of radiological protection by highlighting the potential co-existence of dose limits and dose constraints to manage planned exposure situations with reference levels to manage existing exposure situations (cosmic radiation, radon, contaminated sites, etc.), or even emergency exposure situations in the case of an accident.

This task will continue in the coming years in parallel with the aim of clarifying the ethical foundations of the system, and the social values that shape the tolerability of radiation risk. Despite all the efforts made in the past decades, it must be recognised that, apart from scientists, experts, and radiation protection professionals, citizens are rarely informed about radiation, and even less about the radiological protection system. 'Risk communication' has globally failed to reduce the gap between experienced professionals and uninformed people, and the relationship of our contemporaries with radioactivity remains largely dominated by the spectre of Hiroshima and Nagasaki, and uncertainty about the effects of low doses fed for decades by the ongoing scientific and social controversy on the effects of radiation. The experience of stakeholder engagement in the last two decades shows that, by developing a narrative about the ethical and social values incorporated in the system, it will be possible to improve understanding and gradually deploy a constructive dialogue on radiation risk and its perception with all concerned parties.

REFERENCES

FMU, 2011. 1st International Expert Symposium in Fukushima, 11–12 September 2011, Fukushima, Japan. Available at: http://www.fmu.ac.jp/radiationhealth/1st_International_Expert_Symposium/index.html (accessed 24 March 2015).

FMU, 2013. International Academic Conference on Radiation Health Risk Management in Fukushima, 25–27 February 2013, Fukushima, Japan. Available at: http://www.fmu.ac.jp/radiationhealth/conference/index.html (accessed 24 March 2015).

ICRP, 1977. Recommendations of the ICRP. ICRP Publication 26. Ann. ICRP 1(3).

ICRP, 1991. 1990 Recommendations of the International Commission on Radiological Protection. ICRP Publication 60. Ann. ICRP 21(1–3).

ICRP, 2007a. The 2007 Recommendations of the International Commission on Radiological Protection. ICRP Publication 103. Ann. ICRP 37(2–4).

ICRP, 2007b. Radiological protection in medicine. ICRP Publication 105. Ann. ICRP 37(6).

ICRP, 2009a. Application of the Commission's recommendations for the protection of people in emergency exposure situations. ICRP Publication 109. Ann. ICRP 39(1).

ICRP, 2009b. Application of the Commission's recommendations to the protection of people living in long-term contaminated areas after a nuclear accident or a radiation emergency. ICRP Publication 111. Ann. ICRP 39(3).

ICRP, 2011. ICRP Strategic Plan 2011–2017. International Commission on Radiological Protection, Ottawa. Available at: http://new.icrp.org/docs/ICRP%20Strategic%20Plan%202011-2017.pdf (accessed 24 March 2015).

ICRP, 2012. Report of ICRP Task Group 84 on Initial Lessons Learned from the Nuclear Power Plant Accident in Japan vis-a-vis the ICRP System of Radiological Protection. International Commission on Radiological Protection, Ottawa. Available at: http://new.icrp.org/docs/ICRP%20TG84%20Summary%20Report.pdf (accessed 24 March 2015).

ICRP, 2013. Radiological protection in geological disposal of long-lived solid radioactive waste. ICRP Publication 122. Ann. ICRP 42(3).

ICRP, 2014a. Protection of the environment under different exposure situations. ICRP Publication 124. Ann. ICRP 43(1).

ICRP, 2014b. Radiological protection in security screening. ICRP Publication 125. Ann. ICRP 43(2).

ICRP, 2014c. Radiological protection against radon exposure. ICRP Publication 126. Ann. ICRP 43(3).

ICRP, 2014d. ICRP and Fukushima. International Commission on Radiological Protection, Ottawa. Available at: http://new.icrp.org/page.asp?id = 188 (accessed 24 March 2015).

ICRP, 2014e. Formal Relations with other Organizations. International Commission on Radiological Protection, Ottawa. Available at: http://new.icrp.org/icrp_group.asp?id = 80 (accessed 24 March 2015).

ICRP, 2015. First Asian Workshop on the Ethical Dimensions of the System of Radiological Protection, 27–28 August 2013, Daejeon, Korea. Available at: http://www.icrp.org/page.asp?id = 237 (accessed 1 April 2015).

Lindell, B., 2001. Logic and ethics in radiation protection. J. Radiol. Prot. 21, 377–380.

Lochard, J., 2012. Application of the Commission's recommendations: the activities of ICRP Committee 4. Ann. ICRP. 41(3/4), 32–44.

Lochard, J., 2013. Stakeholder engagement in regaining decent living conditions after Chernobyl. In: Oughton D., Hansson S.O. (Eds.), Social and Ethical Aspects of Radiation Risk Management. Elsevier Science, pp. 311–332.

Silini, G., 1992. Ethical Issues in Radiation Protection. Associazione Italiana di Protezione contro le Radiazioni Sievert Lecture, September 1992, Bologna, Italy. Available at: http://www.iaea.org/inis/collection/NCLCollectionStore/_Public/36/047/36047968.pdf.

Taylor, L., 1957. The philosophy underlying radiation protection. AJR. Am. J. Roent. 77, 914–919.

SAGE

Annals of the ICRP

Overview of ICRP Committee 5

C-M. Larsson[a], K.A. Higley[b], A. Real[c]

[a]*Australian Radiation Protection and Nuclear Safety Agency, PO Box 655, Miranda, NSW 1490, Australia; e-mail: carl-magnus.larsson@arpansa.gov.au*

[b]*Department of Nuclear Engineering and Radiation Health Physics, Oregon State University, USA*

[c]*CIEMAT, Radiation Protection of the Public and the Environment, Spain*

Abstract–The International Commission on Radiological Protection (ICRP) established Committee 5 in 2005 in response to the need to provide direct demonstration of environmental protection from radiation in accordance with national law and international agreements. The development of the ICRP system for environmental protection was facilitated by research over the previous decades, as well as by ICRP's evaluation of the ethical and philosophical basis for environmental protection as laid out in ICRP *Publication 91*. The 2007 Recommendations (*Publication 103*) incorporated environmental protection as one of the integral elements of the radiation protection system. Over a relatively short time, the system has evolved to incorporate a set of 12 Reference Animals and Plants (RAPs), which is a small enough number to develop comprehensive databases for each RAP, but wide ranging enough to provide some insight into radiation impact and protection against such impact, as appropriate, in terrestrial, freshwater, and marine ecosystems. As necessary, the databases can be used to derive supplementary databases for Representative Organisms typical for a particular exposure situation of concern or under study. The system, to date, details biology of the RAPs (*Publication 108*); outlines transfer factors for estimation of internal concentrations of radionuclides of environmental significance under different situations (*Publication 114*); provides further information (*Publication 108*) on dosimetry, biological effects, and derived consideration reference levels (bands of environmental dose rates where potential detrimental effects may deserve attention); and provides information on application of the system in planned, emergency, and existing exposure situations (*Publication 124*). Currently, a review of experimental determinations of relative biological effectiveness, to guide derivation of specific weighting factors for use in environmental radiation protection if possible and necessary, is being concluded, as is work on improved dosimetry. Further work in this area involves consolidation of databases, recommendations for derivation

This paper does not necessarily reflect the views of the International Commission on Radiological Protection.

of specific databases for Representative Organisms on the basis of the RAP data, and recommendations for application of the system to environmental protection in relation to certain human activities of potential environmental concern. Consideration needs to be made for the wider range of ecosystem effects that may be covered in ecological risk assessments, which incorporate the complete suite of stressors that result from human activity, and their effects, to understand the role of radiation effects in this context.

Keywords: Environmental protection; Ionising radiation; Reference Animals and Plants; Radiation effects; Dosimetry

1. INTRODUCTION

This paper sets out to describe the genesis of the International Commission on Radiological Protection's (ICRP) approach to environmental protection, the way the system has evolved, and how it can be applied under different exposure situations. This paper also outlines planned activities to further consolidate the system, and improve its applicability to decisions on projects of environmental significance. It serves as an introduction to the system; the system is explained in detail in recent ICRP publications, including *Publication 108* (ICRP, 2008), *Publication 114* (ICRP, 2009), and *Publication 124* (ICRP, 2014).

A number of specific elements of the system were covered in the Proceedings of the First ICRP Symposium on the International System of Radiological Protection (Copplestone, 2012; Higley et al., 2012; Larsson, 2012; Pentreath, 2012a; Ulanovsky and Pröhl, 2012).

An outline of the work of Committee 5 was presented at the First ICRP Symposium by the then Chair of Committee 5, Professor Jan Pentreath (2012b). The aim of the present paper is to provide an update relative to the situation at the time of the First Symposium and to introduce the work of ICRP Committee 5 and the system of environmental radiological protection to all readers previously unfamiliar with this topic and system.

2. GENESIS

Environmental protection only became an integral part of the ICRP system for radiological protection with the 2007 Recommendations on radiological protection, outlined in *Publication 103* (ICRP, 2007). The 2007 Recommendations superseded the 1990 Recommendations (ICRP, 1991), in which the environment was primarily considered as the medium through which radiation exposure was delivered to humans. However, the inclusion of protection of the environment in its own right in the 2007 Recommendations did not just happen; it was preceded by significant international scientific developments, as well as the gradual recognition by ICRP of the need to demonstrate compliance with environmental standards and criteria as expressed in national legislation and in international 'instruments', whether binding or incentive. The International Atomic Energy Agency (IAEA, 2002) summarised the obligations for protection of the environment as laid out in international instruments as follows:

Activities within a jurisdiction should

- not cause damage to the environment of other states;
- maintain ecosystems and processes that are essential for the functioning of the biosphere;
- maintain biodiversity; and
- observe the 'principle' of optimum sustainable yield in the use of living natural resources.

A full review of the developments referred to above cannot be made within the constraints of these Proceedings. These involve a series of conferences held in Stockholm, Sweden in 1996; Ottawa, Canada in 1999; and Darwin, Australia in 2002; culminating in the IAEA Conference on Protection of the Environment from the Effects of Ionizing Radiation in Stockholm, Sweden in 2003 (IAEA, 2005a). Of significance is the following quote from the President's report on findings and recommendation from the Stockholm conference (IAEA, 2005a):

> *While accepting that there remain significant gaps in knowledge and that there needs to be continuing research, the conference accepted that there was an adequate knowledge base to proceed and strongly supported the development of a framework for environmental radiation protection. The ICRP and the IAEA have important roles to play, and it is vital that the framework be developed in a consultative and inclusive manner. The conference supported the approach based on the development of reference animals and plants, and it noted that these may also serve as a basis for site specific assessments.*

On the basis of the findings presented at the Stockholm conference, IAEA decided on the 'Plan of activities on the radiation protection of the environment' (IAEA, 2005b), and set up a co-ordination group to promote the plan and monitor progress; this group is still in operation today.

ICRP responded to these international developments by commissioning a task group, chaired by the then ICRP Chair-Elect Dr Lars-Erik Holm, to develop a protection policy and suggest a framework for environmental protection based on scientific and ethical-philosophical principles. This resulted in *Publication 91* (ICRP, 2003), which can be regarded as the 'genesis' of the ICRP system for environmental protection as well as for Committee 5 on environmental radiological protection, which commenced its work in 2005. This is further reviewed below.

3. *PUBLICATION 91* AND THE INCEPTION OF COMMITTEE 5

Publication 91 (ICRP, 2003) can be considered the founding publication for the work of ICRP in environmental protection, for the ICRP system of environmental protection, and for Committee 5 of ICRP, although this would not have been possible if the developments briefly reviewed above had not been ongoing and delivered new insight into this area.

The fundamental issues stemming from *Publication 91* were summarised in the Guest Editorial of *Publication 114* (ICRP, 2009) and by Larsson (2012) and Pentreath (2012b). *Publication 91* explores the applicability of different approaches developed for environmental protection in general, and to environmental radiation protection in particular, while acknowledging that any approach to protection of the environment from radiation needs to be harmonised with the system for human radiation protection. ICRP already foreshadowed in the report that:

- a possible future ICRP system addressing environmental assessment and protection would focus on biota, not on the abiotic component of the environment or on environmental media (soil, air, water, sediment);
- the system should be effect-based so that any reasoning about adequate protection would be derived from firm understanding of harm at different exposure levels; and
- the system should be based on data sets for reference fauna and flora (subsequently termed 'Reference Animals and Plants' (RAPs). The definition of a RAP as subsequently developed by ICRP in *Publication 108* (ICRP, 2008) is:

> *a hypothetical entity, with the assumed basic biological characteristics of a particular type of animal or plant, as described to the generality of the taxonomic level of Family, with defined anatomical, physiological and life history properties, that can be used for the purposes of relating exposure to dose, and dose to effects, for that type of living organism.*

The RAPs approach would be analogous to the use of the Reference Person concept in human radiation protection, and guide the assessment of effects and the derivation of dose rate benchmarks to guide protective actions.

In 2005, ICRP established Committee 5 to take charge of the Commission's work on environmental protection, with the view of reviewing progress in this area with time. Committee 5's task is summarised as follows:

> *Committee 5 is concerned with radiological protection of the environment. It will aim to ensure that the development and application of approaches to environmental protection are compatible with those for radiological protection of man, and with those for protection of the environment from other potential hazards.*

Committee 5 is now entering into its third term. The challenges and outputs are briefly reviewed below.

4. THE EVOLUTION OF A PARALLEL PATHWAY

4.1. Basic elements of the parallel pathway

Committee 5 approached the issue of delivering in accordance with its task by developing a pathway for environmentally informed decision-making that is complementary to, analogous to, and parallel to the pathway of reaching a decision based on the aim to protect humans. This 'parallelism' is illustrated in Fig. 1.

Core to the parallel pathway is the creation of two points of reference. First is the creation of a 'reference entity', analogous to the Reference Person, which serves as a basis for developing the necessary reference data sets that are required in order to either assess environmental consequences or – by back tracking from levels of concern – to define environmental circumstances that may be considered 'safe', in the broad sense of the word, to the environment. These 'entities' are the RAPs. The selection of such RAPs has been reviewed previously (Pentreath, 2012b).

The other reference point in this approach is the definition of derived consideration reference levels (DCRLs), analogous to reference levels that are used in the radiological protection of humans. DCRLs are bands of environmental dose rate, spanning one order of magnitude, that are specific for each RAP and can be defined as (ICRP, 2008):

> *. . . a band of dose rate within which there is likely to be some chance of deleterious effect of ionising radiation occurring to individuals of that type of Reference Animal or Plant, derived from knowledge of defined biological effects for that type of organisms that, when considered together with other relevant information, can be used as a point of reference to optimise the level of effort expended on environmental protection, dependent on the overall management objectives and the exposure situation.*

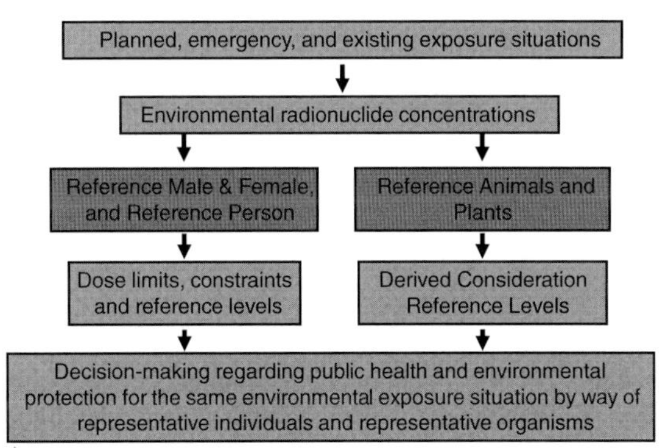

Fig. 1. The parallel pathways of reaching a 'holistic' decision on management of radiation risks; the risks are to human health, to the environment, or both. For further explanations, see the text. Based on *Publication 108* (ICRP, 2008).

Table 1. Broadly defined wildlife groups, general type of environment and derived consideration reference levels (DCRLs) (shaded) for the 12 Reference Animals and Plants (RAPs) currently included in the ICRP system for environmental protection.

Wildlife group	Ecosystem	RAP	DCRL, mGy day^{-1} (shaded)		
			0.1–1	1–10	10–100
Large terrestrial mammals	T	Deer	(shaded)		
Small terrestrial mammals	T	Rat	(shaded)		
Aquatic birds	F, M	Duck	(shaded)		
Large terrestrial plants	T	Pine Tree	(shaded)		
Amphibians	F, T	Frog		(shaded)	
Pelagic fish	F, M	Trout		(shaded)	
Benthic fish	F, M	Flatfish		(shaded)	
Small terrestrial plant	T	Grass		(shaded)	
Seaweeds	M	Brown Seaweed		(shaded)	
Terrestrial insects	T	Bee			(shaded)
Crustacean	F, M	Crab			(shaded)
Terrestrial annelids	T	Earthworm			(shaded)

T, terrestrial environment; F, freshwater environment; M, marine environment.

The DCRLs for the different RAPs are outlined in Table 1, based on *Publication 108* (ICRP, 2008). Note that the DCRL for Reference Brown Seaweed was erroneously quoted as 10–100 mGy day^{-1} in *Publication 108* (ICRP, 2008).

4.2. Transfer, dosimetry, biology, and effects

While Fig. 1 outlines the components of the system for protection of the environment, it does not describe the type of information and data sets necessary to make it work, i.e. how to go from the exposure situation to the (real or postulated) effect – or, from the protection aim and ambition back to an exposure that is commensurate with the aim and ambition. The types of information and data that are necessary can be summarised as follows:

- Biology and ecology; this forms the basis for the selection of the RAPs, the number of which is as low as possible but as large as necessary to provide a reasonable coverage of exposures typical for organisms and environments within terrestrial, freshwater, and marine ecosystems. The biology of the RAPs has been reviewed in *Publication 108* (ICRP, 2008).
- Transfer; this primarily concerns the concentration factors denoting the internal concentration of radionuclides in relation to that of the surrounding medium. *Publication 114* (ICRP, 2009) lists such ratios for the 12 RAPs and for radionuclides of 40 elements based on actual measurements or otherwise inferred.

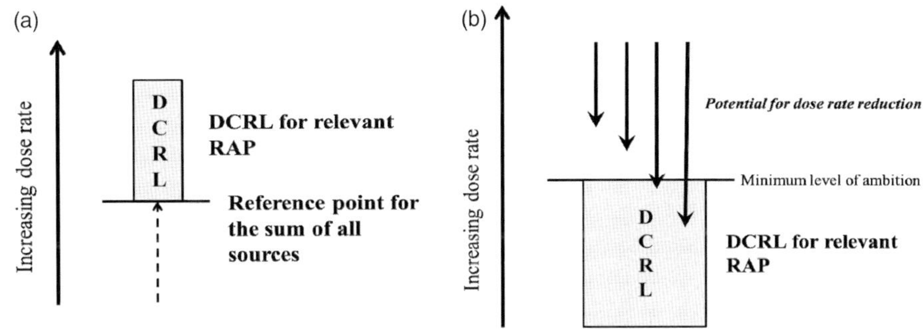

Fig. 2. Application of Reference Animal and Plant (RAP)-specific derived consideration reference levels (DCRLs) to the management of environmental exposures in (a) planned and (b) existing exposure situations. Based on *Publication 124* (ICRP, 2014).

- Dosimetry; dose conversion coefficients for internal and external exposure for geometries and sizes defined for the 12 RAPs are listed in *Publication 108* (ICRP, 2008).
- Effects; effects data, essentially grouped in the umbrella end-point categories of mortality, morbidity, reduced reproductive success, and subtle genetic effects, have been used to derive the DCRLs. The DCRLs and the rationale for their derivation are outlined in *Publication 108* (ICRP, 2008).

4.3. Application in planned, emergency, and existing exposure situations

The application of the ICRP system for environmental protection in planned, emergency, and existing exposure situations has been outlined by Pentreath (2012a) and is elaborated further in *Publication 124* (ICRP, 2013). In *Publication 124*, ICRP recommends the use of DCRLs under all circumstances where there is, or may be, an environmental exposure of significance above the natural background locally experienced by the relevant biota. In planned exposure situations, the lower boundary of the relevant DCRL band should be used as the appropriate reference point for protection. Application of the system under planned exposure situations requires some judgement regarding the size of the area under consideration, and the potential presence of other sources other than the one(s) under consideration in the planning stage. In the case of multiple sources of exposure – for example, from historical discharges or multiple sites – these other sources should be taken into account in comparison with the DCRLs when assessing protection options. If the assessed dose rates are below the appropriate reference point, the level of control is determined by selection of the most reasonable protective action.

For existing exposure situations, it may be difficult, or impractical, to significantly reduce the concentrations or quantities of radioactive material that exist in the affected environment. If dose rates are within the DCRL band, consideration should be given to reduce exposures, assuming that the costs and benefits are such

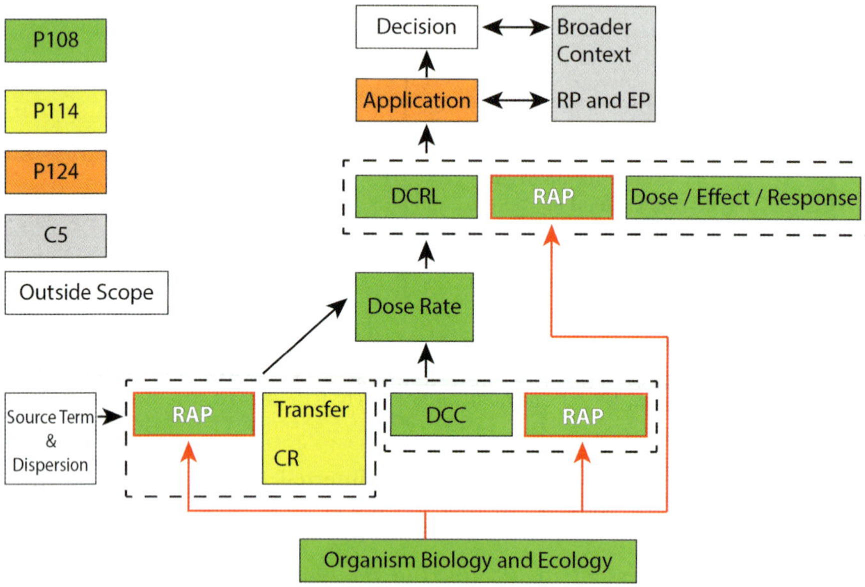

Fig. 3. Schematic representation of assessment of environmental impact using the ICRP system, and where relevant information can be sourced in the suite of ICRP publications. P108, P114 and P124 refer to the ICRP Publications with the same numbers; C5, ICRP Committee 5; DCC, dose conversion coefficient; CR, concentration ratio; DCRL, derived consideration reference level; EP, environmental protection; RAP, reference animals and plants; RP, radiation protection.

that further efforts are warranted. The DCRLs may be used as the criteria for mitigating environmental exposures, just as reference levels are used for mitigating individual exposures for human protection in such situations.

In emergency exposure situations, the levels of exposure may be orders of magnitude greater than the DCRLs. The temporal and spatial patterns of exposure need to be given attention in order to understand the likelihood of effects over the short and long term. The DCRLs may still be considered useful reference points for such assessments, as elaborated further in *Publication 124* (ICRP, 2014).

4.4. Information sources for the system of environmental protection

Fig. 3 outlines how the different components of the ICRP system for environmental protection relate to each other and how they are covered in the ICRP publications. The scheme outlines how source term or dispersion data (observed or modelled) can be progressed through the different steps in the assessment process in order to estimate the likelihood and/or magnitude of environmental effects, which will underpin the appropriate decisions on environmental management. The system can also be used in a reverse manner, so that, on the basis of a desired environmental protection outcome, the appropriate constraints or limits can be set on the discharges into the environment.

5. FUTURE CHALLENGES

During the third term, Committee 5 will work to consolidate the system for radiation protection of the environment by broadening and consolidating the scientific basis that underpins the system, and by drawing on the experience gained in its application in the three exposure situations considered by ICRP.

5.1. Consolidation of the system

Since *Publications 108* and *114*, new data on transfer factors, dosimetry, and radiation-induced effects on several RAPs have become available. Thus, one of the tasks for the next term will be to compile and analyse the new data available, and to update the existing databases. The consolidation of the databases will facilitate the attempt to generalise across different types of animals and plants, i.e. moving from the RAPs to the Representative Organisms (see below) in specific exposure scenarios. The updated databases will also permit evaluation of the uncertainties associated with transfer factors, dosimetry, and radiation-induced effects, and their relative importance in assessments and setting protection goals.

Use of the new dosimetric information and methods will contribute to shedding further light on some of the gaps in current knowledge, including the lack of information on: (1) the concentration of relevant radionuclides (alpha and low-energy electron emitters) in the tissues or organs of interest of the RAPs; (2) the precise geometry of important organs, such as the gonads, for animals of body mass of approximately 1 kg or more [Reference Deer (adult), Reference Duck, Reference Trout, Reference Flatfish, and Reference Crab] exposed to photons; and (3) the dosimetry of terrestrial plants, particularly for Reference Pine Tree. The voxel phantoms developed for several RAPS is an example of the new methodologies that can contribute to fill some of the gaps in dosimetric knowledge.

Regarding the transfer factors, as stated in *Publication 114* (ICRP, 2009), there is a need to further broaden the transfer database for some RAPs and their life stages, as well as for transfer values for a number of RAP/element combinations.

With regard to the radiation-induced effects in animals and plants, all the information available needs to be analysed as a whole in order to develop methodologies that will help to solve the complex issue of extrapolating from: (1) high acute doses/dose rates of low-linear energy transfer radiations to lower doses/dose rates; (2) one organism to another; (3) effects in the individual organism to possible impacts at population and community levels; and (4) laboratory conditions to field conditions.

5.2. Application to real organisms in real environments

The RAPs, and their associated databases, can be used as the default in a variety of situations where an assessment of the significance of environmental exposures is needed to support decisions on environmental management. However, as pointed out by Pentreath (2012b) and elaborated further in *Publication 124* (ICRP, 2014), in situations where there are reasons to expect that a scenario of environmental significance will or can evolve, the ambition should be to make the assessments for real organisms in real environments. A 'real' organism may well be one of the RAPs, and existing databases can be used without further consideration. In other circumstances, the Representative Organism may not be well represented by one of the RAPs, and the differences will need to be assessed. In this case, ICRP has recognised the need to define Representative Organisms, typical for a particular exposure scenario in a particular ecological context.

Due to the vast variety of potential Representative Organisms, there may be considerable differences between the chosen or necessary Representative Organisms and the set of 12 RAPs. If the set of RAPs does not include all or any of the animal or plant types requiring protection, there will be differences in terms of:

- biology, such as life span or life cycle;
- dosimetry, because of size, shape or location; and
- response to radiation at similar dose rates or total dose.

Nevertheless, the reference databases for the RAPs illustrate the types of data that need to be compiled to understand the complexity of an exposure scenario, and suggest the extent to which existing data can be extrapolated to the Representative Organisms. Committee 5 intends to focus on the methodology to identify Representative Organisms and how to link them to the RAP databases in this third term of its deliberations, in order to provide practical recommendations on how to deal with real scenarios of environmental significance and concern.

6. CONCLUDING REMARKS

Notwithstanding the continuing effort to broaden and consolidate the databases underpinning the ICRP system for environmental protection (an effort that is shared with the system for protection of humans), the system is robust and providing valuable guidance to understanding, and protection from, environmental consequences of radiation (Copplestone, 2012; Telleria et al., 2015). This substantiates the statement made during the IAEA Stockholm conference mentioned above (IAEA, 2005a). While under most circumstances, it is clear that the environmental consequences are most likely to be very small (again, a shared feature with radiation protection of humans), the experiences from the aftermath of the Great East-Japan earthquake and tsunami, with acute releases of large amounts of radioactivity and ongoing releases years after the event, illustrate the need to consider environmental exposures both during the planning stage and in the event of an emergency and its legacies. While this was an extreme and rare event, the ICRP system for environmental protection adds a dimension to the decision-making process that allows the decisions to be made on the basis of a holistic understanding of consequences for health and the environment. It is the ambition of Committee 5 to improve the support for such well-founded decisions through its work during the third term.

REFERENCES

Copplestone, D., 2012. Application of radiological protection measures to meet different environmental protection criteria. Ann. ICRP 41(3/4), 263–274.

Higley, K.A., Kocher, D.C., Real, A.G., et al., 2012. Relative biological effectiveness and radiation weighting factors in the context of animals and plants. Ann. ICRP 41(3/4), 233–245.

IAEA, 2002. Ethical Considerations in Protecting the Environment from the Effects of Ionizing Radiation. IAEA-TECDOC-1270. International Atomic Energy Agency, Vienna.

IAEA, 2005a. Protection of the Environment from the Effects of Ionizing Radiation. Proceedings of an International Conference, 6–10 October 2003, Stockholm, Sweden.

IAEA, 2005b. Plan of Activities on the Radiation Protection of the Environment. International Atomic Energy Agency, Vienna. Available at: http://gnssn.iaea.org/RTWS/cgrpe/Shared%20Documents/Other%20Documents/Plan%20of%20Activities%20on%20RPE%20(2005).pdf (last accessed 2 March 2015).

ICRP, 1991. 1990 Recommendations of the International Commission on Radiological Protection. ICRP Publication 60. Ann. ICRP 21(1–3).

ICRP, 2003. A framework for assessing the impact of ionising radiation on non-human species. ICRP Publication 91. Ann. ICRP 33(3).

ICRP, 2007. The 2007 Recommendations of the International Commission on Radiological Protection. ICRP Publication 103. Ann. ICRP 37(2–4).

ICRP, 2008. Environmental protection: the concept and use of reference animals and plants. ICRP Publication 108. Ann. ICRP 38(4–6).

ICRP, 2009. Environmental protection: transfer parameters for reference animals and plants. ICRP Publication 114. Ann. ICRP 39(6).

ICRP, 2014. Protection of the environment under different exposure situations. ICRP Publication 124. Ann. ICRP 43(1).

Larsson, C.-M., 2012. Biological basis for protection of the environment. Ann. ICRP 41(3/4), 208–217.

Pentreath, J., 2012a. Clarifying and simplifying the management of environmental exposures under different exposure situations. Ann. ICRP 41(3/4), 246–255.

Pentreath, J., 2012b. Radiation and protection of the environment: the work of Committee 5. Ann. ICRP 41(3/4), 45–56.

Telleria, D., Cabianca, T., Pröhl, G., et al., 2015. Use of the ICRP system for the protection of marine ecosystems. Ann. ICRP 44.

Ulanovsky, A., Pröhl, G., 2012. Dosimetry for reference animals and plants: current state and prospects. Ann. ICRP 41(3/4), 218–232.

SAGE

ICRP
Annals of the ICRP

Radiobiology of tissue reactions

W. Dörr

Department of Radiation Oncology and Christian Doppler Laboratory for Medical Radiation Research for Radiooncology, Comprehensive Cancer Centre, Medical University Vienna/ Vienna General Hospital, Waehringer Guertel 18-20, A-1090 Vienna, Austria; e-mail: wolfgang.doerr@meduniwien.ac.at

Abstract–Tissue effects of radiation exposure are observed in virtually all normal tissues, with interactions when several organs are involved. Early reactions occur in turnover tissues, where proliferative impairment results in hypoplasia; late reactions, based on combined parenchymal, vascular, and connective tissue changes, result in loss of function within the exposed volume; consequential late effects develop through interactions between early and late effects in the same organ; and very late effects are dominated by vascular sequelae. Invariably, involvement of the immune system is observed. Importantly, latent times of late effects are inversely dependent on the biologically equieffective dose. Each tissue component and – importantly – each individual symptom/endpoint displays a specific dose–effect relationship. Equieffective doses are modulated by exposure conditions: in particular, dose-rate reduction – down to chronic levels – and dose fractionation impact on late responding tissues, while overall exposure time predominantly affects early (and consequential late) reactions. Consequences of partial organ exposure are related to tissue architecture. In 'tubular' organs (gastrointestinal tract, but also vasculature), punctual exposure affects function in downstream compartments. In 'parallel' organs, such as liver or lungs, only exposure of a significant (organ-dependent) fraction of the total volume results in clinical consequences. Forthcoming studies must address biomarkers of the individual risk for tissue reactions, and strategies to prevent/mitigate tissue effects after exposure.

Keywords: Tissue reactions; Radiopathology; Dose fractionation; Dose rate; Volume effect; Repopulation; Biomarker; Intervention

This paper does not necessarily reflect the views of the International Commission on Radiological Protection.

1. INTRODUCTION

Tissue effects of ionising radiation can be observed in virtually all tissues. This article focuses on 'deterministic' tissue responses, and does not consider radiation carcinogenesis or genetic effects after germ cell exposure. One prominent attribute of 'deterministic' effects is that they only occur if a certain, albeit – for some symptoms – small, threshold dose is reached. With regard to their time course, early effects are seen within the first weeks after an acute radiation exposure, while chronic, late radiation sequelae manifest after months to many years. In some instances, the extent, i.e. severity and/or duration, of the early effects has an impact on the risk for the manifestation of late symptoms in the same tissue or organ (consequential late effects). A number of factors have been identified with regard to the exposure characteristics that significantly influence the radiation tolerance of normal tissues, summarised as the 'Rs' of radiation exposure and, particularly, radiotherapy. These include:

- intrinsic *R*adiosensitivity (Steel, 1993);
- *R*ecovery (Withers, 1975);
- *R*epopulation (Withers, 1975);
- *R*edistribution (Withers, 1975);
- *R*eoxygenation (Withers, 1975);
- i*R*radiated volume (Dörr and Van der Kogel, 2009);
- long-term *R*estoration (Dörr and Stewart, 2009); and
- molecular *R*adiopathology (Dörr, 2009c).

These parameters, besides the general pathological principles of tissue effects, will be characterised briefly in this section, as far as they apply to normal tissue effects of irradiation. Redistribution or cell cycle effects in general, based on the varying cellular radiosensitivity over the different cell cycle phases, is a phenomenon that is mainly observed in in-vitro systems. However, the relevance of this factor for tissue effects in vivo, with a variety of factors regulating cell cycle progression, must be doubted. Furthermore, reoxygenation, describing the improvement of the oxygen status of tumours during fractionated radiotherapy, is irrelevant with regard to normal tissue effects. In normal tissues, only a small and constant fraction, if any, of the radiobiologically relevant (tissue-specific target/stem) cells is subject to local hypoxia. Long-term restoration, impacting on tissue tolerance at long intervals of months to years after initial exposure, is also not the subject of the present considerations, but has been described in detail elsewhere (Dörr and Stewart, 2009).

It needs to be emphasised that these radiobiological factors that determine the biological effect of a certain radiation exposure scenario are specific for individual symptoms or endpoints of normal tissue reactions, rather than for the organs as anatomical entities. As an important consequence, these radiobiological factors may impact on different symptoms related to one organ in different ways.

2. PATHOBIOLOGICAL PRINCIPLES OF TISSUE EFFECTS

Based on their time course, early reactions in tissues are defined as those observed within weeks after the (onset of) radiation exposure, e.g. within the first 90 days after the start of radiotherapy. Any symptom that is first diagnosed at a later time is considered a late response, with typical latent times of months to many years and sometimes decades for very late effects, e.g. in the urinary bladder, or even more pronounced in the cardiovascular system. In some organs and tissues, consequential late effects develop based on the clinical manifestation, i.e. severity and/or duration of early effects in the same tissue.

It must be emphasised that tissue effects in many instances are not based on the consequences of radiation exposure of an individual organ, but on the interaction of radiation effects in physiologically 'connected' organs, such as the lung and the heart after thoracic exposure, or the systemic contribution of the immune system to individual organ effects.

2.1. Early radiation effects

Typical early radiation effects are found in turnover tissues (Dörr, 2009a; Dörr and Schmidt, 2014), where physiologically permanent cell loss from the differentiated, post-mitotic compartments of the tissue is well balanced by proliferation in the germinal parts of the tissue. Typical examples are the haematopoietic system, with cell production in the bone marrow, or the epidermis and epithelia of the upper and lower gastrointestinal tract. Radiation exposure of such tissue systems impacts on the cell production rate, while the normal cell loss rate remains unchanged, at least over a wide range of doses. As a consequence of the proliferative impairment, the reduced cellular supply to the differentiated tissue layers results in progressive hypoplasia and, eventually, in complete cell depletion. Regeneration, based on surviving proliferative cells within the exposed volume, or migrating proliferative cells from unexposed tissue or even the circulation, is usually complete. The tissue 'parenchymal', hypoplastic-regenerative response is regularly accompanied, or even preceded, by local vasculoconnective (e.g. vasodilation, vascular leakage, erythema) and immunological reactions, such as macrophage activation (Dörr, 2009a; Dörr and Schmidt, 2014), which also represent the basis of related pain. The interaction between the latter with the parenchymal effects, at molecular level, is subject to current radiobiological investigations.

2.2. Late radiation effects

The pathogenesis of late (chronic) radiation effects is more complex and includes the organ parenchyma, the vasculature and connective tissue components, and, in many instances, a major contribution of the local immune system, mainly macrophages (Dörr et al., 2009b). The response of parenchymal cells is organ specific. Exposure of a certain fraction of these cells may cause cell loss and (slowly progressing) parenchymal hypoplasia, which can occur in slow-turnover parenchymal tissues such as airway epithelia or urothelium, according to the mechanisms described above. Cell loss may also cause recruitment of non-proliferating cells into the cell cycle, and consequent loss of these cells via mitotic death or through other cell death pathways.

Radiation exposure of mitotic fibroblasts triggers their early differentiation into post-mitotic fibrocytes. The consequence is a substantial increase in the synthesis and deposition of collagens, as a basis for the development of tissue fibrosis (Rodemann and Bamberg, 1995; Yarnold and Brotons, 2010).

In the vasculature, radiation exposure results in changes in, or loss of, endothelial cell function, and thus 'leaky' vessels, which progresses into thrombi formation and vascular occlusion, loss of capillaries, or pathological dilation of small vessels, depicted as telangiectasia (Fajardo et al., 2001; Dörr, 2009b). All these changes markedly impact on the blood supply within the exposed volume, as well as downstream tissue compartments, which further promotes the development of progressive loss of tissue function. For very late radiation effects, e.g. in the cardiovascular system, the vascular component appears to be the dominating pathophysiological factor.

Macrophages, present at the time of exposure or recruited into the exposed volume, contribute to the tissue changes by chronic production of reactive oxygen species, and synthesis and release of a number of signalling molecules, with transforming growth factor ß as a promoter of tissue fibroses being one of the most prominent (Hakenjos et al., 2000). They also significantly trigger the chronic inflammatory changes regularly found in association with late radiation sequelae (Rubin et al., 1995; Bentzen, 2006; Yarnold and Brotons, 2010).

2.3. Consequential late effects

Consequential late effects develop in situations where the early radiation responses are associated with breakdown and loss of a physiological protective barrier against mechanical or physical stress (Dörr and Hendry, 2001; Dörr, 2009b). This is, in particular, found for surface epithelia of the upper and lower digestive tract (oral cavity, oesophagus, small and large intestine, rectum) with a combination of mechanical and chemical influences, the epithelium of the urinary bladder (chemical stress), the epidermis at localisations with major mechanical wear and tear, and also in the lung. With regard to time course and clinical manifestation, consequential late effects are similar to generic chronic radiation reactions. Their radiobiological behaviour, however, corresponds to that of the early responding tissue component, e.g. with a less pronounced fractionation effect and a pronounced effect of the overall treatment time (see below).

3. INTRINSIC RADIOSENSITIVITY – TISSUE TOLERANCE

The radiobiological target/stem cell concept postulates that the radiation tolerance of any organ or tissue is defined by the number and intrinsic sensitivity of the tissue-specific target cells (Dörr and Schmidt, 2014; Dörr, 2009b). Specific target cell markers, however, are lacking for the vast majority of normal tissues; hence, this concept must be considered as hypothetical but valid. Detailed information on the tolerance of the major organs to radiation exposure was reviewed in Dörr (2009b), and Dörr and Schmidt (2014). Again, it needs to be emphasised that 'tolerance' applies to individual symptoms or endpoints within an organ, rather than an organ per se. In this context, the tolerance dose is defined as the maximum radiation dose that is associated with a minimum probability (usually of a few percent) of the specific endpoint under consideration (Dörr and Schmidt, 2014).

4. RECOVERY – FRACTIONATION/DOSE-RATE EFFECT

Recovery describes the observation that administration of a certain total radiation dose in separate fractions rather than a single acute exposure decreases the incidence of tissue effects. The same phenomenon is observed when a tissue is exposed to a given total dose at a significantly reduced dose rate, starting from below approximately 1 Gy min^{-1} (Van der Kogel, 2009) down to chronic levels. In turn, the total dose required to induce a certain tissue effect (i.e. the tolerance dose) increases if the exposure occurs in smaller dose fractions or at smaller dose rates. One underlying mechanism is the restoration of the cellular integrity of the target cells during the overall exposure time. This is frequently termed 'repair', which, however, indicates the involvement of DNA repair. Although it is undoubted that DNA repair has a significant role, other mechanisms contribute to the phenomenon of recovery in tissues, e.g. the continued, potentially altered metabolic activity of lethally damaged cells (Dörr and Schmidt, 2014).

Recovery has been quantitated for numerous endpoints in preclinical investigations and clinical studies over the last decades (Bentzen and Joiner, 2009; Dörr, 2009b; Joiner and Bentzen, 2009). In general, the fractionation effect is low, but still significant, for early normal tissue reactions, and pronounced for late tissue endpoints. The linear-quadratic model is currently accepted to describe the relationship between total equieffective doses and dose per fraction or dose rate, respectively, in radiotherapy and other exposure scenarios (Bentzen et al., 2012). It is applied to estimate the toxicity of a given exposure after changes in dose per fraction or dose rate and total dose (Thames and Hendry, 1987; Bentzen and Joiner, 2009; Joiner and Bentzen, 2009; Dörr and Schmidt, 2014). This formalism was initially based on cell survival concepts, but must today just be considered as the most appropriate mathematical formula to fit fractionation/dose-rate data at the tissue level without any biological background (Bentzen et al., 2012). In this formalism, an α/β-value describes the shift of dose–effect curves to higher equieffective doses with a decrease in dose per fraction/dose rate. The lower the α/β-value, the more pronounced the effect of dose fractionation/dose rate, i.e. the increase in tolerance with a reduction in dose per fraction/dose rate. In general, late responding tissues are sensitive to changes in dose per fraction (low α/β-value, <6 Gy), while early responding tissues display a minor, but still significant, fractionation effect (Thames and Hendry, 1987; Bentzen and Joiner, 2009; Joiner and Bentzen, 2009; Dörr and Schmidt, 2014). The consequential component of late effects (see above), however, may not significantly benefit from reduced doses per fraction/dose rates.

The linear-quadratic formalism may underestimate the biological effect of a certain total dose at doses per fraction <1.0 Gy, due to 'low-dose hyper-radiosensitivity' (Marples et al., 1997; Joiner et al., 2001), as well as overestimate the effect of high doses per fraction >10 Gy (Joiner, 2009; Bentzen et al., 2012).

5. REPOPULATION – THE TIME FACTOR

Repopulation in normal tissues, generally defined as an increase in radiation tolerance with increasing overall treatment time (Dörr, 2009a; Dörr and Schmidt, 2014), is observed for typical early radiation effects that occur in turnover tissues (see above). This phenomenon is based on a tissue regeneration response that appears to be initiated by the tissue changes induced during protracted radiation exposure. The complex biological mechanisms include a profound reorganisation of the proliferative structure, both at the stem/target cell and the total tissue level. Clinical and experimental observations of normal tissue effects regarding variations in overall treatment time (Dörr, 1997, 2003, 2009a; Dörr and Schmidt, 2014) consistently illustrate that:

- the exposure effect is compensated with increasing overall treatment time, once the repopulation processes have become effective;
- the rate at which this compensation occurs is in the range of up to 5×2 Gy week^{-1} (e.g. in human oral mucosa); and
- the rate at which tissue cell depletion is observed decreases substantially after the onset of repopulation.

The biological mechanisms underlying normal tissue repopulation have been summarised as the '3As' (Dörr, 1997, 2003, 2009a; Dörr and Schmidt, 2014): Asymmetry loss; Acceleration of stem/target cell divisions; and Abortive divisions of doomed cells.

The target/stem cell hypothesis postulates that the radiation tolerance of (turnover) tissues is defined by the number of tissue-specific target cells. Therefore, compensation of the expected decrease in radiation tolerance expected from continued exposure, that is observed once repopulation has started, must be based on generation of new target cells to replace those sterilised by irradiation. Physiologically, target cells, on average, divide asymmetrically, i.e. into one new stem and one transit/differentiating daughter cell, thus maintaining the stem cell number. Additional production of new target cells, in contrast, requires symmetrical divisions into two stem cell daughters (asymmetrical loss).

The rate of dose compensation during repopulation in oral mucosa has been found to be in the range of 5×0.5 to 5×1 2 Gy-fractions week^{-1} (Dörr, 1997, 2003, 2009a). Based on a 2-Gy-surviving fraction of the stem cells of approximately 0.5, this requires five symmetrical divisions within 7 days, corresponding to a cell cycle time of 1.4 days, which is substantially shorter than the physiological cell cycle time (acceleration of stem cell divisions).

Effective repopulation is associated with a constant, or only a slight reduction of, overall cell number (Dörr and Kummermehr, 1990; Dörr, 2003, 2009a), although differentiation and cell loss continue at their physiological rate (Dörr et al., 1996). Moreover, assessment of overall cell production revealed a largely unchanged activity (Dörr et al., 1994; Dörr, 1997). However, the vast majority of stem cells are sterilised before repopulation becomes effective. In consequence, generation of the original number of (overall) cells to counteract the cell loss would require the remaining stem cells to proliferate with an unlikely short cycle time of only a few hours (Dörr et al., 1994; Dörr, 1997, 2009a). Hence, cells must be generated from other sources than the surviving stem cells. 'Doomed' (or 'sterilised') stem cells do have a residual proliferative capacity in the form of abortive divisions, which result in daughter cells that can undergo near normal differentiation (Dörr et al., 1996). This mechanism significantly contributes to the compensation of the ongoing physiological cell loss (Dörr et al., 1994; Dörr, 1997, 2009a).

6. EFFECT OF THE EXPOSED VOLUME

In radiotherapy, major advances in radiation physics during the last decades have resulted in a progressive conformation of high-dose volume to the macro- and microscopic tumour tissue. In consequence, the volumes of normal tissues exposed to significant doses were significantly reduced, and dose distribution within these volumes became inhomogeneous. This implies that the effect of exposure of the fractional volume of a normal tissue to certain doses, rather than the dose to the entire tissue or organ, needs to be considered. Moreover, accidental radiation exposures most frequently occur in a localised form. Consequently, the irradiated volume of an organ must be taken into account as an important parameter that determines the clinical consequences of tissue exposures (Dörr and Van der Kogel, 2009; Dörr and Schmidt, 2014).

The exposed volume can affect the quality and consequences of the clinical manifestation of tissue effects, even if the tissue sensitivity and the pathological changes per unit volume are identical, e.g. in the case of skin ulceration. In contrast, the pathological changes underlying identical clinical symptoms and consequences can be dependent on different dose-volume parameters and mechanisms, as has been demonstrated for the lung (Novakova-Jiresova et al., 2007). With regard to detailed conclusions on the volume effect in individual normal tissues, recent reviews by Dörr and Van der Kogel (2009) or the Quantitative Analysis of Normal Tissue Effects in the Clinic (QUANTEC) initiative (Marks et al., 2010) are recommended.

6.1. Volume effects and tissue architecture

The concept of functional subunits (FSUs), proposed by Withers et al. (1988), defines an FSU as the largest tissue sub-volume that can be restored from a single surviving stem/clonogenic cell. The number of FSUs sterilised by a certain exposure is hence dependent on the parameters influencing clonogenic cell survival. While structural damage is related to the individual FSU, the clinical manifestation is a consequence of the 'anatomical' arrangement of the FSUs, which can be either in parallel or in a series. In tissues with parallel structure, the FSUs are functioning independently of each other. Thus, clinical consequences of the exposure only become manifest if the number of eliminated FSUs, i.e. the fractional tissue volume, reaches a (tissue-specific) threshold. Here, the exposure constraints should refer to a threshold organ fractional volume, which is not to be exceeded (Dörr and Van der Kogel, 2009; Dörr and Schmidt, 2014). Examples for (predominantly) parallel organs are lung, kidney, and liver.

In organs with a serial ('tubular') organisation, e.g. spinal cord, intestine, or oesophagus, inactivation of one single FSU results in clinically manifest effects in the downstream tissue compartments in a binary response pattern (Dörr and Van der Kogel, 2009; Dörr and Schmidt, 2014). However, the FSUs are not simply arranged as a bunch or in a chain in any real organs. The vasculature in any organ, for example, represents a serial tissue component.

6.2. Normal tissue effect models and the QUANTEC initiative

Mathematical models describing normal tissue effect probabilities (in radiation oncology: normal tissue complication probabilities) include estimates of tissue-specific tolerance doses and fractionation parameters, such as the α/β-values and halftimes of recovery. For tolerance doses, the compilation by Emami et al. (1991) has frequently been applied in the past. More recently, the results of the QUANTEC analyses, reported in the *International Journal of Radiation Oncology Biology Physics*, summarised the available information on dose-volume-effect relationships for a variety of organs and tissues (Bentzen et al., 2010b, Marks et al., 2010), including the clinical significance of various endpoints, dose-volume tolerance data, and risk factors. For further analyses of specific tissue and organ tolerances, the reader is strongly recommended to screen the current literature.

7. MOLECULAR RADIOPATHOLOGY – BIOMARKERS AND INTERVENTIONAL STRATEGIES

7.1. Molecular and 'tissular' radiopathology

After radiation exposure of tissues, a whole orchestra of events, which may be summarised as 'damage processing' (Dörr, 2009c; Dörr and Schmidt, 2014), is seen way before any clinical changes become manifest. The cascades are initiated by the induction of free radicals and acute oxidative stress. These result in changes in the activity of transcription factors, and thus in the modification of various intracellular and extracellular signalling pathways. Such changes can be demonstrated in all tissue components (parenchyma, fibroblasts, vasculature) and also in activated – directly by exposure or due to the abovementioned changes – macrophages and other immune cells. The combination of all these events induces unspecific as well as tissue-specific changes at the cellular/histological level (e.g. cell death, differentiation or proliferation, DNA damage response, chronic oxidative stress, and many others). The integrated response eventually results in the known, pathological changes described earlier. Detailed knowledge of the molecular/tissular radiopathology processes is a major prerequisite for the identification of early biomarkers of the risk for the development of subsequent macroscopic tissue effects, and also for the development of biology-based interventional strategies for the modification of such effects.

7.2. Tissue effect biomarkers

Early radiopathology-based indicators of exposure-related tissue effects (effect biomarkers) can facilitate the stratification of individuals, already timely after their radiation exposure according to their individual risk for the manifestation of (severe) clinical consequences. Major activities in this field, particularly in relation to radiation dose/dose distributions, have only recently been initiated (Bentzen et al., 2010a). Such biomarkers need to be identified according to a precise and detailed knowledge of the individual molecular and cellular process cascades involved in the eventual clinical manifestation of specific exposure endpoints. Some promising examples are inflammatory and immune response markers, e.g. in the urinary bladder and the intestine (Hille et al., 2009; Varela et al., 2009; Gibson and Bowen, 2011; Henson and Ang, 2012).

7.3. Interventional strategies

Any interventions in the manifestation of normal tissue effects may be described as recommended at a National Cancer Institute workshop on normal tissue protection (Stone et al., 2004) according to their timing in relation to the radiation exposure as:

- prophylaxis/protection: pre-exposure;
- mitigation: during or shortly after exposure, before clinically manifest symptoms occur (i.e. during the latent time); and
- treatment/management: in the symptomatic phase.

Interventions in the damage processing pathways can, in principle, be performed at each of the levels of the event cascade described above. Some prominent principles and examples have been systematically summarised by Dörr (2009c), and Dörr and Schmidt (2014). In general, although a variety of such interventional strategies have been suggested, most of these approaches are highly experimental, and few have entered advanced preclinical studies. Very few approaches have been translated into clinical studies to date.

REFERENCES

Bentzen, S.M., 2006. Preventing or reducing late side effects of radiation therapy: radiobiology meets molecular pathology. Nat. Rev. Cancer 6, 702–713.

Bentzen, S.M., Joiner, M.C., 2009. The linear-quadratic approach in clinical practice. In: Joiner, M., Van der Kogel, A. (Eds.), Basic Clinical Radiobiology. 4th edn. Hodder Arnold, London, pp. 120–134.

Bentzen, S.M., Parliament, M., Deasy, J.O., et al., 2010a. Biomarkers and surrogate endpoints for normal-tissue effects of radiation therapy: the importance of dose-volume effects. Int. J. Radiat. Oncol. Biol. Physics 76 (Suppl.), S145–S150.

Bentzen, S.M., Constine, L.S., Deasy, J.O., et al., 2010b. Quantitative analyses of normal tissue effects in the clinic (QUANTEC): an introduction to the scientific issues. Int. J. Radiat. Oncol. Biol. Physics 76 (Suppl.), S3–S9.

Bentzen, S.M., Dörr, W., Gahbauer R., et al., 2012. Bioeffect modeling and equieffective dose concepts in radiation oncology – terminology, quantities and units. Radiother. Oncol. 105, 266–268.

Dörr, W., 1997. Three A's of repopulation during fractionated irradiation in squamous epithelia: asymmetry loss, acceleration of stem-cell divisions and abortive divisions. Int. J. Radiat. Biol. 72, 635–643.

Dörr, W., 2003. Modulation of repopulation processes in oral mucosa: experimental results. Int. J. Radiat. Biol. 79, 531–537.

Dörr, W., 2009a. Time factors in normal tissue responses to irradiation. In: Joiner, M., Van der Kogel, A. (Eds.), Basic Clinical Radiobiology. 4th edn. Hodder Arnold, London, pp. 149–157.

Dörr, W., 2009b. Pathogenesis of normal tissue side effects. In: Joiner, M., Van der Kogel, A. (Eds.), Basic Clinical Radiobiology. 4th edn. Hodder Arnold, London, pp. 169–190.

Dörr, W., 2009c. Biological response modifiers: normal tissues. In: Joiner, M., Van der Kogel, A. (Eds.), Basic Clinical Radiobiology. 4th edn. Hodder Arnold, London, pp. 301–315.

Dörr, W., Kummermehr, J., 1990. Accelerated repopulation of mouse tongue epithelium during fractionated irradiations or following single doses. Radiother. Oncol. 17, 249–259.

Dörr, W., Emmendörfer, H., Haide, E., et al., 1994. Proliferation equivalent of 'accelerated repopulation' in mouse oral mucosa. Int. J. Radiat. Biol. 66, 157–167.

Dörr, W., Emmendörfer, H., Weber-Frisch, M., 1996. Tissue kinetics in mouse tongue mucosa during daily fractionated radiotherapy. Cell Prolif. 29, 495–504.

Dörr, W., Hendry, J.H., 2001. Consequential late effects in normal tissues. Radiother. Oncol. 61, 223–231.

Dörr, W., Stewart, F.A., 2009. Retreatment tolerance of normal tissues. In: Joiner, M., Van der Kogel, A. (Eds.), Basic Clinical Radiobiology. 4th Hodder Arnold, London, pp. 259–270.

Dörr, W., Van der Kogel, A.J., 2009. The volume effect in radiotherapy. In: Joiner, M., Van der Kogel, A. (Eds.), Basic Clinical Radiobiology. 4th edn. Hodder Arnold, London, pp. 191–206.

Dörr, W., Schmidt, M., 2014. Normal tissue radiobiology. In: Brahme, A., Hendry, J.H. (Eds.), Comprehensive Biomedical Physics, Vol. 7. Elsevier, Amsterdam, Chapter 7.05, pp. 75–95.

Emami, B., Lyman, J., Brown, A., et al., 1991. Tolerance of normal tissue to therapeutic irradiation. Int. J. Radiat. Oncol. Biol. Phys. 21, 109–122.

Fajardo, L.F., Berthrong, M., Anderson, R.E., 2001. Radiation Pathology. Oxford University Press, New York.

Gibson, R.J., Bowen, J.M., 2011. Biomarkers of regimen-related mucosal injury. Cancer Treat. Rev. 37, 487–493.

Hakenjos, L., Bamberg, M., Rodemann, H.P., 2000. TGFbeta1-mediated alterations of rat lung fibroblast differentiation resulting in the radiation-induced fibrotic phenotype. Int. J. Radiat. Biol. 76, 503–509.

Henson, C.C., Ang, Y.S., 2012. Biomarkers of normal tissue toxicity after pelvic radiotherapy. Curr. Opin. Support Palliat. Care 6, 33–40.

Hille, A., Rave-Fränk, M., Christiansen, H., et al., 2009. Faecal calprotectin and lactoferrin values during irradiation of prostate cancer correlate with chronic radiation proctitis: results of a prospective study. Scand. J. Gastroenterol. 44, 939–946.

Joiner, M.C., 2009. Quantifying cell kill and cell survival. In: Joiner, M., Van der Kogel, A. (Eds.), Basic Clinical Radiobiology. 4th edn. Hodder Arnold, London, pp. 48–55.

Joiner, M.C., Marples, B., Lambin, P., et al., 2001. Low-dose hypersensitivity: current status and possible mechanisms. Int. J. Radiat. Oncol. Biol. Physics 49, 379–389.

Joiner, M.C., Bentzen, S.M., 2009. Fractionation: the linear-quadratic approach. In: Joiner, M.,Van der Kogel, A. (Eds.), Basic Clinical Radiobiology. 4th edn. Hodder Arnold, London, pp. 102–119.

Marks, L.B., Ten Haken, R.K., Martel, M.K., 2010. Guest editor's introduction to QUANTEC: a users' guide. Int. J. Radiat. Oncol. Biol. Physics 76 (Suppl.), S1–S2.

Marples, B., Lambin, P., Skov, K.A., et al., 1997. Low dose hyper-radiosensitivity and increased radioresistance in mammalian cells. Int. J. Radiat. Biol. 71, 721–735.

Novakova-Jiresova, A., van Luijk, P., van Goor, H., et al., 2007. Changes in expression of injury after irradiation of increasing volumes in rat lung. Int. J. Radiat. Oncol. Biol. Physics 67, 1510–1518.

Rodemann, H.P., Bamberg, M., 1995. Cellular basis of radiation-induced fibrosis. Radiother. Oncol. 35, 83–90.

Rubin, P., Johnston, C.J., Williams, J.P., et al., 1995. A perpetual cascade of cytokines post irradiation leads to pulmonary fibrosis. Int. J. Radiat. Oncol. Biol. Physics 33, 99–109.

Steel, G.G. (Ed.), 1993. Basic Clinical Radiobiology for Radiation Oncologists. Edward Arnold Publications, London.

Stone, H.B., Moulder, J.E., Coleman, C.N., et al., 2004. Models for evaluating agents intended for the prophylaxis, mitigation and treatment of radiation injuries. Report of an NCI Workshop, 3–4 December, 2003. Radiat. Res. 162, 711–728.

Thames, H.D., Hendry, J.H., 1987. Fractionation in Radiotherapy. Taylor and Francis, London.

Van der Kogel, A.J., 2009. The dose-rate effect. In: Joiner, M., Van der Kogel, A. (Eds), Basic Clinical Radiobiology. 4th Hodder Arnold, London, pp. 158–168.

Varela, E., Antolín, M., Guarner, F., et al., 2009. Faecal DNA and calprotectin as biomarkers of acute intestinal toxicity in patients undergoing pelvic radiotherapy. Aliment. Pharmacol. Ther. 30, 175–185.

Withers, H.R., 1975. The four Rs of radiotherapy. In: Lett, J.T., Adler, H. (Eds.), Advances in Radiation Biology. 5th Academic Press, New York, pp. 241–271.

Withers, H.R., Taylor, J.M., Maciejewski, B., 1988. Treatment volume and tissue tolerance. Int. J. Radiat. Oncol. Biol. Physics 14, 751–759.

Yarnold, J., Brotons, M.C., 2010. Pathogenetic mechanisms in radiation fibrosis. Radiother. Oncol. 97, 149–161.

SAGE

ICRP
Annals of the ICRP

Threshold doses and circulatory disease risks

J.H. Hendry
Christie Medical Physics and Engineering, Christie Hospital, Manchester M20 4BX, UK; e-mail: jhendry2002uk@yahoo.com

Abstract–Tissue reactions (deterministic effects) become manifest either early or late after doses above a threshold dose, which is the basis for recommended dose limits for avoiding such effects. Threshold doses have been defined for comparative purposes at 1% incidence of an effect, although the choice of incidence level may be scenario-dependent in practice. Latency time before manifestation is related to cell turnover rates and tissue complexity. In general, threshold doses become lower for longer follow-up times because of the slow progression of injury before manifestation, particularly after lower doses. Radiosensitive individuals may contribute to low threshold doses, which would provide a safety margin for the majority of a population. A threshold dose of 0.5 Gy was proposed for radiation-induced circulatory disease, after acute or chronic exposures, in the International Commission on Radiological Protection *Publication 118*. However, more recent meta-analyses of low-dose population studies suggest that, if a linear dose-incidence is assumed, the risk of some types of circulatory disease after doses <0.5 Gy or <10 mGy day^{-1} may be positive and similar to that for induced cancer. Animal studies show that doses >2 Gy induce the expression of inflammatory and thrombotic molecules in endothelial cells. This causes progressive loss of capillaries in the heart and leads to reduced perfusion, myocardial cell death, and fibrosis. However, doses <1 Gy inhibit both inflammatory cell adhesion to endothelial cells and the development of atherosclerosis in mice. Different mechanisms of injury at low and high doses preclude the simple extrapolation of risk on a linear-quadratic basis from acute to chronic exposures.

Keywords: Radiation biology; Circulatory disease; Tissue reactions; Deterministic effects; Threshold doses

This paper does not necessarily reflect the views of the International Commission on Radiological Protection.

1. THRESHOLD DOSE

A threshold dose for a given effect can be defined as a dose below which the effect does not occur. That dose is often difficult to determine, but is the basis for recommended dose limits pertaining to tissue reactions. One way in which epidemiological evidence for a threshold can be assessed is by examination of the lowest dose at which a statistically significant positive dose–response can be detected. This is subject to uncertainties due to constraints on sample sizes and to the particular model used to fit the data. In the present context of tissue reactions, the 'threshold dose' is defined as estimated dose for 1% incidence (ED_1), denoting the amount of radiation that is required to cause a specific, observable effect in only 1% of individuals exposed to radiation (ICRP, 2007, 2012). Although ED_1 is not a true threshold in the sense of the effect not occurring at all, it was considered appropriate for comparative purposes among tissue types. The use of a smaller level than ED_1 would entail a greater extrapolation of response frequencies to even lower doses, with concomitant greater uncertainties attached to the value. The use of a higher level would have less uncertainty in the value, but it would be even further from the 'true' threshold. When these threshold doses are used for protection purposes, the context should be considered carefully. An incidence of 1% or even more of a serious tissue reaction may be an acceptable risk in the radiotherapy of a life-threatening tumour in a cohort of 100–1000 cancer patients, but unacceptable in a fit worker or public population of 10,000 to millions unless the injury is readily correctable such as in the case of induced cataract by lens replacement.

The estimation of ED_1 may be complicated by substantial baseline levels of specific tissue effects or diseases that develop with ageing in the absence of radiation exposure, e.g. cataracts and circulatory disease. In these cases, ED_1 refers to effects just starting to rise above the baseline levels in unirradiated, age-matched individuals. In the case of circulatory disease, it refers to a dose that would increase the already high natural incidence or mortality by only 1%. ED_1 does not imply that no biological effects occur at lower doses; it merely defines the dose above which a specified effect becomes clinically apparent in a small increasing percentage of individuals. The incidence then rises with increasing dose to form a sigmoid dose–response relationship.

Note that the frequency of highly radiosensitive individuals in the general population is considered to be considerably below 1%, and such a population may contribute to a low threshold dose. Also, children appear to be more sensitive than adults to the induction of circulatory disease (e.g. Tukenova et al., 2010). In addition, various biological response modifiers given after exposure can provide dose-modifying factors of 1.1–1.5 for various types of tissue reactions occurring in experimental animal systems (ICRP, 2012). It is hoped that, in the future, some of these modifiers may be found to be helpful in increasing threshold doses in human populations.

2. LATENCY INTERVALS AND RADIATION RISKS

Early tissue reactions after radiation exposure generally have a well-defined latency and expression period that is related to the turnover time of the tissue. In contrast, the majority of late reactions progress and increase in frequency with increasing time. In this late phase, the first reactions to manifest are found after higher doses, and more reactions occur later after lower doses. Hence, threshold doses for specific levels of late tissue reactions are not absolute, and may decrease with increasing follow-up time as more reactions appear after the lower doses. Thus, the thresholds should be quoted as pertaining to a specified time after exposure. A review of many different clinical datasets demonstrated that the development of the incidence of late normal tissue injury occurred with approximately exponential kinetics after high radiation doses, which could be quantified as the percentage of patients at risk of developing a specific effect per year (Jung et al., 2001). This annual percentage risk remained relatively constant with increasing time after irradiation regarding a specific late effect, but it varied between tissues, e.g. 5% year^{-1} for dermal injury, and 12–14% year^{-1} for injury in bladder and ileum after pre-operative radiotherapy for rectal cancer. Most of the data reviewed showed annual increases in incidence up to 10 years maximum follow-up, but some studies showed an increasing incidence of late injury to 20, 25, and 30 years follow-up for some tissues. This indicates that very long follow-up times are needed to assess the accumulated expression of injury over a lifetime.

Also, the exponential kinetics implied that a random process might be involved in the occurrence of late radiation reactions (Jung et al., 2001). In addition, a study of patient-related vs stochastic components of variability for late blood vessel telangiectasia in paired bilateral irradiated areas of skin found that for a given dose-fractionation schedule, patient-related factors explained 81–90% of the patient-to-patient variation in telangiectasia level. The remaining 10–19% was explained by stochastic effects (Safwat et al., 2002). Hence, there is some evidence, albeit quite limited to date, for a stochastic component in tissue responses to radiation. This was also alluded to in the UK Health Protection Agency report on circulatory disease after radiation exposure (HPA, 2010):

> *Were the involvement of a stochastic process to be demonstrated convincingly, it would have significant implications with respect to radiation risk coefficients. However, atherosclerotic disease is a multifunctional disorder and all aspects of its biology need to be considered in relation to causal factors. We do not consider that the available evidence justifies consideration of a stochastic component as being established, although it remains as a possibility. Clearly, further work is needed to establish whether or not radiation can induce transformation of smooth muscle cells to a plaque-type phenotype, whether this induction is a stochastic process, and whether it plays a significant role in atherogenic development. (HPA, 2010).*

3. LOW DOSES AND CHRONIC DOSE RATES

Although higher accumulated total doses would generally be expected to be tolerated after protracted/chronic exposures than after acute exposures, similar threshold doses would be expected if the risk at doses up to the threshold dose was governed by single-hit irreparable injury, with no split-dose repair, slow repair, or cell repopulation effects differentially involved at very low dose levels or low dose rates. Alternatively, the mechanisms of injury may be different as a function of dose and dose-delivery pattern. In that case, it would probably be fortuitous if threshold doses were found to be similar, and statistical uncertainties may also be reflected in this conclusion.

4. CIRCULATORY DISEASE

The most recent analysis of circulatory disease mortality in the atomic bomb survivors showed a near-curvilinear dose–response curve at 20 years follow-up (since exposure in 1945) and a threshold dose consistent with the above-mentioned value of 0.5 Gy (Fig. 1). However, between 20 and 58 years follow-up, the dose–response curve became near-linear with no threshold, consistent with the prediction from biological evidence and human data for blood vessel telangiectasia (Turesson, 1989) that threshold doses may decrease with increasing time after exposure. It is recognised that this is a general conclusion, and the various types of circulatory disease have different shapes of dose–response curve (Takahashi et al., 2013). Also of note is that the general shapes of the dose–response curves in terms of excess relative risk (ERR) for circulatory disease mortality in the two follow-up periods were similar to those for mortality from all solid cancers, with the values of ERR for all circulatory diseases being four to five times less than the values for all solid cancers over the range of doses up to 3 Gy (Fig. 1).

A systematic review and meta-analysis of circulatory disease mortality from low-level radiation was carried out for 10 medical, occupational, and atomic bomb survivor studies, where doses were <0.5 Gy or <10 mGy day^{-1} (Little et al., 2012). A short 5-year latency followed by a constant ERR with further time after exposure, and a linear-no-threshold response, were assumed. The excess population risk for all circulatory diseases combined was 2.5–8.5% Sv^{-1}, which was of the same order as the cancer risk of 4.2–5.6% Sv^{-1} for these populations, indicating their similar importance at these low doses.

Animal studies have contributed knowledge in terms of dose–effect relationships and mechanisms of injury. For example, several studies have been performed using genetically modified ApoE–/– mice, which have a five-fold increased level of cholesterol and a predisposition to atherosclerosis. A single dose of 4 Gy or 20 × 2-Gy fractions delivered over 4 weeks accelerated atherosclerosis with an inflammatory thrombotic plaque phenotype (Hoving et al., 2008). Also, doses of 0.025–0.5 Gy, at an early or late stage of disease, impacted variously on the development of atherosclerosis (Mitchel et al., 2011). In addition, cardiac exposure to 0.2 Gy induced

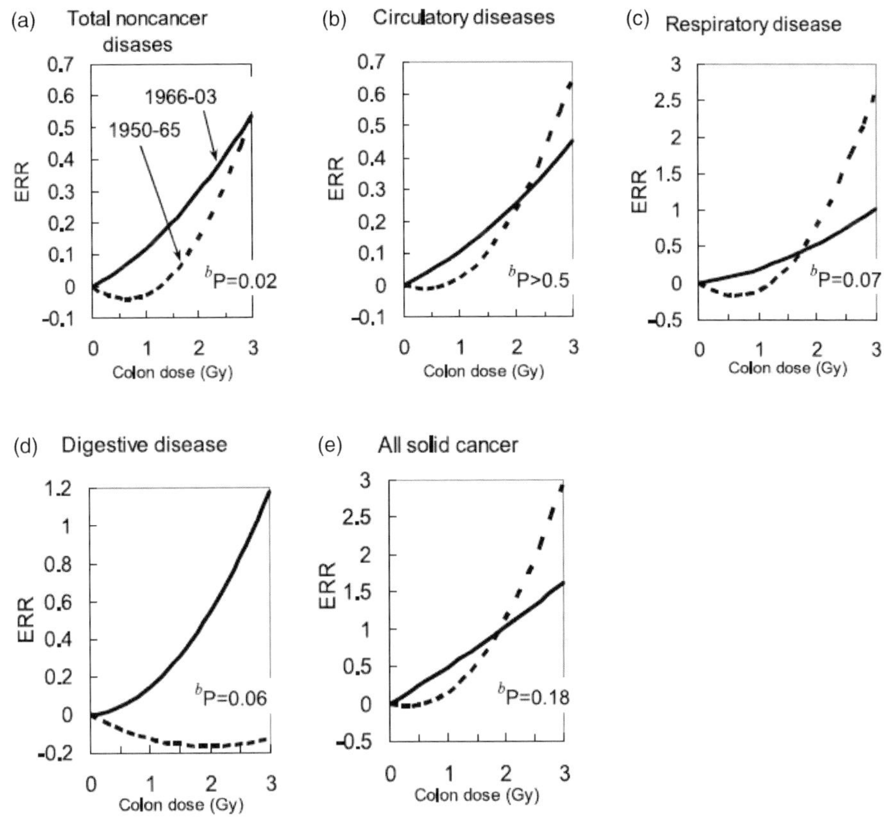

Figure 1. Dose–response curves for the early period (1950–1965, dashed curves) and the later period (1966–2003) for circulatory diseases in the atomic bomb survivors. ^{b}P values denote the significance of the difference between the two curves. ERR, excess relative risk. Source: Ozasa et al. (2012), figure reprinted with permission from *Radiation Research* and Dr Kotaro Ozasa.

significant physiological, histopathological, cellular, and molecular heart alterations, with mild functional impairment and early pro-inflammatory polarisation of macrophages (Monceau et al., 2013). Another model system is the stroke-prone spontaneously hypertensive rat, considered to be a good model of cerebrovascular diseases caused by severe hypertension and arteriosclerosis because the overall vascular changes in the brain and other organs are consistent with those observed in malignant hypertension (Takahashi et al., 2013).

A succinct summary of the mechanisms operating as a function of dose was given by Stewart (2012):

> *In large arteries, doses of ≥8 Gy, combined with elevated cholesterol, initiate atherosclerosis and predispose to the formation of inflammatory, unstable lesions, which are prone to*

rupture and may cause a fatal heart attack or stroke. Doses of ≥2 Gy induce the expression of inflammatory and thrombotic molecules in endothelial cells. In the heart, this causes progressive loss of capillaries and eventually leads to reduced perfusion, myocardial cell death, and fibrosis. Doses <1 Gy inhibit inflammatory cell adhesion to endothelial cells and inhibit the development of atherosclerosis in mice. It seems likely that mechanisms other than accelerated atherosclerosis are responsible for cardiovascular effects after low total-body exposures of radiation (e.g. impaired T-cell immunity or persistent increase in systemic cytokines).

Research in this topic is continuing and notable is the large European-Union-funded project on radiation-induced circulatory disease (www.procardio.eu), which follows a previous project (www.cardiorisk.eu). This support is helping to increase knowledge in this area.

Circulatory disease was recognised some years ago as one of the important non-cancer diseases induced by radiation, and several review articles were published covering epidemiology, dose–response relationships, and potential mechanisms of the effects (e.g. Hendry et al., 2008; UNSCEAR, 2008; Darby et al., 2010; HPA, 2010; ICRP, 2012). In 2013, with the accrual of more information on this subject, some new review articles were published. These cover, for example, epidemiology of circulatory effects after low-dose exposures (Little, 2013), circulatory disease induced by higher doses received from radiotherapeutic exposures and strategies for intervention (Stewart et al., 2013), a workshop report from the Radiation Effects Research Foundation (Japan) on many aspects of radiation-induced circulatory disease including directions for future research (Takahashi et al., 2013), and a review by consultant experts to the International Atomic Energy Agency also aimed at formulating new research questions in this area (Wondergem et al., 2013). Clearly, the topic of radiation-induced circulatory disease has become recognised as very important in protection and medical scenarios.

REFERENCES

Darby, S.C., Cutter, D.J., Boerma, M., et al., 2010. Radiation-related heart disease: current knowledge and future prospects. Int. J. Radiat. Oncol. Biol. Physics 76, 656–665.

Hendry, J.H., Akahoshi, M., Wang, L.S., et al., 2008. Radiation-induced cardiovascular injury. Radiat. Environ. Biophys 47, 189–193.

Hoving, S., Heeneman, S., Gijbels, M.J., et al., 2008. Single-dose and fractionated irradiation promote initiation and progression of atherosclerosis and induce an inflammatory plaque phenotype in ApoE(–/–) mice. Int. J. Radiat. Oncol. Biol. Physics 71, 848–857.

HPA, 2010. Circulatory Disease Risk. Report of the Independent Advisory Group on Ionising Radiation. Documents of the Health Protection Agency, RCE-16. Health Protection Agency, Chilton.

ICRP, 2007. The 2007 Recommendations of the International Commission on Radiological Protection. Publication 103. Ann. ICRP 37(2–4).

ICRP, 2012. ICRP statement on tissue reactions, and early and late effects of radiation in normal tissues and organs – threshold doses for tissue reactions in a radiation protection context. Publication 118. Ann. ICRP 41(1/2).

Jung, H., Beck-Bornholdt, H.P., Svoboda, V., et al., 2001. Quantification of late complications after radiation therapy. Radiother. Oncol 61, 233–246.

Little, M.P., 2013. A review of non-cancer effects, especially circulatory and ocular diseases. Radiat. Environ. Biophys 52, 435–449.

Little, M.P., Azizova, T.V., Bazyka, D., et al., 2012. Systematic review and meta-analysis of circulatory disease from exposure to low-level ionizing radiation and estimates of potential population mortality risks. Environ. Health Perspect 120, 1503–1511.

Mitchel, R.E., Hasu, M., Bugden, M., et al., 2011. Low-dose radiation exposure and atherosclerosis in ApoE–/– mice. Radiat. Res 175, 665–676.

Monceau, V., Meziani, L., Strup-Perrot, C., et al., 2013. Enhanced sensitivity to low dose irradiation of ApoE–/– mice mediated by early pro-inflammatory profile and delayed activation of the TGFb1 cascade involved in fibrogenesis. PLOS One 8, e57052.

Ozasa, K., Shimizu, Y., Suyama, A., et al., 2012. Studies of the mortality of atomic bomb survivors, Report 14, 1950–2003: an overview of cancer and noncancer diseases. Radiat. Res 177, 229–243.

Safwat, A., Bentzen, S.M., Hendry, J.H., et al., 2002. Deterministic rather than stochastic factors explain most of the variation in the expression of skin telangiectasia after radiotherapy. Int. J. Radiat. Oncol. Biol. Physics 52, 198–204.

Stewart, F.A., 2012. Mechanisms and dose–response relationships for radiation-induced cardiovascular disease. Ann. ICRP 41(3/4): 72–79.

Stewart, F.A., Seemann, I., Hoving, S., et al., 2013. Understanding radiation-induced cardiovascular damage and strategies for intervention. Clin. Oncol 25, 617–624.

Takahashi, I., Ohishi, W., Mettler, F.A., et al., 2013. A report from the 2013 International Workshop: Radiation and Cardiovascular Disease, Hiroshima, Japan. J. Radiol. Protect 33, 869–880.

Tukenova, M., Guibout, C., Oberlin, O., et al., 2010. Role of cancer treatment in long-term overall and cardiovascular mortality after childhood cancer. J. Clin. Oncol 28, 1308–1315.

Turesson, I., 1989. The progression rate of late radiation effects in normal tissue and its impact on dose–response relationships. Radiother. Oncol 15, 217–226.

UNSCEAR, 2008. United Nations Scientific Committee on the Effects of Atomic Radiation. Volume 1, Effects of Ionising Radiation. 2006 Report to the General Assembly, Annex B: Epidemiological Evaluation of Cardiovascular Disease and Other Non-Cancer Diseases Following Radiation Exposure. United Nations, New York.

Wondergem, J., Boerma, M., Kodama, K., et al., 2013. Cardiovascular effects after low-dose exposure and radiotherapy: what research is needed? Radiat. Environ. Biophys. 52, 425–434.

SAGE
Annals of the ICRP

General tissue reactions and implications for radiation protection

S. Miyazaki[a], C. Hill[b]

[a]*Atomic Energy Research Institute, Kinki University, Osaka, Japan; Email: smiyazaki@neltd.co.jp*
[b]*Stoller Nuclear News (SN3) Corporation, USA*

Abstract–Non-cancer effects and risks at low doses from ionising radiation are controversial topics within the field of radiation protection. These issues are discussed in International Commission on Radiological Protection (ICRP) *Publication 118*, 'ICRP statement on tissue reactions'. Both non-cancer effects and risks are expected to become increasingly important to the system of radiation protection. Before this can happen, several factors must be considered: thorough characterisation of the relationship between dose and risk; verification of the biological mechanisms for any noted excess risk; and adjustment of noted excess risks through the use of a detriment factor. It is difficult to differentiate the relatively small risks associated with radiation from other risk factors in the low-dose region of the dose–response curve. Several recent papers have indicated the possibility of a non-linear dose-response relationship for non-cancer effects. In addition, there are still many uncertainties associated with the biological mechanisms for non-cancer effects. Finally, it is essential to consider the incorporation of detriment into a well-defined system of radiological protection. Given the recent interest in non-cancer effects, it is essential to facilitate discussions in order to define dose limits more clearly within the existing system of radiation protection for both cancer and non-cancer effects.

Keywords: Tissue reactions; Radiation protection; Detriment; Risk; Epidemiology

1. INTRODUCTION

Several publications by the International Commission on Radiological Protection (ICRP) outline risk assessment for cancer in great detail. Most recently, this was addressed in *Publication 103* (ICRP, 2007). In *Publication 103*, cancer risks are determined through epidemiological analysis, and adjusted to account for reduced quality of life and length of life lost should a cancer occur.

In 2012, an extensive update to the recommendations regarding deterministic effects (now tissue reactions) was issued in *Publication 118* (ICRP, 2012). One of the biggest differences between *Publication 118* and prior ICRP publications regarding tissue reactions is the treatment of cardiovascular diseases. *Publication 118* estimates threshold doses for circulatory disease effects using three important concepts: epidemiological analysis to determine the shape of the dose–response curve; detriment; and mechanisms for damage. The epidemiological treatment of non-cancer data is similar to that of the cancer risk estimations in *Publication 103*, although accounting for severity or detriment differs significantly from the cancer case.

Recent evidence suggests that uncertainties exist between the risk of cardiovascular disease and the dose–response curve, calling an assumed linear relationship (as in *Publication 118*) into question. Health effects for many non-cancer diseases appear minimal to negligible in the lower regions of the dose–response curve, although this is not necessarily reflected in the recommendations. The shape of the dose–response curve is a crucial point when discussing risks associated with non-cancer effects, and should be clarified. In addition, current estimations for threshold doses of non-cancer diseases from ionising radiation make use of an incidence value, although this implies that a specific degree of harm has occurred. In previous ICRP publications on non-stochastic or deterministic effects, the degree of harm (or severity of an effect) was indicated by cell death. No similar explanation for severity is provided in *Publication 118*.

ICRP publications such as *Publications 103* and *118* have had a major effect on radiological protection systems worldwide. For this reason, it is important to develop a radiological protection system based on scientific information and to develop consistency in the treatment of low-dose cancer and non-cancer effects. This paper examines areas where additional consideration may be needed for the assessment of non-cancer risks, such as the shape of the dose–response curve, and the concept of severity or detriment. The assessment of circulatory disease risks is emphasised due to recent ICRP recommendations for significantly reduced threshold doses in this area, and their potential impact on the nuclear power and medical industries. The issues discussed, however, may also apply to other types of non-cancer effects.

2. EPIDEMIOLOGICAL ANALYSIS (DOSE–RESPONSE CURVE)

Estimated excess risks for 'all circulatory disease' and cataracts in *Publication 118* are based on an assumed linear dose–response relationship. This type of analysis is based primarily on epidemiological analyses. Clarification of the shape of these dose–response curves is essential for developing a fair representation of threshold dose for circulatory disease.

2.1. Variations in dose-response for circulatory disease subtypes

While broad categories for circulatory disease (e.g. 'all circulatory disease', 'heart disease', or 'stroke') seem to indicate a positive dose–response relationship, closer inspection reveals some discrepancies. When examining the dose–response relationship for circulatory disease subtypes, the shape of the curve and even the apparent positive dose-response relationship may be questionable. For example, Ozasa et al.

(2012) and Suzuki (2012) examined excess risks calculated using a linear model fit to atomic bomb life span study (LSS) data for well-defined subtypes of heart disease over discrete time intervals. The data show no significant excess risk for ischaemic heart disease over the subject time periods. In contrast, a significant excess risk is indicated for hypertensive heart disease for most time periods. In neither case is a linear curve necessarily indicated when the excess risks are plotted against absorbed dose.

Both studies suggest that a seeming linear dose–response relationship for the broad category of all cardiovascular disease may be an artefact of the combination of disease subtypes and time periods. In fact, certain well-defined disease subtypes tend to indicate non-linear responses with a potential threshold (e.g. rheumatic and hypertensive heart diseases). Suzuki (2012) demonstrated that in several studies pertaining to circulatory disease risks at low doses, the apparent dose–response relationship can vary when evaluated against additional factors (e.g. population, irradiated volume, etc.). ICRP suggested for non-cancer risks lean heavily on the LSS data. While these data may indicate a potential linear dose–response relationship for circulatory disease when taken as a whole, they may not support an argument for linearity when disease subtypes are examined individually.

Although the data from the two studies illustrate some uncertainties in the apparent positive dose–response relationship, the data also show the importance of considering the overall effect for lifetime data. The data for later time periods may indicate that there is some trend towards a positive dose–response relationship in later time periods for ischaemic heart disease, as might be expected for a late manifesting effect. However, in short, the analyses by Ozasa et al. (2012) and Suzuki (2012) demonstrated that a non-linear dose–response relationship with possible threshold may be a more accurate model for cardiovascular disease.

Like Suzuki (2012), the National Council on Radiation Protection and Measurements (NCRP) noted that only certain subcategories of heart disease demonstrate excess risks (NCRP, 2012). NCRP concluded that evidence supporting a reduced threshold for cardiovascular disease in the range of 0 to 0.5 Gy is weak (NCRP, 2012). Specifically, NCRP stated the following (NCRP, 2012).

- In a review of several studies on the circulatory disease risks following low-dose radiation in medical settings, 'inconsistent results...limit the confidence in establishing a single value for a low-dose threshold for cardiac effects following therapeutic radiations' (p. 217).
- Similarly, a review of worker studies on the association between heart disease and radiation dose did not provide consistent evidence for a positive dose–response relationship, nor was it possible to exclude effects on the same order of magnitude as reported for the LSS cohort.
- NCRP supports the 2008 (United Nations Scientific Committee on the Effects of Atomic Radiation) conclusion by UNSCEAR that there is little evidence to 'establish a causal relation following heart doses <1 to 2 Gy' (p. 222).

2.1.1. Questions of linearity

Several other studies have illustrated an apparent non-linear dose-response relationship for circulatory disease risks. For example, Schöllnberger et al. (2012) used a multi-model inference method whereby various dose–response models were combined mathematically so that risk estimates could be based on several possible dose–response models. Using this approach on data for cerebrovascular disease yielded a weak dose–response relationship below approximately 0.6 Gy with a risk estimate approximately one-third of that estimated using a linear non-threshold model. Similarly, for circulatory disease, the combined model approach indicated zero risk below approximately 2.2 Gy. Takahashi et al. (2012) analysed stroke incidence data from the atomic bomb Adult Health Study, which also suggests a potential threshold for stroke.

2.1.2. Possible treatments for data

While some circulatory disease subtypes indicate little to no dose–response relationship in the low-dose region (or, at the very least, a non-linear dose–response) relationship, it must be acknowledged that other types of circulatory disease clearly show effects that should be accounted for. In this regard, the treatment of the data is important. For example, (a) circulatory disease data could be aggregated according to the affected organ or tissue (e.g. brain vs heart); or (b) specific subtypes of disease could be examined individually, possibly even with separate methods or adjustments for detriment.

The first method would have the advantage of providing some further differentiation and possibly some alleviation of the aforementioned 'artefact' of linearity from blending multiple disease subtypes, as there seem to be significant differences between the dose–response relationships for heart disease vs stroke. A disadvantage of this method would be that the artefact effect would not be eliminated completely, as some heart disease subtypes demonstrate positive dose–response relationships while others do not. Option (b), above, would eliminate any question of an artificially imposed dose–response relationship, although examining several possible disease subtypes begs the questions as to which subtypes to analyse and how to incorporate the risks from several different diseases (some of which may show excess risks whereas others do not). Another consideration is the fact that, in general, when data are divided into subtypes, larger uncertainties can arise with the use of smaller sample sizes.

Ultimately, if the shape of the dose–response curve is not truly linear, an assumed linear non-threshold relationship provides a possibly overly conservative estimate of risk. In addition, there is not sufficient evidence to maintain that low-dose/dose-rate exposures cause significant non-cancer health effects. Recent studies have emphasised the importance of including Adult Health Study data and the heterogeneity amongst disease subtypes when determining excess risks for non-cancer effects. Such findings indicate that further research into the dose–response relationship may be warranted.

3. DETRIMENT

This section reviews the application of detriment to the calculation of cancer risks and the potential importance of this concept to current estimations of non-cancer risks.

3.1. Detriment and cancer risks

Stochastic (e.g. cancer) risks are modelled such that the probability of occurrence increases with dose. Severity for stochastic risks is expressed by adjusting the probability of occurrence using factors to account for various components of detriment. According to *Publication 103*, the primary components of cancer detriment are the probability of attributable, fatal cancer, the weighted probability of attributable, non-fatal cancer, and the length of life lost if harm occurs (ICRP, 2007). The process for cancer risk detriment adjustment is shown in Fig. 1.

From here, dose limits are estimated based on a variety of considerations in addition to the calculated fatal cancer risk estimate.

3.2. Severity and non-cancer risks

For deterministic (non-cancer) effects, both the probability and the severity of an effect increase above some threshold doses. Traditionally [i.e. in *Publication 41* (ICRP, 1984)], the Commission has based the concept of severity for deterministic effects on cell death following irradiation. That is, the severity of the effect (and thus clinical manifestation of an effect) of radiation on a particular organ or system is determined primarily by the rates of cell repopulation and repair of sublethal damage in the affected organs/systems. *Publication 41* describes iso-effect models developed for use in radiotherapeutic research and clinical oncology to 'quantify the relationship between the severity of tissue damage, the total dose, dose per exposure, number of exposures, and overall duration of exposure' (p. 10). These models are based on the concept of cell killing to describe dose-related damage (i.e. severity). In *Publication 41*, ICRP estimates the equivalent highly fractionated dose that would produce the same effect (i.e. clinical manifestation as a measure of severity) as a radiotherapeutic dose in 1–5% of patients. Threshold doses are determined for both acute and highly fractionated exposures.

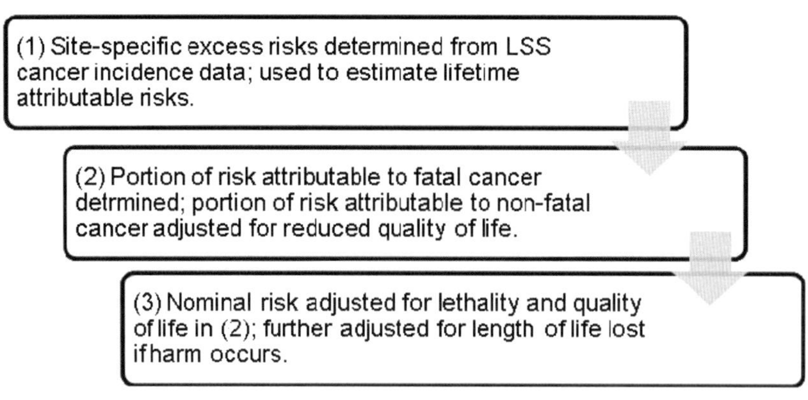

Figure 1. Detriment adjustment for cancer risks. LSS, Life Span Study.

In *Publication 118*, ICRP defines threshold dose as that dose estimated to result in the incidence of a defined tissue reaction in 1% of the population. The use of an incidence value, however, implies a method to account for severity (such as cell death), as this implies a specific degree of harm has occurred. In addition, a clear definition of the highly sensitive population represented by the 1% incidence value has not been provided. An appropriate incidence value is still under discussion, and some have suggested that it should be as high as 5% (AGIR, 2013).

The concept of severity plays an important role in both cancer and non-cancer risk estimations. In *Publication 118*, however, a concrete definition of severity is not provided; severity is not modelled as a function of cell death in iso-effect models, nor is severity accounted for in the estimate of excess risks through the use of detriment factors. This is a departure from the traditional treatment of non-cancer risk estimations. Further discussion is needed to clarify non-cancer risk assessments, namely in the areas of the use of an incidence value and a potential metric to account for severity. Defining detriment for non-cancer effects may be a potential method to harmonise risk assessments for cancer and non-cancer effects.

4. MECHANISMS

The understanding of the mechanisms behind non-cancer effects will also contribute to a better interpretation of the dose–response relationship, although the study of biological mechanisms for radiation-induced cardiovascular effects is still nascent. There is even some indication that mechanisms for damage may be different at low doses compared with high doses (ICRP, 2012). Little et al. (2012) acknowledged that postulated mechanisms for cardiovascular disease are still not well understood due to several factors. For one, target tissues for circulatory system effects at doses less than approximately 0.5 Gy 'remain uncertain' (Little et al., 2012: 17). Secondly, although certain animal models (such as the ApoE (Apolipoprotein E) mouse) are used to study cardiovascular disease effects, there are no current animal models for cardiovascular disease that can be used confidently to translate results from animal studies to human application.

Recent research indicates that the mechanisms for both cancer and non-cancer effects likely include non-targeted effects (genomic instability and bystander effects) and inflammatory processes. Such processes have been primarily observed in in-vitro systems at low doses. However, if the fundamental mechanisms for radiation-induced damage are similar for cancer and non-cancer effects, it is worth exploring whether differences between the methodologies for cancer and non-cancer risk determinations can be resolved. Research into non-targeted effects is still relatively new for both cancer and non-cancer effects, although ongoing research into mechanisms of response could contribute to a better interpretation of dose–response curves for circulatory diseases and may even provide a basis for the future harmonisation of cancer and non-cancer risk estimates. At the very least, emerging results indicating that non-targeted effects may play a role in circulatory disease induction following exposure to ionising radiation suggest that much work is needed to reconcile epidemiological data with radiobiology at low doses.

Efforts to understand underlying biological mechanisms for cardiovascular disease are complicated by the existence of confounding factors and noises. Of particular interest in the current framework of radiological protection is the effect of stress on the development of cardiovascular disease. While the risks of physical health effects directly caused by ionising radiation decrease according to reductions in exposure dose, the same cannot necessarily be said of psychological trauma. Striving to quantify the separate contributions of ionising radiation and stress to the development of cardiovascular disease, while continuing to inform the public on the scientific results, is particularly important in light of the 2011 accident at the Fukushima Daiichi nuclear power plant. Eventually, it may even be possible to develop a radiological protection system that accounts for the social aspect of risks as well as physiological damage.

5. CONCLUSIONS

Due to the fact that recommendations and publications from ICRP carry great weight within the radiation protection community, discussions regarding the treatment of epidemiological data, shape of the dose–response curve, and definitions of severity and detriment should be ongoing. Potential benefits could also be realised by exploring the harmonisation of cancer and non-cancer risk estimates. For non-cancer risks, specifically, this will involve examination of the importance of including data from the Adult Health Study, recognising heterogeneity among disease subtypes, exploring the appropriateness of an incidence value, creating metrics to account for severity and detriment, and further researching mechanisms for damage.

The radiation protection community has begun to recognise the importance of this topic, even proposing at the 2012 Science and Values Workshop in Tokyo, Japan, an ICRP working group to address non-cancer detriment. This topic was also mentioned at a February 2013 workshop on 'Radiation and cardiovascular disease' hosted by the Radiation Effects Research Foundation. As further insights are made into the mechanisms of cancer and non-cancer effects from ionising radiation, it is important to explore potential paths forward towards definitions of severity and detriment, and the potential harmonisation of risk assessments.

REFERENCES

AGIR, 2013. Human Radiosensitivity. RCE-21. Health Protection Agency, Chilton.

ICRP, 1984. Non-stochastic effects of ionising radiation. ICRP Publication 41. Ann. ICRP 14(3).

ICRP, 2007. 2007 Recommendations of the International Commission on Radiological Protection. ICRP Publication 103. Ann. ICRP 37(2–4).

ICRP, 2012. ICRP statement on tissue reactions/early and late effects of radiation in normal tissues and organs – threshold doses for tissue reactions in a radiation protection context. ICRP Publication 118. Ann. ICRP 41(1/2).

Little, M., Azizova, T., Bazyka, D., et al., 2012. Systematic review and meta-analysis of circulatory disease from exposure to low-level ionizing radiation and estimates of potential population mortality risks. Environ Health Perspect 120(11), 1503–1511.

NCRP, 2012. Uncertainties in the Estimation of Radiation Risks and Probability of Disease Causation. NCRP Report No. 171. National Council on Radiation Protection and Measurements, Bethesda, MD.

Ozasa, K., Shimizu, Y., Takahashi, I., et al., 2012. Non-cancer Effects in the Life Span Study (Cardiovascular Diseases). Presentation at the 2012 OECD/NEA Science and Values Workshop, 6–8 November 2012, Tokyo, Japan.

Schöllnberger, H., Kaiser, J.C., Jacob, P., et al., 2012. Dose-responses from multi-model inference for the non-cancer disease mortality of atomic bomb survivors. Radiat. Environ. Biophys. 51, 165–178.

Suzuki, G., 2012. Review of Epidemiological Studies of Non-cancer Diseases. Presentation at the 2012 OECD/NEA Science and Values Workshop, 6–8 November 2012, Tokyo, Japan.

Takahashi, I., Abbott, R., Ohshita, T., et al., 2012. A prospective follow-up study of the association of radiation exposure with fatal and non-fatal stroke among atomic bomb survivors in Hiroshima and Nagasaki (1980–2003). BMJ Open 2, e000654.

SAGE

ICRP
Annals of the ICRP

The lens of the eye: exposures in the UK medical sector and mechanistic studies of radiation effects

S.D. Bouffler[a], S. Peters[a], P. Gilvin[a], K. Slack[a], E. Markiewicz[b], R.A. Quinlan[b], J. Gillan[a], M. Coster[a], S. Barnard[a], K. Rothkamm[a], E. Ainsbury[a]

[a]*Centre for Radiation, Chemical and Environmental Hazards, Public Health England, Chilton, Didcot, OX11 0RQ, UK; e-mail: simon.bouffler@phe.gov.uk*
[b]*Biophysical Sciences Institute, Durham University, UK*

Abstract–The recommendation from the International Commission on Radiological Protection that the occupational equivalent dose limit for the lens of the eye should be reduced to 20 mSv year^{-1}, averaged over 5 years with no year exceeding 50 mSv, has stimulated a discussion on the practicalities of implementation of this revised dose limit, and the most appropriate risk and protection framework to adopt. This brief paper provides an overview of some of the drivers behind the move to a lower recommended dose limit. The issue of implementation in the medical sector in the UK has been addressed through a small-scale survey of doses to the lens of the eye amongst interventional cardiologists and radiologists. In addition, a mechanistic study of early and late post-irradiation changes in the lens of the eye in in-vivo-exposed mice is outlined. Surveys and studies such as those described can contribute to a deeper understanding of fundamental and practical issues, and therefore contribute to a robust evidence base for ensuring adequate protection of the eye while avoiding undesirable restrictions to working practices.

Keywords: Eye lens; Radiation; Cataract; Exposure; Medical; Mechanisms

This paper does not necessarily reflect the views of the International Commission on Radiological Protection.

1. INTRODUCTION

The International Commission on Radiological Protection (ICRP) published its report on tissue reactions in 2012 (ICRP, 2012). This report is an extensive and comprehensive review of the effects of ionising radiation on tissues; it also considers the dose thresholds, below which the incidence of any given tissue reaction occurs in less than 1% of an exposed group. In many cases, there was little indication that threshold doses were much different from those previously recognised. However, evidence had accumulated from epidemiological studies indicating that the threshold dose for effects on the lens of the eye had been overestimated. Previously, absorbed dose thresholds for cataract were considered to be approximately 2 Gy for acute exposures and on the order of 4–5 Gy for fractionated and protracted exposures; the newer evidence suggested that the threshold dose was around 0.5 Gy for both acute and protracted/fractionated exposures. On the basis of this new evidence, ICRP considered the dose limit for the lens of the eye and, prior to *Publication 118* (ICRP, 2012), issued a statement on tissue reactions (ICRP, 2011) which recommended that the dose limit for occupational exposure should be reduced from 150 mSv $year^{-1}$ to 20 mSv $year^{-1}$ averaged over 5 years, with no individual year exceeding 50 mSv.

The conclusion that the threshold dose for the lens of the eye had previously been overestimated was based largely on analysis of epidemiological studies of cataract and lens opacities, particularly those of Japanese atomic bomb survivors who had extended post-exposure follow-up periods, and Chernobyl liquidators in whom exposures were protracted. These studies have been considered in recent reviews (Ainsbury et al., 2009; Bouffler et al., 2012) which also concluded that the absorbed dose threshold for cataract is around 0.5 Gy in the case of acute exposures, with protraction or fractionation of exposure having little effect on the value. A particularly important study is that of Neriishi et al. (2007), which reports on the prevalence of cataracts requiring surgical intervention among atomic bomb survivors. This study reports a 0.5-Gy threshold dose for significant excess relative risk; the confidence intervals on the 0.5-Gy threshold value included 0 Gy, indicating that a non-threshold relationship might apply. This study therefore indicates that radiation exposure is associated with lens opacities that impair vision, not just smaller pre-cataractous lesions that have a minimal effect on vision.

ICRP's recommendation of a reduced dose limit for the lens of the eye has caused considerable concern in some occupational areas, particularly interventional radiology and cardiology, where compliance with a new dose limit may be difficult and may require modifications to working practices (Englefield, 2011; Martin, 2011; Broughton et al., 2013).

In considering potential mechanisms of radiation-associated lens cataract, Bouffler et al. (2012) observed that the dose–response slope for cataract and lens opacities reported in the more recent epidemiological investigations is not as steep as conventionally considered to apply to tissue reactions, and several studies have indicated that incidence can be described by a non-threshold dose–response relationship. This suggests that effects on the lens of the eye may be governed by processes other than cell killing or functional inactivation, as generally accepted to be important for the induction of tissue reactions. Ionising radiation produces a characteristic posterior subcapsular cataract that is distinct from, for instance, a nuclear cataract that would be typical of an age-related cataract phenotype (Little, 2013). Therefore, the mechanistic detail associated with the development of posterior subcapsular cataract and the involvement of the underlying cortex (Merriam and Worgul, 1983) will likely involve specific events that are associated with cataractogenesis in general. Oxidative stress clearly plays an important role in the development of cataract, and reduction in anti-oxidant capacity is associated with age-related cataract. The incidence of cataract is strongly age-related, and it has been suggested that radiation acts in cataractogenesis by accelerating the ageing process, which in itself is likely to include an effect of oxidative stress (Pendergrass et al., 2010). There is good evidence for a role of genetic factors in determining cataract sensitivity (Churchill and Graw, 2011; He et al., 2013). In experimental models, genes that play important roles in suppression of tumourigenesis are also demonstrated to affect cataract formation (e.g. Wiley et al., 2011), and DNA-repair-related genes are reported to affect radiation-induced cataract (Kleiman et al., 2007). These observations lend weight to the argument that the process of cataract formation following radiation exposure has similarity with mechanisms of stochastic effects. However, further mechanistic investigations are required to define the mechanisms that lead to cataract following ionising radiation exposure.

On the basis of the brief discussion above, two main issues relevant to the revised ICRP recommendation for a reduced occupational dose limit for the eye stand out as requiring further attention: (1) investigation of exposures in the medical sector to identify any difficulties there may be in compliance with a reduced dose limit; and (2) further investigation of the mechanisms contributing to radiation cataractogenesis to help clarify how radiation causes cataract, as well as whether radiation cataractogenesis is best treated as a stochastic effect or a tissue reaction for low dose risk estimation. Two studies that consider these issues are summarised below.

2. SMALL-SCALE SURVEY OF EYE DOSES IN THE UK MEDICAL SECTOR

As noted above, concerns have been expressed that doses to the lens of the eye received by those working in the medical sector may be high enough to present some difficulties in implementation of the recommended 20-mSv occupational equivalent dose limit for the eye. In particular, concern has been noted in the interventional cardiology and radiology fields, as highlighted by Vañó et al. (2009). To address this issue, a small-scale study of doses to the lens of the eye was undertaken recently in three UK radiology and cardiology departments (Ainsbury et al., 2014).

In brief, this survey targeted three busy departments offering a full range of radiology services including computed tomography, fluoroscopy, mammography, magnetic resonance imaging, nuclear medicine, ultrasound, and x ray. Within

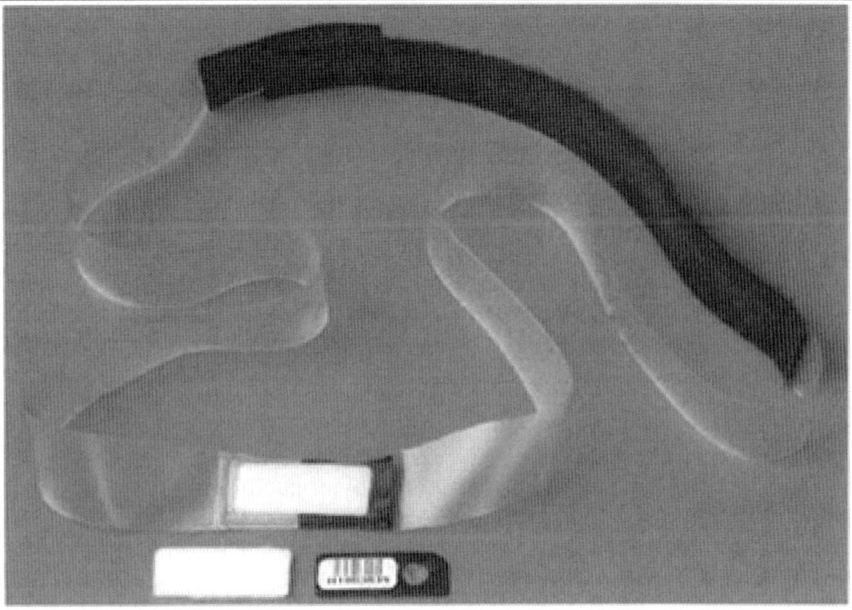

Fig. 1. Illustration of Public Health England's eye lens dosimeter (Gilvin et al., 2013).

these centres, consultant and registrar interventional cardiologists and radiologists were the focus as these clinicians were predicted to be among the most exposed. Radiation doses to the lens of the eye were assessed over a 4-week period using the Public Health England (PHE) eye lens dosimeter (Gilvin et al., 2013). This dosimeter is supplied as a headband worn just above the eye, and measures the $H_p(3)$ operational quantity using a tissue equivalent thermoluminescence dosimeter behind a filter equivalent to a 3-mm thickness of tissue (Fig. 1).

Sixty-eight dosimeters were provided to the hospitals along with instructions for use and a questionnaire. The questionnaire aimed to establish:

- if dosimeter use followed the supplied instructions;
- job title of user;
- types and numbers of procedures undertaken during the survey period; and
- personal protective equipment (PPE) used.

Hospital-based radiation protection specialists distributed the dosimeters to hospital staff, collected the dosimeters and questionnaires, and returned them to PHE for analysis. Of the 68 distributed dosimeters, 61 were returned along with 58 completed questionnaires.

Full results are presented elsewhere (Ainsbury et al., 2013); however, to summarise:

- Doses received by staff carrying out a full range of interventional procedures, including coronary angiogram, cardiac catheterisation, transcatheter aortic valve implantation, angioplasty, and pulmonary vein isolation, were monitored.
- Over 1000 procedures were carried out by the monitored staff.
- A median of 15 procedures were carried out per monitored individual, with a wide range of 1–70 on the median value.
- Reported use of PPE was good, although only nine individuals reported the use of lead glasses.
- Thirteen of the returned dosimeters recorded doses above the detection limit of 0.15 mSv.
- The median recorded dose over the 61 dosimeters was below the detection limit of 0.15 mSv.
- Two returned dosimeters recorded doses of potential concern (1.60 and 1.58 mSv, respectively).

If the two staff with recorded doses of 1.60 and 1.58 mSv over the 4-week period continued to be exposed at the same level over 1 whole year, doses just above the revised dose limit (equivalent yearly doses of 20.80 and 20.54, respectively) could accumulate. In these cases, one participant reported that lead glasses were not worn, while the other participant reported that lead glasses were used but the dosimeter was worn outside the glasses. This indicated that doses to the eye in these two cases would be below 20 mSv year^{-1} if doses were recorded behind properly worn lead glasses that can provide a 5–10-fold reduction in dose (Koukorava et al., 2011).

Overall, this small-scale, three-centre survey of doses to the lens of the eye in the UK suggests that compliance with the ICRP recommended dose limit (ICRP, 2012) should not be difficult where there is good uptake and correct use of PPE. However, the survey has limitations in the number and types of staff surveyed, and the number of centres covered. It was also notable that no obvious relationship was observed between the number of procedures carried out and recorded doses to the lens of the eye. Therefore, there appear to be individual factors that affect eye exposures and, although not observed in this survey, there may be variation in practice and therefore eye exposures in different hospitals. Nonetheless, the doses recorded in this survey are comparable with those reported elsewhere (Donadille et al., 2011; Efstathopoulos et al., 2011).

3. MECHANISTIC STUDY OF RADIATION CATARACTOGENESIS

Given the uncertainty on the risk of radiation cataract at low doses and the need for a sound basis for risk extrapolation, mechanistic studies can contribute to providing a robust scientific basis for the protection of humans against radiation-associated cataract. With this aim in mind, a study of the dose–response for radiation cataract and early lens changes has been initiated. The main part of the study uses C57BL/6 J mice exposed to 20 mGy–2 Gy whole-body x rays. Eye lenses are removed at times between 1 hour and 10 months following irradiation. Multiple endpoints are under investigation in anterior marginal lens epithelial cells (i.e. the region of the lens where lens fibres are formed), including DNA damage assessed by the presence of gamma-H2AX staining, cell proliferation, cellular senescence, cell cycle arrest, and lens fibre morphogenesis. Additional tissues are being stored from the mice exposed in the study to allow investigation of comparative radiosensitivity in a range of tissues (including blood, skin, and whiskers).

This collaborative study between PHE and Durham University is at an early stage; experimental assays are being refined but it is clear that the mouse lens provides an accessible system for the study of early- and late-phase effects related to radiation exposure and cataract formation. An important aspect of the study is sampling over an extended timeframe, so that early changes can be related to later changes more evidently involved in cataract formation. In the longer term, the availability of well-defined experimental mouse models with a wide range of genetic conditions will help to characterise relevant pathways and susceptibility factors that contribute to cataract formation at low doses.

REFERENCES

Ainsbury, E.A., Bouffler, S.D., Dörr W, et al., 2009. Radiation cataractogenesis: a review of recent studies. Radiat. Res. 172, 1–9.

Ainsbury, E.A., Bouffler, S., Cocker, M., et al., 2014. Public Health England survey of eye lens doses in the UK medical sector. J. Radiol. Protect. 34, 15–29.

Bouffler, S., Ainsbury, E., Gilvin, P., et al., 2012. Radiation-induced cataracts: the Health Protection Agency's response to the ICRP statement on tissue reactions and recommendation on the dose limit for the eye lens. J. Radiol. Protect. 32, 479–488.

Broughton, J., Cantone, M.C., Ginjaume, M., et al., 2013. Report of task group on the implications of the implementation of the ICRP recommendations for a revised dose limit to the lens of the eye. J. Radiol. Protect. 33, 855–868.

Churchill, A., Graw, J., 2011. Clinical and experimental advances in congenital and paediatric cataracts. Phil. Trans. R. Soc. Lond. 366, 1234–1249.

Donadille, L., Carinou, E., Brodecki, M., et al., 2011. Staff eye lens and extremity exposure in interventional cardiology: results of the ORAMED project. Radiat. Measur. 46, 1203–1209.

Efstathopoulos, E.P., Pantos, I., Andreou, M., et al., 2011. Occupational radiation doses to the extremities and the eyes in interventional radiology and cardiology procedures. Br. J. Radiol. 84, 70–77.

Englefield, C., 2011. Is the new ICRP eye dose limit justified? J. Radiol. Protect. 31, 499–500.

Gilvin, P.J., Baker, S.T., Gibbens, N.J., et al., 2013. Type testing of a head band dosemeter for measuring eye lens dose in terms of Hp(3). Radiat. Protect. Dosimetry 157, 430–436.

He, S., Li, X., Chan, N., et al., 2013. Epigenetic mechanisms in ocular disease. Molec. Vision 19, 665–674.

ICRP, 2011. Statement on Tissue Reactions. International Commission on Radiological Protection, Ottawa. Available at: http://www.icrp.org/page.asp?id = 123 (last accessed 14 January 2015).

ICRP, 2012. Statement on tissue reactions/early and late effects of radiation in normal tissues and organs - threshold doses for tissue reactions in a radiation protection context. ICRP Publication 118. Ann. ICRP 41(1/2).

Kleiman, N.J., David, J., Elliston, C.D., et al., 2007. Mrad9 and atm haploinsufficiency enhance spontaneous and X-ray-induced cataractogenesis in mice. Radiat. Res. 168, 567–573.

Koukorava, C., Carinou, E., Ferrari, P., et al., 2011. Study of the parameters affecting operator doses in interventional radiology using Monte Carlo simulations. Radiat. Measur. 46, 1216–1222.

Little, M.P., 2013. A review of non-cancer effects, especially circulatory and ocular diseases. Radiat. Environ. Biophysics 52, 435–449.

Martin, C.J., 2011. A 20 mSv dose limit for the eye: sense or no sense? J. Radiol. Protect. 31, 385–387.

Merriam, G.R. Jr, Worgul, B.V., 1983. Experimental radiation cataract – its clinical relevance. Bull. N.Y. Acad. Med. 59, 372–392.

Neriishi, K., Nakashima, E., Minamoto, A., et al., 2007. Post-operative cataract cases among atomic bomb survivors: radiation dose response and threshold. Radiat. Res. 168, 404–408.

Pendergrass, W., Zitnik, G., Tsai, R., et al., 2010. X-ray induced cataract is preceded by LEC loss, and coincident with accumulation of cortical DNA, and ROS; similarities with age-related cataracts. Molec. Vision 16, 1496–1513.

Vañó, E., Ubeda, C., Leyton, F., et al., 2009. Staff radiation doses in interventional cardiology: correlation with patient exposure. Pediatr. Cardiol. 30, 409–413.

Wiley, L.A., Rajagopal, R., Dattilo, L.K., et al., 2011. The tumor suppressor gene Trp53 protects the mouse lens against posterior subcapsular cataracts and the BMP receptor Acvr1 acts as a tumour suppressor in the lens. Dis. Model. Mech. 4, 484–495.

SAGE

ICRP
Annals of the ICRP

Dosimetric models of the eye and lens of the eye and their use in assessing dose coefficients for ocular exposures

W.E. Bolch[a], G. Dietze[b], N. Petoussi-Henss[c], M. Zankl[c]

[a]*University of Florida, Gainesville, FL, USA; e-mail: wbolch@ufl.edu*
[b]*Braunschweig, Germany*
[c]*Helmholtz Zentrum Munchen, Germany*

Abstract–Based upon recent epidemiological studies of ocular exposure, the Main Commission of the International Commission on Radiological Protection (ICRP) in ICRP *Publication 118* states that the threshold dose for radiation-induced cataracts is now considered to be approximately 0.5 Gy for both acute and fractionated exposures. Consequently, a reduction was also recommended for the occupational annual equivalent dose to the lens of the eye from 150 mSv to 20 mSv, averaged over defined periods of 5 years. To support ocular dose assessment and optimisation, Committee 2 included Annex F within ICRP *Publication 116*. Annex F provides dose coefficients – absorbed dose per particle fluence – for photon, electron, and neutron irradiation of the eye and lens of the eye using two dosimetric models. The first approach uses the reference adult male and female voxel phantoms of ICRP *Publication 110*. The second approach uses the stylised eye model of Behrens et al., which itself is based on ocular dimensional data given in Charles and Brown. This article will review the data and models of Annex F with particular emphasis on how these models treat tissue regions thought to be associated with stem cells at risk.

Keywords: Ocular exposure; Eye lens; Cataract; Dose conversion coefficient

1. INTRODUCTION

Radiation exposure to the eye and lens of the eye may occur within many occupational and medical settings (Ainsbury et al., 2009). Ocular exposures can be a routine facet of radiological protection within any facility where spatially broad

This paper does not necessarily reflect the views of the International Commission on Radiological Protection.

fields of gamma rays, neutrons, or charged particles may arise. Occupational exposures to the lens of the eye may occur in nuclear power plants, in research accelerator laboratories, or in industrial irradiation facilities. Exposures of the eye of medical patients may result during computed tomography (CT) examinations of the head or during neurovascular interventional procedures (Stein et al., 2010). These same medical radiation sources may also result in ocular exposures to physicians, nurses, and radiological technologists. Low-energy photon fields are produced through machine scatter and leakage to varying degrees across all forms of diagnostic imaging – radiography, diagnostic fluoroscopy, CT, and nuclear medicine. For the latter, the patients themselves serve as an additional radiation source. Scatter fields of higher energy photons may be experienced within areas of external beam radiation therapy. Some of the highest occupational ocular exposures in medicine involve physicians, radiologists, and cardiologists performing fluoroscopically guided interventions and, more recently, CT angiography (Vañó et al., 2010). Unique exposures to the lens of the eye from high linear energy transfer (LET) radiations occur during space exploration activities. Aircraft crew are also subject to neutron and charged-particle radiation fields following cosmic ray interactions with the aircraft hull. In these scenarios, the lens of the eye, owing to its superficial location within the body, may receive absorbed doses that can potentially exceed the threshold for tissue reactions (Rafnsson et al., 2005).

The primary radiobiological concern during ocular irradiation is the induction of one or more radiogenic cataracts. This paper will cover the International Commission on Radiological Protection (ICRP) recommendations for dose thresholds and dose limits for ocular exposure, review dosimetric models for assessment of absorbed dose to the eye and lens of the eye using Monte Carlo radiation transport methods, and conclude with a review of the most recent reference dose conversion coefficients for photon, electron, and neutron ocular exposures as given in *Publication 116* (ICRP, 2010).

2. ICRP RECOMMENDATIONS FOR DOSE THRESHOLDS AND DOSE LIMITS FOR OCULAR EXPOSURES

The most recent recommendations by the Main Commission of ICRP were published in 2007 in *Publication 103* (ICRP, 2007). In this document, the Commission recommended an annual dose limit for planned exposure situations set at 20 mSv effective dose, with an additional annual limit on the equivalent dose to the lens of the eye of 150 mSv. It is noted that the lens of the eye is thought to be subject to tissue reactions alone (formerly deterministic effects), and thus the lens of the eye is not a constituent tissue of the effective dose, even in its remainder term. For members of the general public, the annual limits on effective dose and equivalent dose to the lens of the eye are 1 mSv and 15 mSv year^{-1}, respectively. The Commission, however, explicitly noted that: 'this limit [to the lens of the eye] is currently being reviewed by an ICRP Task Group', inferring that the dose limit could be subject to change given current reviews of the radiobiology and radiation epidemiology literature.

While the parameter values used in the calculation of effective dose – specifically w_R and w_T – were updated in *Publication 103* (ICRP, 2007) over those given previously in *Publication 60* (ICRP, 1991), the numeral limits for annual effective dose and annual equivalent dose to the lens of the eye remained unchanged from the 1990 Recommendations. In both cases, the annual limit for equivalent dose to the lens of the eye is based, in part, on assumed values of the threshold estimate of acute absorbed dose for 1% incidence of morbidity after whole-body gamma ray exposure. For cataract formation (visual impairment), the assumed dose threshold was thought to be approximately 1.5 Gy with a time-to-effect of several years (see Table A.3.4 of *Publication 103*). The primary references for this dose threshold were *Publication 41* (ICRP, 1984) and Edwards and Lloyd (1996).

In 2011, the Main Commission of ICRP issued its Statement on Tissue Reactions as Part 1 of *Publication 118* (ICRP, 2012). Part 2 of that report included a comprehensive review of early and late effects of radiation in normal tissues and organs including the lens of the eye. A wide variety of human epidemiological studies were reviewed that supported or questioned a lower (or even zero) threshold model for radiation induction of cataracts. Studies that supported a lower threshold covered ocular exposures to patients undergoing diagnostic procedures or radiotherapy, astronauts, atomic bomb survivors, residents in highly contaminated buildings, nuclear power plant workers, first responders to the Chernobyl accident, and workers in various medical exposure fields. Studies that questioned a lowering of the dose threshold for radiogenic cataracts included additional reviews of ocular exposures in diagnostic medicine, radiotherapy, and nuclear power plants.

Following this review, the Commission concluded that recent epidemiological evidence suggests that for some tissue reactions, particularly those with very late manifestation, threshold doses are lower than previously assumed. For the lens of the eye, the Commission, in their 2011 Statement (ICRP, 2012), indicates that the threshold in absorbed dose for cataract induction is now considered to be not higher than 0.5 Gy, which is lower by more than a factor of 3 from that stated in *Publication 60* and *Publication 103* (ICRP, 1991, 2007). Resultantly, for occupational exposures in planned exposure situations, the Commission now recommends an equivalent dose limit to the lens of the eye of 20 mSv year^{-1}, averaged over defined periods of 5 years, with no single year exceeding 50 mSv. This nominal dose limit is thus a factor of 7.5 lower than that given previously in its 1990 and 2007 Recommendations (ICRP, 1991, 2007). The 2011 Statement (ICRP, 2012) did not specifically address changes in the annual dose limits for ocular exposures to members of the general public.

3. USE OF OPERATIONAL QUANTITIES FOR ASSESSING DOSE TO THE LENS OF THE EYE

The revised ICRP annual limit of 20 mSv on the radiation exposure to the lens of the eye is based on the radiation protection quantity, equivalent dose. This quantity is not directly measurable within the exposed individual, and thus cannot be used directly for radiation monitoring. ICRP thus adopts various operational quantities as defined by the International Commission on Radiation Units and Measurements (ICRU) for demonstration of compliance with regulations on occupational exposures and for dose optimisation. Three operational quantities are formally defined: the ambient dose equivalent $H^*(d)$; the directional dose equivalent $H'(d,\Omega)$; and the personal dose equivalent $H_p(d)$ (ICRU, 1993). The depth parameter d is given as 0.07 mm, 3 mm, or 10 mm (in tissue-equivalent material) for monitoring exposures of the skin, lens of the eye, or effective dose, respectively. Both the ambient and directional dose equivalent are applied to area monitoring, while the personal dose equivalent is reserved for individual monitoring of radiation workers (via personal dosimeters). While the quantity $H^*(10)$ is formally assigned for area monitoring in the control of effective dose, all three quantities – $H^*(3)$, $H'(3,\Omega)$, and $H_p(3)$ – have been used or proposed for control and monitoring of dose to the lens of the eye in relation to the ICRP dose limits. Very few physical dosimeters, however, have been constructed and/or calibrated to assess these operational quantities properly for $d = 3$ mm. In both *Publication 103* (ICRP, 2007) and *Publication 116* (ICRP, 2010), the Commission notes that dosimeters designed to measure $H'(0.07,\Omega)$ and $H_p(0.07)$ for skin dose monitoring may also be applied for the purpose of monitoring exposure of the lens of the eye. With increased interest in dose assessment for the lens of the eye, given the Commission's downward revision of its annual dose limit, many investigators are putting new emphasis on devices that measure dose equivalent at the nominal 3-mm depth in tissue.

4. USE OF DOSE COEFFICIENTS FOR ASSESSING DOSE TO THE LENS OF THE EYE

An alternative approach to dose assessment for the lens of the eye is through the use of numerical computation of particle fluence coupled with reference external fluence-to-dose conversion coefficients. A dose conversion coefficient for external radiation fields is defined as the tissue-equivalent dose per particle fluence (for example, in units of pGy cm^{-2}). Values of the dose conversion coefficient for the lens of the eye as a function of particle energy may then be compared to values of $H^*(3)/\Phi$, $H'(3,\Omega)/\Phi$, and $H_p(3)/\Phi$ in determining the appropriateness of the operational quantities in monitoring whether or not annual limits have been met or exceeded. Dose conversion coefficients for the lens of the eye have recently been compiled in *Publication 116* using geometric models of the eye and lens of the eye, and are reviewed briefly below (ICRP, 2010).

5. DOSIMETRIC MODELS FOR DOSIMETRY OF THE LENS OF THE EYE

Current reference values for dose conversion coefficients for the lens of the eye adopted by the Commission in *Publication 116* (ICRP, 2010) were developed using two geometric models of the eye. The first is the voxelised representation of the head, eye, and lens of the eye within the *Publication 110* reference voxel phantoms of the reference adult male and reference adult female (ICRP, 2009). The second is a geometric stylised eye model taken from the studies by Behrens et al. (2009). This stylised eye model was used to supplement dose conversion coefficients for the lens of the eye derived from the *Publication 110* phantoms (ICRP, 2009), where the lower voxel resolution within the later did not permit adequate spatial resolution of energy deposition within the lens of the eye, particularly at low-incident particle energies. The Behrens et al. (2009) model was, in turn, based upon an early and seminal study by Charles and Brown (1975) of ocular dimensions and the location of cells at risk.

5.1. Geometric model of Charles and Brown (1975)

Charles and Brown (1975) sought to provide guidance on improved dosimeter design for monitoring dose to the lens of the eye. Their main thesis was that the radiosensitive cells responsible for radiogenic cataract induction were not distributed uniformly across the whole lens of the eye and, thus, their specific location should be quantified and used as the basis for improved dosimeter design. Figs 1 and 2 show the relevant structures of the eye and lens of the eye reviewed in this study.

5.2. Location of cells at risk

The authors noted that Goldman and Liechti (1938) were perhaps the first to demonstrate that radiation damage to the equatorial region of the lens was a prerequisite to radiogenic cataract induction. The lens is a transparent elastic structure that is biconvex in shape, with the rounded junction of the two surfaces referred to as the 'equator'. Goldman and Liechti used narrow x-ray beams to irradiate selective portions of rabbit lens, and found that cataract formation occurred only within the

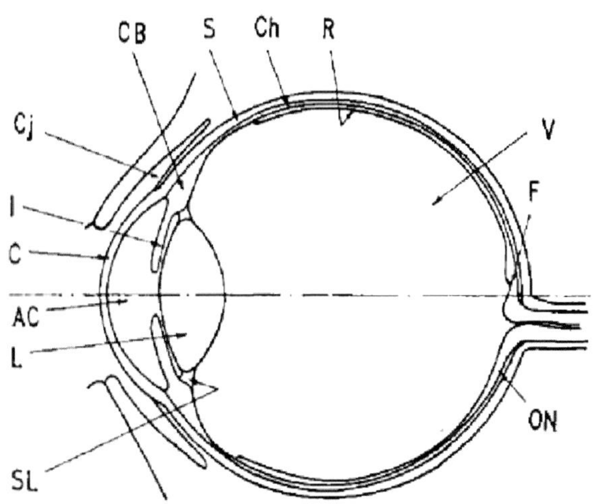

Fig. 1. The eye in its sagittal cross-section. AC, anterior chamber; CB, ciliary body; ON, optic nerve; Cj, conjunctiva; SL, suspensory ligament of lens; Ch, choroid; V, vitreous body; C, cornea; F, fovea; I, iris; L, lens; R, retina; S, sclera. Source, Fig. 1 of Charles and Brown (1975).

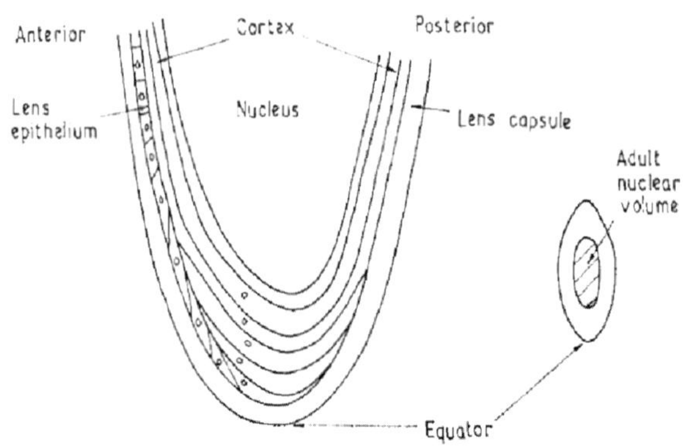

Fig. 2. Transverse section of the lens. Source: Fig. 2 of Charles and Brown (1975).

equatorial region of the irradiated lens. In a subsequent study by Alter and Leinfelder (1953), quadrantal exposure of the rabbit lens with x rays produced partial cataracts only in the exposed section, and no cataracts were produced when only the central region of the lens was irradiated. Von Sallman et al. (1955) further demonstrated that the central area of the lens was relatively radio-resistant to deuterons irradiation, whereas the equatorial region of the lens was highly radiosensitive with regard to lens opacities. These and other experiments reviewed by Charles and Brown provided evidence that the damaged cells of the germinative epithelium in the equatorial region of the lens were the precursors for radiation-induced cataract formation.

5.3. Data of cells at risk

The question of the depth of the cells at risk was further addressed in Charles and Brown (1975) through the development of geometric models of the normal (emmetropic) eye and their extremes as given in the myopic (shorted-sighted) eye and the hypermetropic (long-sighted) eye. Dimensional data were gathered on nine parameters: (1) the radius of curvature of the anterior surface of the cornea; (2) the radius of curvature of the anterior surface of the lens; (3) the radius of curvature of the posterior surface of the lens; (4) the radius of curvature of the anterior surface of the sclera; (5) the depth of the anterior chamber along the optical axis; (6) the thickness of the cornea along the optical axis; (7) the thickness of the lens along the optical axis; (8) the diameter of the cornea; and (9) the equatorial diameter of the lens. Figs 3 and 4 show examples of the acquired data gathered from prior studies by Lowe and Clark (1973a,b) and Brown (1973, 1974a,b). Fig. 5 displays the final 'reference' eye model constructed by Charles and Brown. The extremes in ocular dimension – represented by the young myope and the older hypermetrope – are shown in Fig. 6. Data from Fig. 5 were used extensively in the chapter on eye anatomy given in *Publication 89* (ICRP, 2002).

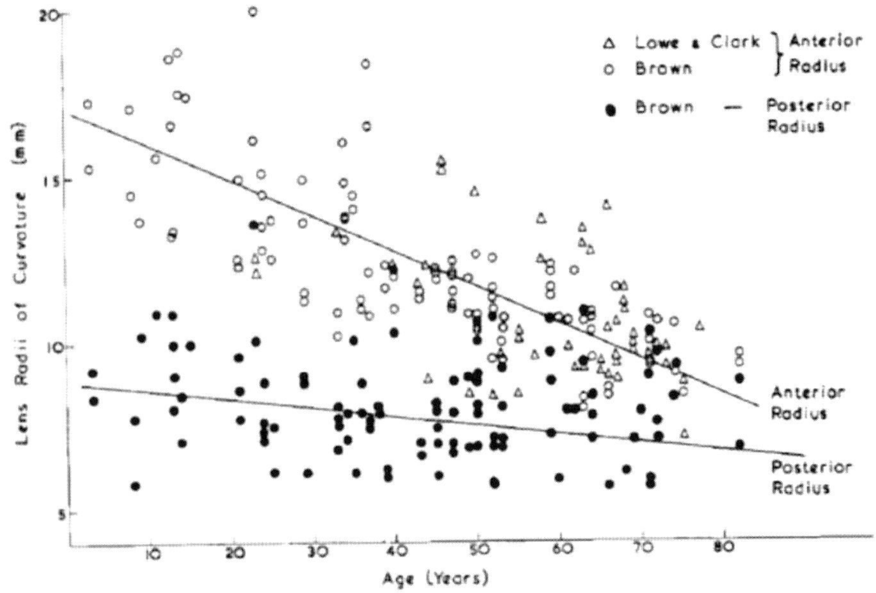

Fig. 3. Radii of curvature of the lens as a function of subject age. Source: Fig. 4 from Charles and Brown (1975).

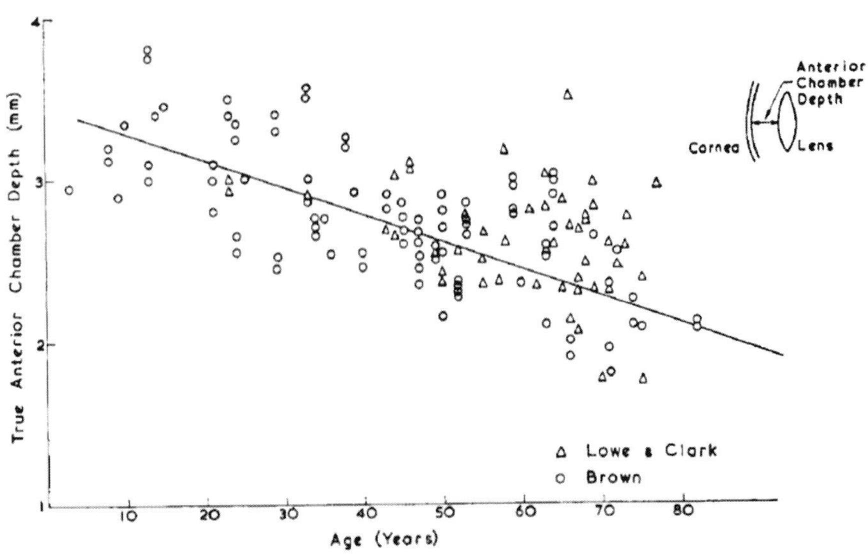

Fig. 4. Anterior chamber depth as a function of subject age. Source: Fig. 4 from Charles and Brown (1975).

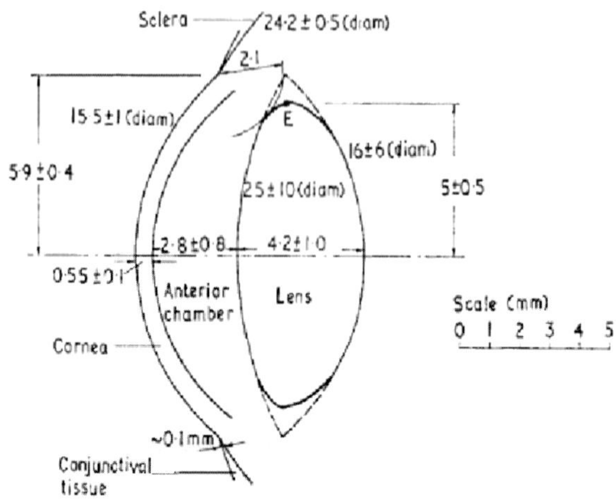

Fig. 5. Mean ocular dimensions for the unaccommodated eye. The indicated tolerances represent the limits found in normal adults aged 20–65 years. Source: Fig. 8 from Charles and Brown (1975).

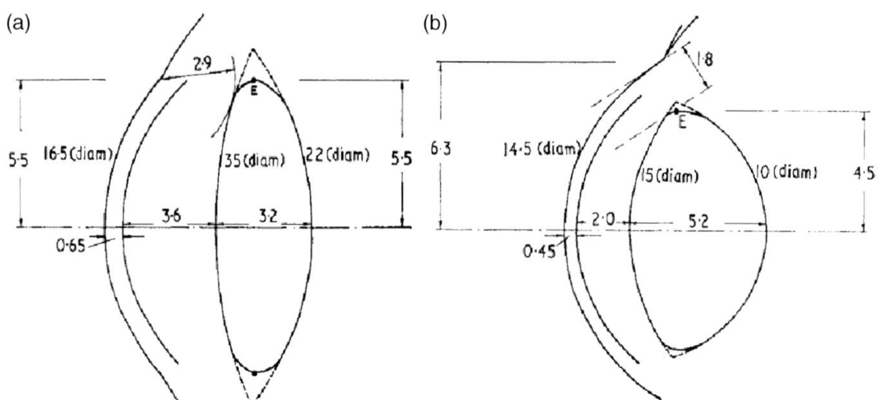

Fig. 6. Mean ocular dimensions for the deep lying lens of the young myope (a) and for the shallow lying lens of the older hypermetrope (b). Source: Figs 9 and 10 from Charles and Brown (1975).

In their review, given the dimension models of the eye shown above, Charles and Brown (1975) came to the following conclusions as relevant to radiological protection of the eye. First, because the curvature of the anterior cornea is always greater than that of the anterior surface of the lens, the equatorial portion of the lens – wherein lie the cells at risk – is positioned at a depth shallower to the eye surface than at any other region of the lens. Second, the depth of the lens equator varies very little over the extremes of eye geometry in the normal adult population. The depth of the equator as measured to the sclera–cornea junction is 2.6 ± 0.6 mm, and the depth measured in the anterior–posterior (AP) direction is 3.1 ± 0.3 mm. The lower and upper limits correspond to older and younger age groups, respectively. Third, the minimum depth of any part of the lens is indicated by drawing an arc from the cornea–sclera junction, which just touches the lens. This corresponds to a value of 2.3 ± 0.4 mm and is associated with a lens region just anterior to the equator.

In their recommendations for dosimeter design, the authors note that a planar dosimeter that measures tissue dose at a depth of approximately 3 mm should give 'a reasonable overall measure of the mean equatorial dose for a variety of eye and irradiation geometries likely to be met during the lifetime's exposure experience at radiation protection levels'. Furthermore, an ideal dosimeter would integrate the tissue dose between depths of 2.5 and 3.5 mm as this 1-mm thickness covers the approximate extent of the equatorial regions of the lens over which the cells at risk lie.

5.4. Geometric eye model of Behrens et al. (2009) for electron dosimetry

In a study by Behrens et al. (2009), a new three-dimensional (3D) geometric, stylised model of the eye was developed for the purpose of computing dose conversion coefficients for externally incident mono-energetic electrons using Monte Carlo radiation transport methods. The model was constructed, in large part, based upon dimensional data taken from Charles and Brown (1975). Figs. 7 and 8 show the Behrens et al. eye model in its 2D schematic and 3D pictorial views, respectively. The model includes an analysis of the elemental composition and mass density of each of the coloured regions of Fig. 8. Mass densities (units of g cm^{-3}) were assigned as 1.09 for the eye lid, 1.06 for the lens (both sensitive and insensitive regions), 1.00 for the aqueous and vitreous humour, and 1.076 for the cornea. Data sources included *Publication 23* (ICRP, 1975) and Report 56 (ICRU, 1997).

The model shown in Figs. 7 and 8 results in a tissue-equivalent depth of the sensitive volume of the lens between 2.80 mm and 3.82 mm (mean 3.36 mm) and a sensitive volume thickness of between 0 mm and 0.595 mm (mean 0.495 mm). The mass of the entire lens and its sensitive volume are given at 229.1 mg and 39.2 mg, respectively. The authors note that the mean depth of 3.36 mm is quite similar to the nominal value of 3 mm in the definition of the operational quantities, directional dose equivalent $H'(3,\Omega)$ and personal dose equivalent $H_p(3)$ for ocular exposure monitoring and regulatory compliance.

In Behrens et al. (2009), the eye model was used to assess dose conversion coefficients for mono-energetic electrons normally incident upon the eye and at three off-axis polar angles – 15°, 30°, and 45° – and for three directions – side, bottom, and top of the eye. The range of electron energies considered were 100 keV to 12 MeV. All Monte Carlo simulations were performed using EGSnrc Version 4. In this study, the geometry of the surrounding head and tissue structures were not considered in that: (1) electrons at angles of incidence less than 45° were considered as most relevant to occupational and medical exposures; and (2) electrons of energies below several MeV

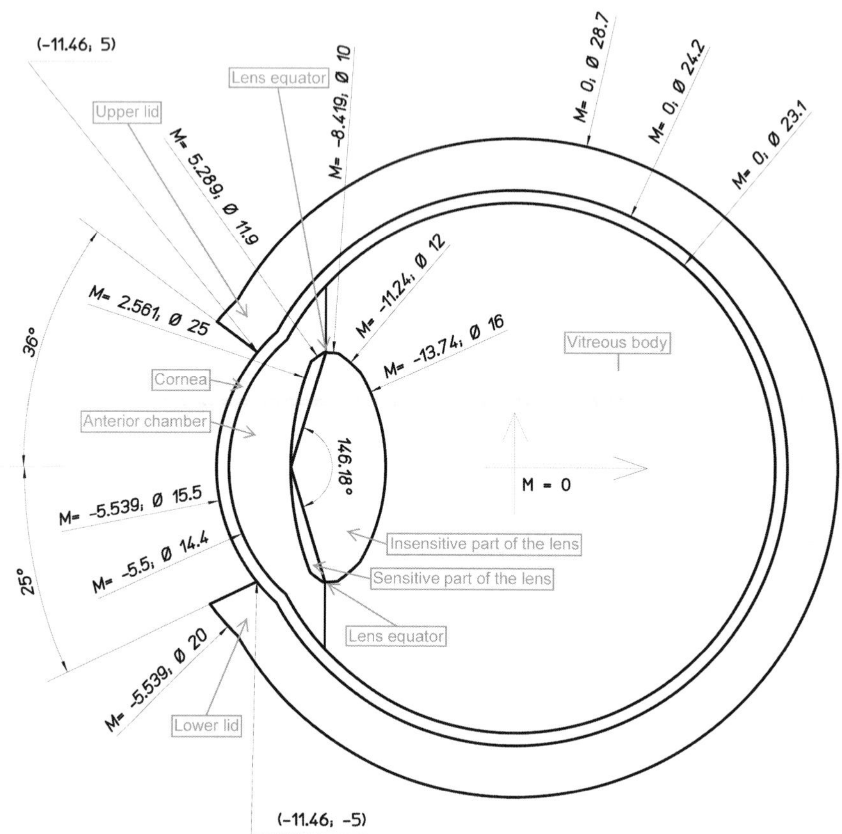

Fig. 7. Eye geometry used in the study by Behrens et al. (2009). All dimensions are in millimetres and M denotes the x-position of the centre of the eye sphere. Source: Fig. 1 of Behrens et al. (2009).

have small tissue ranges and thus would not reach the lens of the eye at larger incident angles. Virtually all beta particles of relevance to radiological protection have endpoint energies below approximately 3.5 MeV.

5.5. Geometric eye model of Behrens and Dietze (2011) for photon dosimetry

In 2011, the original stylised eye model used for Monte Carlo radiation transport simulation of external electron ocular irradiation was extended to the study of photon exposures of the eye and lens of the eye. In this study, Behrens and Dietze (2011) used the geometric model of Behrens et al. (2009) described above, but considered the influence of scatter and attenuation of incident photons striking the exposed individual and their influence on ocular tissue dose. Accordingly, a left and right stylised eye model was inserted within a larger gender-averaged whole-body anatomic model as defined by Kramer et al. (1982) (described in detail within Appendix A of their paper).

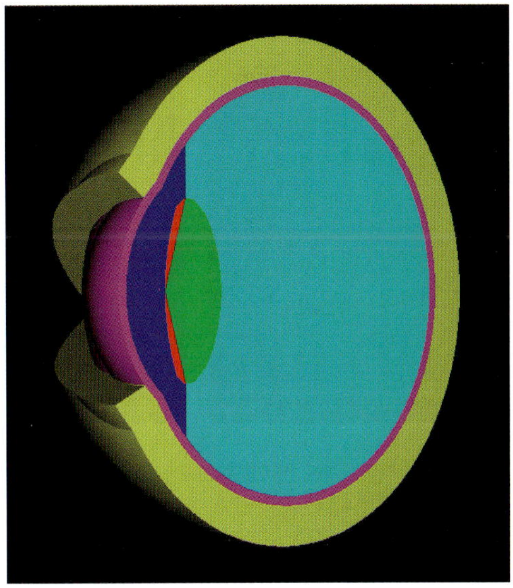

Fig. 8. Three-dimensional view of the eye geometric model of Behrens et al. The following colour code is used: yellow: skin of eye lid; red: sensitive region of lens of the eye; green: insensitive region of lens of the eye; dark blue: aqueous humour; light blue: vitreous humour; magenta: cornea. Source: Fig. 2 of Behrens et al. (2009).

The resulting model is shown in Fig. 9. Mono-energetic photons of energies ranging from 5 keV to 10 MeV were considered under standard ICRP exposure geometries: AP; posterior–anterior (PA); left lateral; average of left and right lateral (LAT); and rotational (ROT) incidence. All Monte Carlo simulations were performed using EGSnrc Version 4 with full secondary charged particle transport. The dose conversion coefficients to the sensitive volume of the lens and the entire lens volume were valuated.

5.6. Geometric eye model of Manger et al. (2012) for neutron dosimetry

In 2012, the previous studies of electron and photon eye dosimetry were further extended to neutron ocular exposures. In a study by Manger et al. (2012), the stylised eye model of Behrens et al. (2009) was again adopted and inserted into a larger model of the head and neck. In this study, the latter model selected was that of the modified Oak Ridge National Laboratory stylised phantom as reported by Han et al. (2006). Dose conversion coefficients to the whole lens of the eye and to its sensitive layer were reported for discrete mono-energetic neutron energies between 10^{-9} and 20 MeV for AP, RLAT (right lateral), and ROT orientations. All simulations were performed using MCNPX Version 2.6.0.

5.7. Eye models in the ICRP reference phantoms

Publication 116 (ICRP, 2010) represents the very first application of the *Publication 110* reference phantoms – both the adult male and the adult female – in the calculation

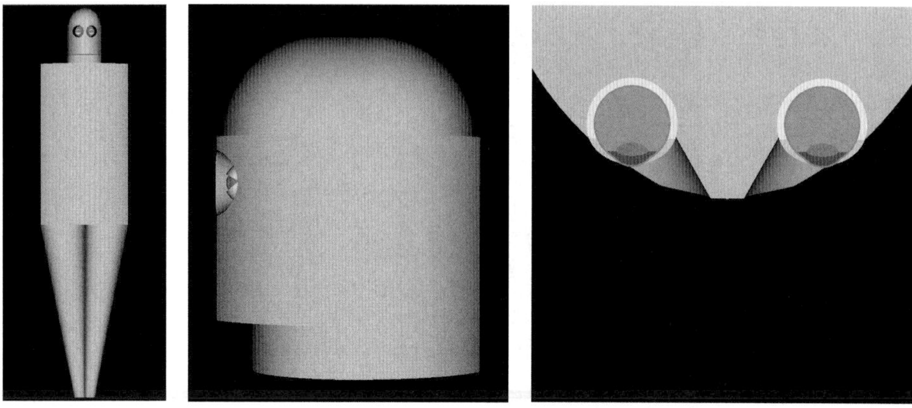

Fig. 9. View of the geometry used in the Monte Carlo simulations studies of Behrens and Dietze for external photon irradiation of the eye and lens of the eye. Source: Fig. 1 of Behrens and Dietze (2011).

of reference dose conversion coefficients within ICRP Committee 2. These ICRP reference phantoms are consistent with the information given in *Publication 89* (ICRP, 2002) on the reference anatomical parameters for both male and female adults and were constructed through modification of the voxel models 'Golem' and 'Laura' (Zankl and Wittmann, 2001; Zankl et al., 2005) of two individuals whose body height and mass closely resembled the reference data. Each phantom is represented as a 3D array of cuboid voxels, arranged in columns, rows, and slices. Each entry in the array identifies the organ or tissue to which the corresponding voxel belongs. The male reference computational phantom consists of 1.95 million tissue voxels (excluding voxels representing the surrounding vacuum), each with a slice thickness (corresponding to the voxel height) of 8.0 mm and an in-plane resolution (i.e. voxel width and depth) of 2.137 mm, corresponding to a voxel volume of 36.54 mm^3. The female reference computational phantom consists of 3.89 million tissue voxels, each with a slice thickness of 4.84 mm and an in-plane resolution of 1.775 mm, corresponding to a voxel volume of 15.25 mm^3. The number of individually segmented structures is 136 in each phantom, and 53 different tissue compositions have been assigned to them.

In the *Publication 110* reference male phantom, the right/left eye and lens of the eye are approximately 7.3 g and approximately 0.19 g, respectively (ICRP, 2009). At a mass density of 1.05 g cm^{-3}, these tissue regions correspond to 190 voxels for each male eye, but only approximately 5 voxels for each male lens. Correspondingly, in the *Publication 110* reference female phantom, the right/left eye and lens of the eye have masses of approximately 7.3 g and approximately 0.20 g, respectively, which correspond to voxel counts of approximately 456 voxels for each eye and 12 voxels for each lens. Transaxial views of both models are shown in Fig. 10. For moderately to highly penetrating external radiations, reasonable estimates of dose to the lens of the eye can be expected from Monte Carlo radiation transport within the *Publication 110* reference phantoms. For low-energy penetrating radiations, the coarse nature of the voxel geometry in these

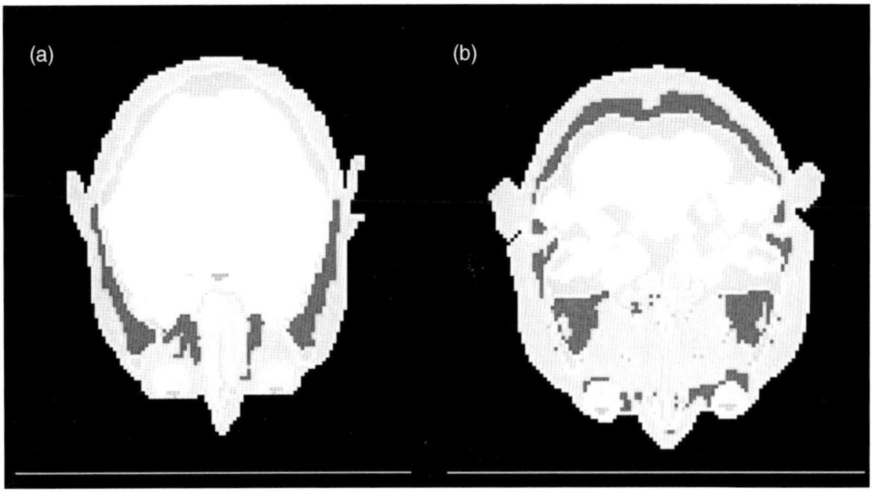

Fig. 10. Transaxial views through the *Publication 110* reference computational phantoms [male (a) and female (b)] showing the voxelised representation of the eye and lens of the eye.

phantoms will not fully capture rapidly changing dose gradients for external ocular irradiations. Furthermore, it is not possible to model the sensitive volume of the lens of the eye at these voxel dimensions. It was for these reasons that reference dose conversion coefficients in *Publication 116* (ICRP, 2010) were supplemented with values determined using the various stylised eye models presented above.

6. DOSE COEFFICIENTS FROM ANNEX F OF *PUBLICATION 116*

Annex F of *Publication 116* (ICRP, 2010) presents the Commission's recommended reference dose conversion coefficients for ocular exposures given in terms of absorbed dose to the lens of the eye per particle fluence (pGy cm^2). These reference conversion coefficients were defined either based upon values using the stylised eye model or from the voxelised model of the lens of the eye within the *Publication 110* reference computational phantoms (ICRP, 2009). Reference dose conversion factors for the lens of the eye are given for external photons (10 keV to 10 GeV), electrons (10 keV to 10 GeV) and neutrons (1 meV to 10 GeV) in tabular form within Annex F of *Publication 116*. A more extensive listing of these conversion coefficients is given in the electronic annex of *Publication 116* for photons, electrons, neutrons, positrons, protons, muons, pions, and He ions.

6.1. Photon irradiation of the eye

Data for external photon irradiation of the lens of the eye are shown in Fig. 11 for the AP, LAT, and ROT irradiation geometries. Data points show separately the dose conversion coefficients for the reference phantoms (gender averaged), the total lens within the stylised eye model and the sensitive volume of the lens of the eye within the stylised eye model. Solid curves give the final values of ICRP reference conversion

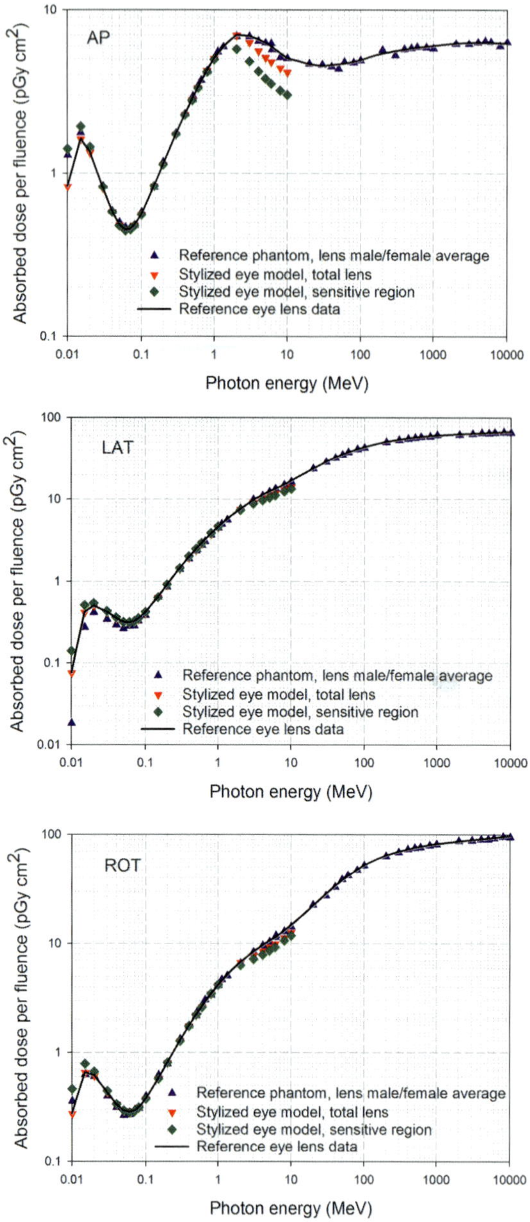

Fig. 11. Absorbed doses for the lens of the eye per photon fluence (pGy cm^2) for anterior–posterior (AP), average of left and right lateral (LAT), and rotational (ROT) irradiation as calculated with the male and female reference computational phantoms and with the stylised eye model of Behrens and Dietze (2011). The solid curves show the *Publication 116* reference values. Source: Fig. F.3 of ICRP (2010).

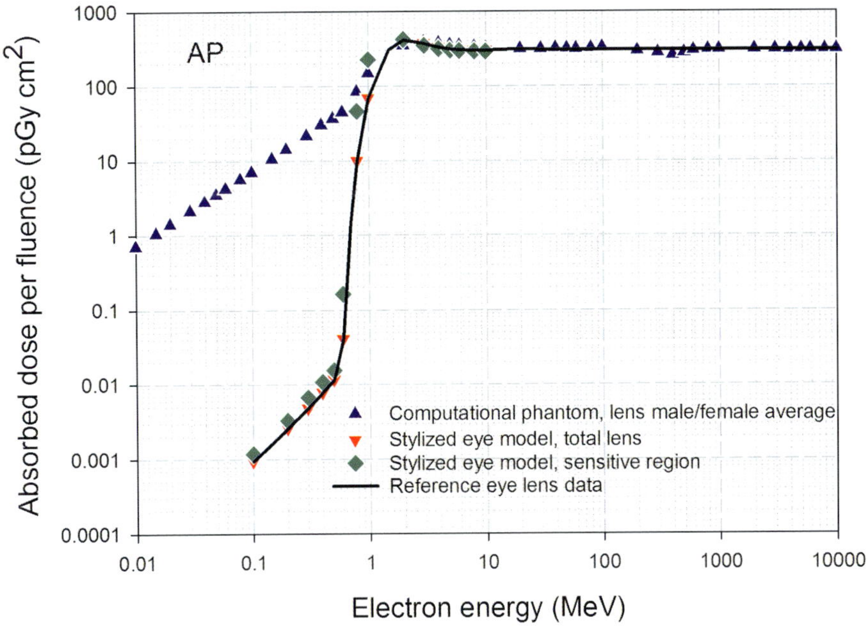

Fig. 12. Absorbed dose for the lens of the eye per electron fluence (pGy cm^2) for anterior–posterior (AP), as calculated with the male and female reference computational phantoms and the stylised eye model of Behrens et al. (2009). The solid curves show the *Publication 116* reference values. Source: Fig. F.4 of ICRP (2010).

coefficients for external photons as a function of photon energy. The graphs within Fig. 11 indicate that, with the exception of photon energies below approximately 20 keV for the AP geometry and 50 keV for other geometries, there is good agreement across the various eye models – voxel and stylised. Above 2 MeV, and in particular for AP irradiation, the data for the stylised model have a tendency to be less than those from the *Publication 110* reference phantoms (ICRP, 2009). It can also be seen that absorbed doses in the sensitive region of the lens of the eye are reasonably represented by the reference values at photon energies up to 1 MeV.

Following a conservative approach, the data from the stylised model was used in *Publication 116* (ICRP, 2010) as reference dose conversion coefficients for AP, PA, LAT, and ROT geometries for energies up to 2 MeV. For all other situations, the reference conversion coefficients for the lens of the eye were evaluated as their gender-averaged value from the *Publication 110* male and female computational phantoms (ICRP, 2009).

6.2. Electron irradiation of the eye

Fig. 12 shows a similar comparison of model results for external electron irradiation of the eye and lens of the eye under AP irradiation. Below 1 MeV, the lens dose for the stylised eye model is significantly lower than that given by the lower resolution eye model of the reference computational phantoms. This is due to the fact that at least one voxel in the reference phantom is in direct contact with the phantom's exterior (vacuum in the Monte Carlo simulations), which is not equivalent to that in the real ocular anatomy whereby the externally incident electrons would have to first traverse both the cornea and anterior chamber of the eye prior to reaching the lens. At energies between 1 and 10 MeV, however, there is good agreement between the model results.

Among the irradiation geometries considered in *Publication 116* (ICRP, 2010), the only significant dose to the lens of the eye for electrons of energies below 1 MeV are those normally incident under AP irradiation. Data for AP irradiation can be considered as conservative upper estimates for the other geometries below 1 MeV electron energy. For isotropic (ISO) irradiation and energies above 1 MeV, and for PA irradiations at all energies, no significant differences between the dose conversion coefficients from voxel and stylised models were evident. For ISO geometry and electrons with energies lower than 1 MeV, the same dose conversion coefficients as for AP geometry were adopted. For all other situations, reference data were evaluated as their gender-averaged values from the reference computational phantoms.

6.3. Neutron irradiation of the eye

Fig. 13 shows a further comparison of model results for external neutron irradiation of the eye and lens of the eye under the AP, LAT, and ROT irradiation geometries. Data points show separately the dose conversion coefficients for the reference phantoms (gender averaged), the total lens within the stylised eye model, and the sensitive volume of the lens of the eye within the stylised eye model. Solid curves give the final values of ICRP reference conversion coefficients for external neutrons as a function of neutron energy. Manger et al. (2012) concluded that the dose to the sensitive region of the lens of the eye was similar to that of the total lens over all neutron energies and exposure geometries, with the greatest differences being 13% at the lowest incident neutron energy. Regardless of the orientation of the exposure, no significant difference was found in the absorbed dose in the sensitive volume and total volume of the lens of the eye.

Reference values for dose conversion coefficients for externally incident neutrons are given in *Publication 116* (ICRP, 2010) as taken from the stylised model of Manger et al. (2012) for AP, LAT and ROT geometries and energies up to 4 MeV. All other reference conversion coefficients were assigned as their gender-averaged values seen in the male and female reference computational phantoms.

6.4. Relationship to the operational quantities for the lens of the eye

An important analysis presented in Chapter 5 of *Publication 116* (ICRP, 2010) is the comparison of the new reference dose conversion coefficients for external radiation exposure with the various operational quantities used for either area or individual monitoring of workers exposed to these fields. Such comparisons are shown in Fig. 14 for photon irradiation of the lens of the eye under various exposure situations (AP, PA, LAT, ROT, and ISO). In Fig. 14, it is clear that the ambient dose equivalent

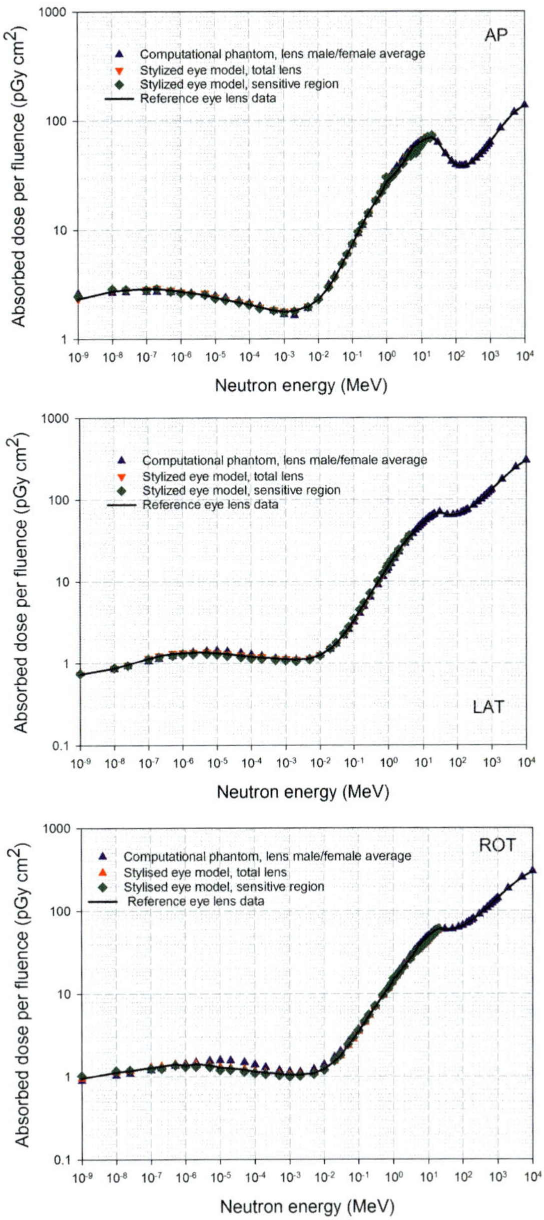

Fig. 13. Absorbed doses for the lens of the eye per neutron fluence (pGy cm^2) for anterior–posterior (AP), average of left and right lateral (LAT), and rotational (ROT) irradiation as calculated with the male and female reference computational phantoms and with the stylised eye model of Manger et al. (2012). The solid curves show the *Publication 116* reference values. Source: Fig. F.5 of ICRP (2010).

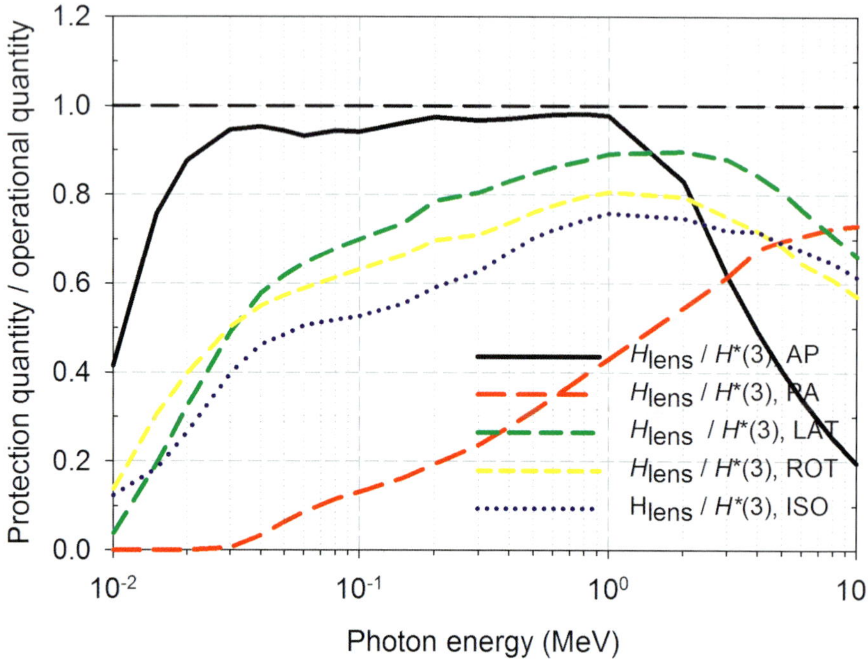

Fig. 14. Ratios of equivalent dose to the lens of the eye (*Publication 116*) for various exposure situations (AP, anterior–posterior; PA, posterior–anterior (PA); LAT, average of left and right lateral; ROT, rotational; ISO, isotropic) to the ambient dose equivalent at 3-mm depth $H^*(3)$ from *Publication 74* for externally incident photons. Source: Fig. 5.8 of ICRP (2010).

evaluated at a tissue-equivalent depth of 3 mm, $H^*(3)$, is a reasonably good estimator of photon dose to the lens of the eye under AP irradiation at energies between 100 keV to 1 MeV and is a very conservative estimator of that same dose at energies below 100 keV, and above 1 MeV. $H^*(3)$ is an even more conservative estimator of dose to the lens of the eye for other irradiation geometries across the energy range considered.

A comparison of dose to the lens of the eye for externally incident mono-energetic electrons and the operational quantities $H^*(3)$ and $H'(3)$ are shown in Fig. 15. At electron energies of 3 MeV and above under AP irradiation, the ambient dose equivalent evaluated at 3-mm tissue-equivalent depth provides a very good estimate of the dose to the lens of the eye and is a conservative estimate at lower energies. For both PA and ISO irradiation geometries, $H^*(3)$ is shown to be a conservative estimator of electron dose to the lens of the eye at energies up to 100 MeV, above which the ambient dose equivalent underestimates the value of dose to the lens of the eye. The authors of *Publication 116* (ICRP, 2010) note that precise data for the personal dose equivalent $H_p(3)$ on a head phantom are not available at present. Furthermore, ICRU is currently evaluating the suitability of all existing operational quantities, their calibration procedures, and appropriate calibration phantoms.

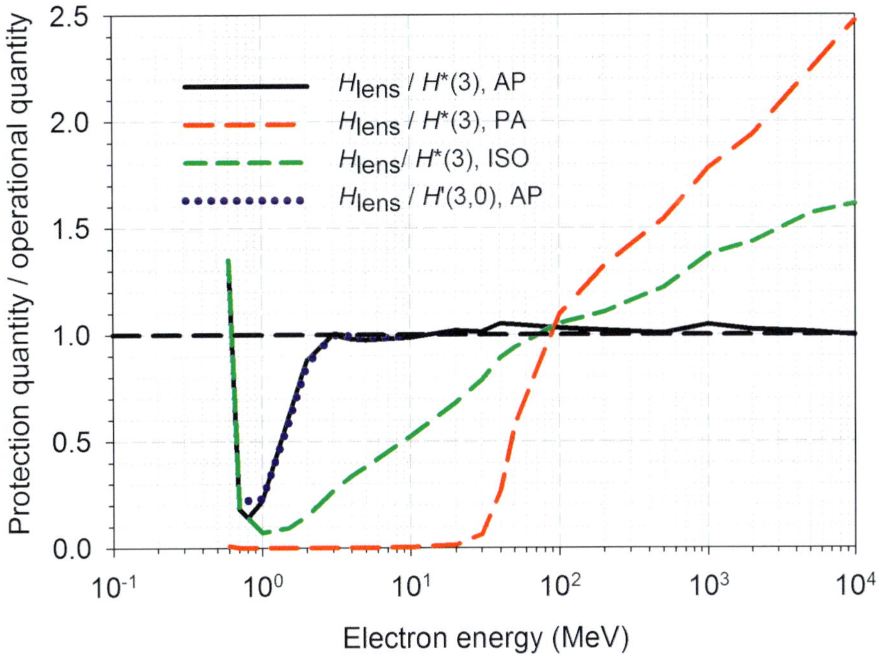

Fig. 15. Ratios of equivalent dose to the lens of the eye [*Publication 116* (ICRP, 2010)] for various exposure situations (AP, anterior–posterior; PA, posterior–anterior (PA); LAT, average of left and right lateral; ROT, rotational; ISO, isotropic) to the directional dose equivalent $H'(3)$ for electrons from *Publication 74* and to the ambient dose equivalent $H^*(3)$ based on data from Ferrari and Pelliccioni (1994). Source: Fig. 5.9 of ICRP (2010).

7. SUMMARY

In the past decade, new and significant information has arisen regarding our understanding of the radiation risk of cataract formation. As a result, in *Publication 118* (ICRP, 2012), ICRP revised downwards the presumed threshold for tissue reactions of the lens of the eye at 1% morbidity incidence from 1.5 Gy to a current value of no more than 0.5 Gy, a reduction by a factor of 3. Accordingly, the annual limit recommended by ICRP for exposures of the lens of the eye has similarly been revised downwards from 150 mSv to 20 mSv, a reduction of a factor of 7.5. Occupational monitoring of eye exposures to patients, physicians, and radiation workers has thus become of increased importance. To assist in computational assessments of dose to the lens of the eye in exposed persons, Annex F of *Publication 116* (ICRP, 2010) provides newly approved reference data on = conversion coefficients for the lens of the eye – absorbed dose to the lens of the eye per particle fluence (pGy cm^2) – for external fields of photons, electrons, and neutrons. These reference conversion coefficients were determined using two data sets – one based on Monte

Carlo simulations with the *Publication 110* reference computational phantoms (ICRP, 2009) and another based upon stylised models of the eye in which the more radiosensitive layer of the lens is included as a target region. Given a priori knowledge of the particle fields at the level of the eye exposure, estimates of lens dose can be assessed using these reference coefficients. Regarding dosimetric monitoring of eye exposures, further work is needed in the development of meaningful operational quantities and their associated calibration phantoms for lens dosimetry, particularly in light of these new and lower annual limits.

REFERENCES

Ainsbury E.A., Bouffler S.D., Dörr W., et al., 2009. Radiation cataractogenesis: a review of recent studies. Radiat. Res. 172, 1–9.

Alter A.J., Leinfelder P.J., 1953. Roentgen-ray cataract – effects of shielding of the lens and ciliary body. AMA. Arch. Ophthalmol. 49, 257–260.

Behrens R., Dietze G., 2011. Dose conversion coefficients for photon exposure of the human eye lens. Phys. Med. Biol. 56, 415–437.

Behrens R., Dietze G. and Zankl M., 2009. Dose conversion coefficients for electron exposure of the human eye lens. Phys. Med. Biol. 54, 4069–4087.

Brown N., 1973. Change in shape and internal form of lens of eye on accommodation. Exp. Eye Res. 15, 441–459.

Brown N., 1974a. Change in lens curvature with age. Exp. Eye Res. 19, 175–183.

Brown N., 1974b. Shape of lens equator. Exp. Eye Res. 19, 571–576.

Charles M.W., Brown N., 1975. Dimensions of the human eye relevant to radiation protection. Phys. Med. Biol. 20, 202–218.

Edwards A.A., Lloyd D.C., 1996. Risk from deterministic effects of ionising radiation. Doc. NRPB. 7, 1–31.

Ferrari A., Pelliccioni M., 1994. Dose equivalents for monoenergetic electrons incident on the ICRU sphere. Radiat. Prot. Dosim. 55, 207–210.

Goldman H., Liechti A., 1938. Experimental studies on the genesis of the x-ray stars. Arch. Ophthalmol. 138, 722.

Han E.Y., Bolch W.E. and Eckerman K.F., 2006. Revisions to the ORNL series of adult and pediatric computational phantoms for use with the MIRD schema. Health Phys. 90, 337–356.

ICRP, 1975. Report on the Task Group on Reference Man. ICRP Publication 23. Pergamon Press, Oxford.

ICRP, 1984. Non-stochastic effects of irradiation. ICRP Publication 41. Ann. ICRP 14(3).

ICRP, 1991. 1990 Recommendations of the International Commission on Radiological Protection. ICRP Publication 60. Ann. ICRP 21(1–3).

ICRP, 1996. Conversion coefficients for use in radiological protection against external radiation. ICRP Publication 74. Ann. ICRP 26 (3/4).

ICRP, 2002. Basic anatomical and physiological data for use in radiological protection: reference values. ICRP Publication 89. Ann. ICRP 32(3/4).

ICRP, 2007. The 2007 Recommendations of the International Commission on Radiological Protection. ICRP Publication 103. Ann. ICRP 37(2–4).

ICRP, 2009. Adult reference computational phantoms. ICRP Publication 110. Ann. ICRP 39(2).

ICRP, 2010. Conversion coefficients for radiological protection quantities for external radiation exposures. ICRP Publication 116. Ann. ICRP 40(2–5).
ICRP, 2012. ICRP statement on tissue reactions/early and late effects of radiation in normal tissues and organs – threshold doses for tissue reactions in a radiation protection context. ICRP Publication 118. Ann. ICRP 41(1/2).
ICRU, 1993. Quantities and Units in Radiation Protection Dosimetry. ICRU Report 51. International Commission on Radiation Units and Measurements, Bethesda, MD.
ICRU, 1997. Dosimetry of External Beta Rays for Radiation Protection. ICRU Report 54. International Commission on Radiation Units and Measurements, Bethesda, MD.
Kramer, R., Zankl, M., Williams, G., et al., 1982. The Calculation of Dose from External Photon Exposures Using Reference Human Phantoms and Monte-Carlo Methods. Part I. The Male (ADAM) and Female (EVA) Adult Mathematical Phantoms. GSF Bericht S-885. GSF, Neuherberg.
Lowe R.F., Clark B.A., 1973a. Posterior corneal curvature – correlations in normal eyes and in eyes involved with primary angel-closure glaucoma. Br. J. Ophthalmol. 57, 464–470.
Lowe R.F., Clark B.A., 1973b. Radius of curvature of anterior lens surface – correlations in normal eyes and in eyes involved with primary angle-closure glaucoma. Br. J. Ophthalmol. 57, 471–474.
Manger R.P., Bellamy M.B. and Eckerman K.F., 2012. Dose conversion coefficients for neutron exposure to the lens of the human eye. Radiat. Prot. Dosim. 148, 507–513.
Rafnsson V., Olafsdottir E., Hrafnkelsson J., et al., 2005. Cosmic radiation increases the risk of nuclear cataract in airline pilots: a population-based case–control study. Arch. Ophthalmol. 123, 1102–1105.
Stein E.G., Haramati L.B., Bellin E., et al., 2010. Radiation exposure from medical imaging in patients with chronic and recurrent conditions. J. Am. Coll. Radiol. 7, 351–359.
Vañó E., Kleiman N.J., Durán A., et al., 2010. Radiation cataract risk in interventional cardiology personnel. Radiat. Res. 174, 490–495.
Von Sallman L., Tobias C.A., Anger H.O., et al., 1955. Effects of high-energy particles, x-rays, and aging of lens epithelium. AMA. Arch. Ophthalmol. 54, 489–514.
Zankl M., Wittmann A., 2001. The adult male voxel model 'Golem' segmented from whole body CT patient data. Radiat. Environ. Biophys. 40, 153–162.
Zankl, M., Becker, J., Fill, U., et al., 2005. GSF male and female adult voxel models representing ICRP Reference Man – the present status. In: The Monte Carlo Method: Versatility Unbounded in a Dynamic Computing World. Chattanooga, TN.

SAGE

Annals of the ICRP

Dose limits to the lens of the eye: International Basic Safety Standards and related guidance

T.J. Boal, M. Pinak

Radiation Safety and Monitoring Section, International Atomic Energy Agency, Vienna, Austria; e-mail: t.boal@iaea.org

Abstract–The International Atomic Energy Agency (IAEA) safety requirements: 'General Safety Requirements Part 3 – Radiation protection and safety of radiation sources: International Basic Safety Standards' (BSS) was approved by the IAEA Board of Governors at its meeting in September 2011, and was issued as General Safety Requirements Part 3 in July 2014. The equivalent dose limit for the lens of the eye for occupational exposure in planned exposure situations was reduced from 150 mSv $year^{-1}$ to 20 mSv $year^{-1}$, averaged over defined periods of 5 years, with no annual dose in a single year exceeding 50 mSv. This reduction in the dose limit for the lens of the eye followed the recommendation of the International Commission on Radiological Protection in its statement on tissue reactions of 21 April 2011. IAEA has developed guidance on the implications of the new dose limit for the lens of the eye. This paper summarises the process that led to the inclusion of the new dose limit for the lens of the eye in the BSS, and the implications of the new dose limit.

Keywords: IAEA; Dose limit; Lens of the eye; ICRP

1. INTRODUCTION

The statute of the International Atomic Energy Agency (IAEA) (IAEA, 1956) authorises IAEA:

> ... to establish or adopt, in consultation and, where appropriate, in collaboration with the competent organs of the United Nations and with the specialized agencies concerned, standards of safety for protection of health and minimization of danger to life and property, ... and to provide for the application of these standards ...

This paper does not necessarily reflect the views of the International Commission on Radiological Protection.

The key safety standard for radiation safety is the IAEA safety requirements publication 'GSR Part 3 – Radiation protection and safety of radiation sources: International Basic Safety Standards' (BSS) (IAEA, 2014). The BSS is a safety requirements document that applies to all facilities and all activities that give rise to radiation risks, and covers all exposure situations and categories of exposure. The BSS was first published in 1962 (IAEA, 1962), and the most recent edition was approved by the IAEA Board of Governors in September 2011.

The requirements in the BSS are governed by the objectives, concepts, and principles of the fundamental safety principles (IAEA, 2006), and draw upon information derived from the experience of states in applying the requirements of the previous edition of the BSS (IAEA, 1996). They also draw upon the extensive research on the health effects of radiation exposure, in particular the findings of the United Nations Scientific Committee on the Effects of Atomic Radiation and the recommendations of the International Commission on Radiological Protection (ICRP). In addition, when the IAEA Board of Governors first approved radiation protection and safety measures in March 1960 (IAEA, 1960), it was stated that 'The Agency's Basic Safety Standards... will be based, to the extent possible, on the recommendations of the International Commission on Radiological Protection (ICRP)'.

2. ICRP RECOMMENDATIONS ON THE DOSE LIMIT FOR THE LENS OF THE EYE

ICRP published its 2007 Recommendations in *Publication 103* (ICRP, 2007). These recommendations reiterated ICRP's three general principles of radiation protection, which concern justification of exposure, optimisation of protection, and application of dose limits. The ICRP recommendations also maintained the numerical values of the dose limits given in its earlier recommendations (ICRP, 1991) and the 1996 edition of the BSS (IAEA, 1996). However, ICRP pointed out in *Publication 103* (ICRP, 2007) that the annual equivalent dose limit for the lens of the eye was currently being reviewed by an ICRP task group, with account being taken of new data on the radiosensitivity of the eye with regard to visual impairment.

In *Publication 118*, ICRP published its review of recent epidemiological evidence and concluded that there were some deterministic effects of radiation exposure, particularly those with very late manifestation, where threshold doses are or might be lower than previously considered (ICRP, 2012). Threshold dose was defined for practical purposes as the dose resulting in 1% incidence of specified tissue or organ reactions (i.e. deterministic effects). For the lens of the eye, the threshold for absorbed dose is now considered to be 0.5 Gy and, on that basis, ICRP has reduced its recommended dose limit for the lens of the eye.

In April 2011, ICRP issued a statement on tissue reactions in which it recommended, for occupational exposure, a dose limit for equivalent dose to the lens of the eye of 20 mSv year^{-1}, averaged over defined periods of 5 years, with no annual dose in a single year exceeding 50 mSv. Previously, this dose limit had been 150 mSv year^{-1}. ICRP did not specifically address the dose limit for the lens of the eye for members of the general public.

3. THE INTERNATIONAL BASIC SAFETY STANDARDS

3.1 Process for including the new dose limit for the lens of the eye in the BSS

The revision process for the 1996 edition of the BSS commenced in early 2007. A technical meeting was held in July 2007, with the participation of representatives of the sponsoring organisations.[1] The technical meeting concluded that the new edition of the BSS should follow, to the extent possible, the new recommendations of ICRP. The IAEA Safety Standards Committees and groups of experts from some of the sponsoring organisations provided comments on draft texts of the revised BSS in 2008 and 2009. A further technical meeting with the participation of the sponsoring organisations was held in December 2009 to discuss a statement on radon that was issued by ICRP in November 2009, and to assess its implications for the new edition.

The draft text of the new edition was submitted for comment to IAEA member states on 28 January 2010 for a 4-month comment period. The text was revised after a review of the comments.The draft text of the revised BSS was approved by the IAEA Safety Standards Committees at their meetings in November and December 2010. The Committees noted that ICRP was to release a statement relating to the dose limit for the lens of the eye, which should be taken into account prior to endorsement by the Commission on Safety Standards of the draft text of the new edition.

ICRP issued a statement on tissue reactions in April 2011. The Commission on Safety Standards endorsed the draft text of the new edition at its meeting in May 2011, and requested that member states and the sponsoring organisations be consulted on the proposed new dose limit for the lens of the eye. This process resulted in the new dose limit for the lens of the eye being included in the revised BSS. The text of the revised BSS was then approved by the IAEA Board of Governors in September 2011, and the 2014 edition was published in July 2014.

3.2 The new dose limit for the lens of the eye in the BSS

The dose limits for occupational exposure of workers over the age of 18 years, occupational exposure of apprentices of 16 to 18 years of age who are being trained for employment involving radiation, and public exposure for equivalent dose to the lens of the eye are presented in Table 1, along with dose limits from the 1996 edition and the 2014 edition of the BSS.

[1]The joint sponsors of the BSS are: European Commission, Food and Agriculture Organization of the United Nations, International Labour Organization, International Atomic Energy Agency, Nuclear Energy Agency of the Organisation for Economic Co-operation and Development, Pan American Health Organization, United Nations Environment Programme, and the World Health Organization.

Table 1. Limits on the equivalent dose to the lens of the eye in the 2014 edition of the International Basic Safety Standards (BSS) and the 1996 edition of the BSS.

Occupational exposure of workers		Occupational exposure of apprentices and students		Public exposure	
>18 years		16–18 years			
2014 edition	1996 edition	2014 edition	1996 edition	2014 edition	1996 edition
20 mSv year^{-1} averaged over 5 consecutive years and 50 mSv in a single year	150 mSv year^{-1}	20 mSv year^{-1}	50 mSv year^{-1}	15 mSv year^{-1}	15 mSv year^{-1}

4. GUIDANCE ON THE IMPLICATIONS OF THE NEW DOSE LIMIT FOR THE LENS OF THE EYE

4.1 Safety guides for the provision of guidance on the new dose limit

Safety guides provide recommendations and guidance on how to comply with the safety requirements. At the time when the 2014 edition of the BSS was approved, the Secretariat was asked to develop guidance as early as possible to assist member states in the observance of the new dose limit.

Guidance in relation to the new dose limit for the lens of the eye is to be provided in two safety guides that are currently being developed, entitled 'Occupational radiation protection' (DS453) and 'Radiation safety in the medical uses of ionising radiation' (DS399). It is foreseen that these two safety guides will be published in 2015–2016. DS453 is a general safety guide that is intended to provide generic guidance for all facilities and activities. DS399 is a specific safety guide that will provide guidance on all aspects of radiation safety for medical facilities, e.g. occupational protection, protection of patients, safety of sources, etc. DS399 will cover specific guidance for the occupational radiation protection of medical practitioners performing interventional procedures that will not be covered in DS453. It is recognised that guidance material is required before these two safety guides are finalised in order to allow member states the opportunity to put appropriate actions and planning in place to address the introduction of the new dose limit for the lens of the eye.

TECDOCs are developed by IAEA to provide explanatory material or interim guidance to support the safety standards. A TECDOC was therefore developed to provide advice on the implications for occupational radiation protection of the new dose limit for the lens of the eye, and to allow comment on detailed recommendations that may be incorporated into the above safety guides.

4.2 Process for development of the TECDOC

As a first step towards developing the TECDOC (IAEA, 2013), a technical meeting was held in October 2012 that was attended by 47 experts from 25 IAEA member states and eight international organisations. Experts from ICRP made presentations on the scientific basis of the new dose limit for the lens of the eye, and on the process for establishing the new dose limit. There were also presentations on the practical issues for implementing the dose limit to the lens of the eye in relation to medical, industrial, and nuclear facilities.

The discussions and the report of the meeting formed the basis of the content of the TECDOC. The text of the TECDOC was drafted in early 2013, and circulated to all participants in the technical meeting for comment. The draft text was provided to the Radiation Safety Standards Committee for information and comment at their meeting in early July 2013.

4.3 Guidance on implications of the new dose limit for the lens of the eye

The guidance in the TECDOC covers the following topics: safety assessment; workers for whom exposure of the lens of the eye might be important; optimisation of protection; monitoring of doses to the lens of the eye due to external radiation; health surveillance; and exposure of emergency workers.

An example of the guidance in the TECDOC is that provided for monitoring of the dose to the lens of the eye. The TECDOC states that the most accurate method for monitoring the equivalent dose to the lens of the eye is to measure Hp(3) with a dosimeter worn as close as possible to the eye, and calibrated on a phantom representative of the head. However, dosimeters designed for Hp(3) are not broadly available. The TECDOC provides guidance on when Hp(0.07) or Hp(10) dosimeters can be used, and when Hp(3) dosimeters have to be used. The TECDOC recommends that routine monitoring of dose to the lens of the eye should be undertaken if annual equivalent dose to the lens is likely to exceed 5 mSv.

REFERENCES

IAEA, 1956. The Statute of the IAEA. International Atomic Energy Agency, Vienna. Available at: http://www.iaea.org/about/statute (last accessed 16 February 2015).

IAEA, 1960. The Agency's Health and Safety Measures. INFCIRC/18. International Atomic Energy Agency, Vienna.

IAEA, 1962. Basic Safety Standards for Radiation Protection. IAEA Safety Series No. 9. International Atomic Energy Agency, Vienna.

IAEA, 1996. International Basic Safety Standards for Protection Against Ionizing Radiation and for the Safety of Radiation Sources. IAEA Safety Series No. 115. International Atomic Energy Agency, Vienna.

IAEA, 2006. Fundamental Safety Principles. IAEA Safety Standards Series No. SF-1. International Atomic Energy Agency, Vienna.

IAEA, 2013. Implications for Occupational Radiation Protection of the New Dose Limit for the Lens of the Eye. IAEA-TECDOC No. 1731. International Atomic Energy Agency, Vienna.

IAEA, 2014. Radiation Protection and Safety of Radiation Sources: International Basic Safety Standards. IAEA Safety Standards Series No. GSR Part 3. International Atomic Energy Agency, Vienna.
ICRP, 1991. 1990 Recommendations of the International Commission on Radiological Protection. ICRP Publication 60. Ann. ICRP 21(1–3).
ICRP, 2007. The 2007 Recommendations of the International Commission on Radiological Protection. ICRP Publication 103. Ann. ICRP 37(2–4).
ICRP, 2012. ICRP statement on tissue reactions/early and late effects of radiation in normal tissues and organs – threshold doses for tissue reactions in a radiation protection context. ICRP Publication 118. Ann. ICRP 41(1/2).

SAGE

ICRP
Annals of the ICRP

Implications in medical imaging of the new ICRP thresholds for tissue reactions

E. Vañó[a], D.L. Miller[b] and L. Dauer[c]

[a]*Radiology Department, Medicine School and San Carlos Hospital, Complutense University, Madrid, 28040 Spain; e-mail: eliseov@med.ucm.es*

[b]*Center for Devices and Radiological Health, Food and Drug Administration, USA*

[c]*Department of Medical Physics, Department of Radiology. Memorial Sloan-Kettering Cancer Center, USA*

Abstract–The International Commission on Radiological Protection (ICRP) statement on tissue reactions, issued by the Commission in April 2011, reviewed epidemiological evidence and suggested that there are some tissue reactions where threshold doses are or may be lower than those previously considered. For the lens of the eye, the threshold is now considered to be 0.5 Gy. The absorbed dose threshold for circulatory disease in the heart and brain may be as low as 0.5 Gy. These values can be reached in some patients during interventional cardiology or neuroradiology procedures. They may also be of concern for repeated computed tomography examinations of the head. The new thresholds should be considered in optimisation strategies for clinical procedures, especially in patients likely to require repeated interventions. The new dose thresholds also affect occupational protection for operators and staff. Some operators do not protect their eyes or their brain adequately. After several years of work without proper protection, the absorbed doses to the lens of the eye and the brain of staff can exceed 0.5 Gy. More research is needed to understand the biological effects of cumulative incident air kerma and the instantaneous air kerma rates currently used in medical imaging. The new thresholds, and the need for specific occupational dosimetry related to lens doses, should be considered in radiation protection programmes, and should be included in the education and training of professionals involved in fluoroscopy guided procedures and computed tomography.

Keywords: Radiological protection; Tissue reactions; Interventional radiology; Interventional cardiology; Neuroradiology; Deterministic effects

This paper does not necessarily reflect the views of the International Commission on Radiological Protection.

1. INTRODUCTION

The International Commission on Radiological Protection (ICRP) statement on tissue reactions, issued by the Commission in April 2011, (ICRP, 2012) reviewed available epidemiological evidence and suggested that there are some tissue reactions where threshold doses are or may be lower than those previously considered.

For the lens of the eye, the threshold is now considered to be 0.5 Gy. The absorbed dose threshold for circulatory disease in the heart and brain may be as low as 0.5 Gy. These values can be reached in some patients during interventional cardiology or neuroradiology procedures (Thierry-Chef et al., 2008; Vañó et al., 2013a; Sánchez et al., 2014). They can also be reached in some staff involved in interventional procedures if radiation protection tools are not used properly (Dauer et al., 2010; Vañó et al., 2010, 2013c; Ciraj-Bjelac et al., 2012).

The new thresholds should be considered during the justification process and especially in optimisation strategies for clinical procedures. This is particularly important for patients who are likely to require repeated interventions, and for staff in interventional laboratories who have a high workload (Picano, 2012; Miller, 2013; Bartal et al., 2014).

This paper presents some examples of organ dose values (heart, brain, and lens of the eye) in patients who have undergone cardiology or neuroradiology interventional procedures or head computed tomography (CT), and in staff working in interventional laboratories.

2. HEART DOSES IN INTERVENTIONAL CARDIOLOGY AND CARDIAC CT

Organ doses can be estimated with PCXMC 2.0 software, which uses Monte Carlo-derived calculations (Tapiovaara and Siiskonen, 2008). In interventional cardiology procedures, the absorbed dose to the heart can be estimated from the cumulative air kerma (CAK) at the 'patient entrance reference point' defined by the International Electrotechnical Commission (IEC, 2010). The estimated absorbed dose to the heart from CAK of 1 Gy, delivered from a posterior–anterior projection with the typical x-ray beam quality and field size used in cardiology, is 250 mGy. When collimation is not used – as is usually the case in interventional cardiology – the estimated absorbed dose to the heart is approximately 20–25% of CAK. For cardiac interventions, CAK > 2.0–2.5 Gy therefore implies an absorbed dose to the heart greater than the 0.5 Gy threshold proposed by the Commission (ICRP, 2012). In order to ensure that organ doses that may exceed the threshold are detected and these patients are followed appropriately, patient dose metrics should be recorded and reported for every patient and every examination.

In a sample of 6557 diagnostic and therapeutic cardiac interventional procedures performed at the San Carlos University Hospital in Madrid (Spain) in 2010 and 2011, the median CAK was 1079 mGy and the third quartile CAK was 1662 mGy. In 17% of these procedures, CAK was >2.0 Gy, and in 10% of these procedures, CAK was >2.5 Gy (Fig. 1). This implies that the organ dose to the heart was

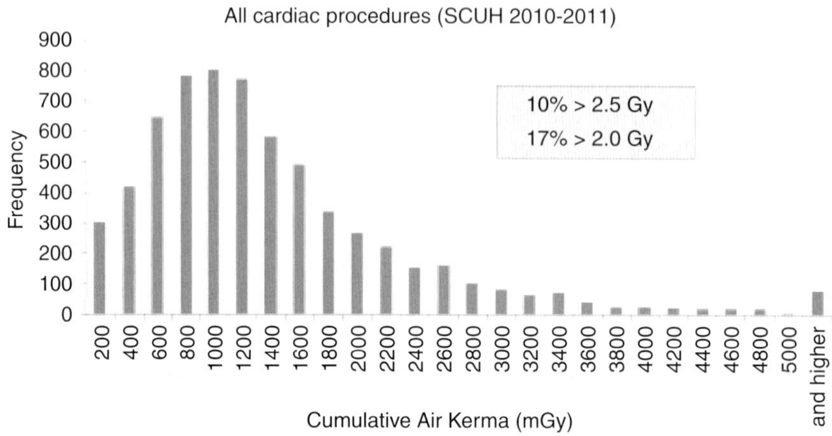

Sample	Median	Mean	SD	Min	Max	3rd quartile
6557	1079	1325	1039	1	12653	1662

Fig. 1. Cumulative air kerma (CAK) at the patient entrance reference point for interventional (diagnostic and therapeutic) cardiac procedures at the San Carlos University Hospital (SCUH) in Madrid, Spain.

approximately 0.5 Gy or greater in 10–17% of these procedures. In a sample of 4128 therapeutic interventional cardiac procedures performed in the same hospital, the median CAK was 1435 mGy and the third quartile CAK was 2076 mGy. Of these procedures, 27% resulted in CAK > 2.0 Gy (i.e. with estimated organ dose to the heart of approximately 0.5 Gy) and 16% resulted in CAK > 2.5 Gy.

Data from the 2008–2009 US nationwide evaluation of x-ray trends (NEXT) survey of cardiac catheterisation have been reported (Miller et al., 2012b). These data were acquired from 171 hospitals in 30 US states. Radiation dose metrics were recorded for diagnostic cardiac catheterisations ($n = 1038$), percutaneous coronary interventions (PCI) ($n = 117$), and combined diagnostic and PCI procedures ($n = 390$). For diagnostic catheterisations, 5.1% resulted in CAK > 2.0 Gy and 3.5% in CAK > 2.5 Gy. For PCI, these percentages were 32.5% and 26.5%, respectively, and for combined procedures, the percentages were 35.2% and 24.9%, respectively.

Absorbed dose to the heart from CT coronary angiography may also be substantial, especially when the patient undergoes multiple examinations. Different acquisition techniques and heart rates determine heart dose in these cases. Matsubara et al. (2011) reported absorbed doses to the heart using a 64 detector-row CT. When helical acquisitions using electrocardiogram modulation of tube current were performed at simulated heart rates of 40, 60, and 90 beats min^{-1}, the absorbed doses to the heart were 178.6, 139.1, and 159.3 mGy, respectively.

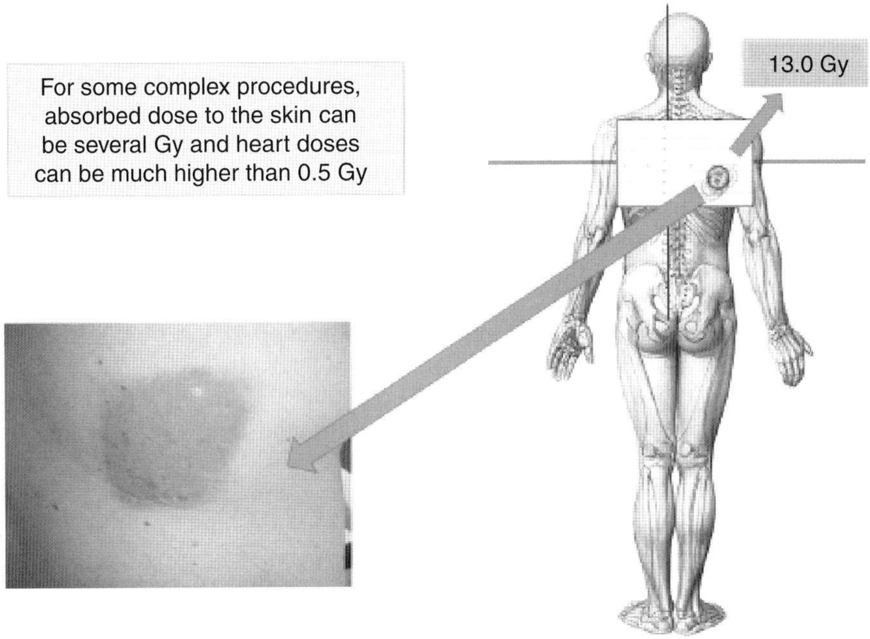

Fig. 2. Example of a skin injury during a complex cardiology procedure (treatment of a chronic total occlusion) (Vañó et al., 2013a). The estimated peak skin dose was 13 Gy.

When skin injuries occur after a cardiac interventional procedure (Fig. 2) (Vañó et al., 2013a), tissue reactions in the cardiovascular system should also be considered as part of the clinical follow-up.

3. BRAIN AND LENS OF THE EYE DOSES IN INTERVENTIONAL NEURORADIOLOGY AND HEAD CT PROCEDURES

Absorbed doses to the brain and cerebrovascular system (including the carotid artery in the neck) delivered to patients during interventional neuroradiology procedures may also be substantial. Clinicians should consider the new threshold for tissue reactions in the cerebrovascular system when they develop optimisation strategies.

In a recent survey at the San Carlos University Hospital in Madrid (Sánchez et al., 2014) the absorbed dose to the brain was estimated using PCXMC 2.0 software (Tapiovaara and Siiskonen, 2008) for 99 diagnostic and therapeutic neuroradiology procedures. In total, 9031 beam projections were processed. Brain doses $\geq$0.5 Gy were observed in 34% of cerebral embolisation procedures. For these procedures, if collimation is not used, CAK of 1000 mGy will result in an estimated absorbed dose to the brain of approximately 250 mGy.

Thierry-Chef et al. (2008) published absorbed doses to the brain for 49 paediatric patients undergoing intracranial embolisation. The average dose to the whole brain was calculated using Monte Carlo software. They noted that the distribution of the absorbed dose within the brain can vary substantially depending on field size and field motion during procedures. The median absorbed dose, averaged over the entire brain, was 440 mGy (range 50–2700 mGy) assuming large uniform frontal and lateral fields. If collimated frontal and lateral fields focused near the centre of the brain were assumed, the median dose was estimated to be 60 mGy (range 8–310 mGy). Average absorbed doses to the brain, when exposed to large uniform fields, were >500 mGy in 21 of the 49 cases (43%). Brain dose is dramatically reduced if the brain is irradiated with narrow non-uniform fields that vary in location, as compared with irradiation by large uniform frontal and lateral fields. If this optimisation strategy was used, all the analysed procedures could be performed with brain doses <0.5 Gy. Collimation and limiting fluoroscopy time and dose rate were the most effective means to minimise dose.

During some CT procedures, especially brain perfusion examinations, brain and lens absorbed doses may approach or exceed the new threshold of 0.5 Gy. Zhang et al. (2012) reported eye doses of 81 to 348 mGy from brain perfusion CT using Monte Carlo simulations. The wide range is due to the different CT scanners and protocols used.

Using Monte Carlo simulation software designed for patient CT dosimetry, Perisinakis et al. (2013) studied the effect of head size and shape, miscentring, and bowtie filter size on peak patient tissue doses from modern brain perfusion 256-slice CT in 106 individual-specific adult head phantoms. The mean peak doses to the lens of the eye, skin, brain, and red bone marrow were 124, 120, 95, and 163 mGy, respectively. They concluded that doses to these tissues from a standard low-dose brain perfusion 256-slice CT protocol are well below the thresholds for the induction of erythema, cataract, cerebrovascular disease, and depression of haematopoiesis. Nonetheless, every effort should be made to optimise the procedure and minimise the dose received by these tissues.

Estimation of brain doses and evaluation of dose-sparing techniques should be part of the optimisation process, particularly for patients who have or will undergo multiple procedures and for paediatric patients. This implies that patient dose metrics should be recorded and reported for every patient and every such examination. If cumulative absorbed brain dose is expected to be >0.5 Gy, the possibility of a tissue reaction in the cerebrovascular system should be considered as part of the informed consent process in addition to the risk of a tissue reaction in the skin or a stochastic effect.

4. OCCUPATIONAL LENS DOSES IN INTERVENTIONAL STAFF

As a consequence of the new threshold for lens effects, the Commission provided a new recommendation for the occupational dose limit to the lens of the eye. The recommended equivalent dose limit for the lens of the eye was reduced from 150 mSv $year^{-1}$

to 20 mSv year^{-1}, averaged over a 5-year period, with no year's dose exceeding 50 mSv (ICRP, 2012). The new International and European Basic Safety Standards incorporate this new occupational limit (European Commission, 2012; IAEA, 2014). The UK Health Protection Agency (HPA) endorses the conclusion reached by the Commission in its 2011 statement (Bouffler et al., 2012). The Health Protection Agency is now part of Public Health England, an executive agency of the Department of Health. The HPA believes that further work is desirable to better quantify the risk at low doses and following protracted exposures, as is research into the mechanistic basis for radiation cataractogenesis. This research will help to inform selection of risk projection models.

The new dose thresholds, especially the 0.5 Gy threshold for the lens of the eye, affect occupational protection for operators and other involved staff. This is particularly important for fluoroscopically guided procedures. Some operators do not protect their eyes or their brain adequately during interventional procedures. After several years of work without proper protection, cumulative absorbed doses to the lens can exceed 0.5 Gy (Vañó et al., 2010, 2013c; Ciraj-Bjelac et al., 2012). During most of the IAEA.RELID (Retrospective Evaluation of Lens Injuries and Dose) (see: https://rpop.iaea.org/rpop/rpop/content/news/relid-cataract-study.htm, accessed on 13 January 2015) surveys of radiation-related lens opacities in interventionalists and staff (Vañó et al., 2010, 2013c; Ciraj-Bjelac et al., 2012), 40–50% of the professionals involved in interventional cardiology procedures who volunteered to be examined had lens opacities (posterior subcapsular opacities). Only 10% of the members of the control groups had such opacities. It should be noted that most of the detected lens opacities were in professionals who had worked for several years without any eye protection.

Many medical specialties outside the imaging departments (e.g. vascular surgery, urology, orthopaedics, gastroenterology, and anaesthetics and pain management) are starting to use or increase their use of fluoroscopically guided procedures in surgical theatres without the radiation protection tools available in standard interventional laboratories (ICRP, 2010; Burns et al., 2013). This may result in occupational doses in these physicians that are higher than the typical values recorded in interventional radiology and cardiology services where protection tools are used regularly. Also, because of the growing use of positron-emission tomography (PET)-CT, and the high workload for PET-CT staff, occupational protection is becoming the focus of greater attention in nuclear medicine installations (Ryan et al., 2013).

Lack of compliance with the regular use of personal dosimetry has been a problem for many years in many countries (Kim et al., 2012; Durán et al., 2013; Bartal et al., 2014). This is one of the reasons for the lack of accurate occupational dose data (and reliable radiation risk estimations) in relevant professional groups. Several investigators are working on Monte Carlo calculations and detection of staff location in the catheterisation laboratory to estimate staff doses during clinical procedures through the use of radiographic and geometric data (Badal et al., 2013). Other authors (Vañó et al., 2013b) used a different method to develop a realistic approach to estimate lens doses in interventional cardiology when personal dosimeters are not used regularly. They concluded that for around 2000 diagnostic, PCI, and valvular procedures, the median scatter dose value per procedure at the C-arm was 0.78, 1.07, and 1.45 mSv, respectively. Lens doses are approximately 50–60% of these values when radiation protection tools are not used. For all of these procedures, the ratio between the scatter dose at the C-arm and the kerma-area product was 10.3–11.3 μSv(Gy cm^2)$^{-1}$.

Newer active personal dosimeters display occupational dose rates in real time inside the interventional fluoroscopy suite, and allow subsequent detailed analyses of staff and patient doses by recording dose at very short intervals. These data permit the development of new optimisation strategies to improve occupational protection (Sánchez et al., 2010). Several medical societies have published or endorsed documents on occupational protection that are also expected to improve radiation safety (Miller et al., 2010, 2012a; Durán et al., 2013). Some users have urged that scatter doses be included in patient dose reports. New ethical issues are emerging in situations where reducing patient dose involves increasing staff doses and vice versa.

Occupational radiological protection is still a challenge in several clinical situations. The Commission has included specific recommendations and advice for occupational protection in most of its publications in recent years (ICRP, 2000, 2007a,b,c, 2009, 2010, 2013a,b).

5. DOSE RATES FOR PATIENTS AND STAFF

One area where more research is needed is an understanding of the biological effects of cumulative incident air kerma and the instantaneous air kerma rates used currently in medical imaging. In *Publication 99* (ICRP, 2005), the Commission stated that its radiation protection philosophy is based on the linear no-threshold theory. According to this theory, total radiation-related cancer risk is proportional to dose at low and moderately low doses (of the order of 200 mGy or less) and dose rates (less than 6 mGy h^{-1}, averaged over the first few hours).

Table 1. Ranges of photon energy, dose per procedure and dose rate (incident air kerma) in medical imaging procedures for patients and staff.

Patients	Range
Photon energy	20–130 keV
Dose (incident air kerma)	1–5000 mGy
Dose rate (from fluoroscopy and radiography)	5–5000 mGy min^{-1}
Staff, if radiation protection tools not used	
Photon energy	20–130 keV
Dose (incident air kerma)	1–5000 μGy
Dose rate (patient/1000)	5–5000 μGy min^{-1}
ICRP (2005), DDREF of 2, if:	
Dose rates less than 6 mGy h^{-1} (0.1 mGy min^{-1})	
Acute exposures of total doses less than 0.2 Gy	

DDREF, dose and dose-rate effectiveness factor.

In medical imaging, patient doses may exceed these values (Table 1). Some interventional procedures may result in CAK of several Gy (Fig. 2). Incident air kerma values of 4–8 mGy at the patient entrance reference point may result from some abdominal digital subtraction angiography images obtained with radiation pulses of 40–80 ms. This is equivalent to instantaneous dose rates of 5 Gy min^{-1}. Operator dose (incident air kerma) at the typical working position is around 1/1000 of incident air kerma at the patient's skin, so the instantaneous dose rate to an unprotected operator may reach approximately 5 mGy min^{-1}. During interventional cardiology procedures, air kerma at the patient's skin of 0.5 mGy frame^{-1} can be reached during cineradiography runs, with radiation pulses of 5 ms. This results in an air kerma rate of 100 mGy s^{-1} (6 Gy min^{-1}) at the patient's skin. This dose rate is much higher than the 6 mGy h^{-1} reported in *Publication 99* (ICRP, 2005).

6. CONCLUSIONS

Radiation protection programmes and optimisation strategies for medical imaging should take into account the new thresholds for tissue reactions. Optimisation strategies should attempt to minimise the possibility of exceeding the thresholds for the heart, brain, and lens of the eye in patients, both for individual high-dose exposures (e.g. complex interventional fluoroscopy procedures) and for multiple moderate-dose exposures (e.g. a patient who has or will undergo repeated brain perfusion CT examinations or interventional fluoroscopy procedures). This implies that patient dose metrics should be recorded and reported for every patient and every such examination. Also, if the dose to the brain, heart, or lens of the eye is predicted to exceed 0.5 Gy, tissue reactions in these organs should be considered in the follow-up of these patients.

These concepts should be included in the education and training of professionals involved in fluoroscopy guided procedures and CT. Attention should also be paid to the proper use of protection tools by staff in interventional fluoroscopy, as absorbed dose to the lens of the eye >0.5 Gy and an increased prevalence of lens opacities has been observed when protection tools are not used or not used properly.

Several Commission documents provide additional advice. These include *Publication 85*, 'Avoidance of radiation injuries from medical interventional procedures' (ICRP, 2000); *Publication 105*, 'Radiological protection in medicine' (ICRP, 2007c); *Publication 113*, 'Education and training in radiological protection for diagnostic and interventional procedures' (ICRP, 2009); *Publication 117*, 'Radiological protection in fluoroscopically guided procedures performed outside the imaging department' (ICRP, 2010); *Publication 120*, 'Radiological protection in cardiology' (ICRP, 2013a); and *Publication 121*, 'Radiological protection in paediatric diagnostic and interventional radiology' (ICRP, 2013b).

REFERENCES

Badal, A., Zafar, F., Dong, H., et al., 2013. A real-time radiation dose monitoring system for patients and staff during interventional fluoroscopy using a GPU accelerated Monte Carlo simulator and an automatic 3D localization system based on a depth camera. Proc. SPIE 8668, Medical Imaging 2013: Physics of Medical Imaging, 866828.

Bartal G., Vañó E., Paulo G., et al., 2014. Management of patient and staff radiation dose in interventional radiology: current concepts. Cardiovasc. Intervent. Radiol. 37, 289–298.

Bouffler S., Ainsbury E., Gilvin P., et al., 2012. Radiation-induced cataracts: the Health Protection Agency's response to the ICRP statement on tissue reactions and recommendation on the dose limit for the eye lens. J. Radiol. Protect. 32, 479–488.

Burns S., Thornton R., Dauer L.T., et al., 2013. Leaded eyeglasses substantially reduce radiation exposure of the surgeon's eyes during acquisition of typical fluoroscopic views of the hip and pelvis. J. Bone Joint Surg. Am. 95, 1307–1311.

Ciraj-Bjelac O., Rehani M., Minamoto A., et al., 2012. Radiation-induced eye lens changes and risk for cataract in interventional cardiology. Cardiology. 123, 168–171.

Dauer L.T., Thornton R.H., Solomon S.B., et al., 2010. Unprotected operator eye lens doses in oncologic interventional radiology are clinically significant: estimation from patient kerma-area product data. J. Vasc. Intervent. Radiol. 21, 1859–1861.

Durán A., Hian S.K., Miller D.L., et al., 2013. Recommendations for occupational radiation protection in interventional cardiology. Catheteriz. Cardiovasc. Intervent. 82, 29–42.

European Commission, 2012. Proposal for a Council Directive Laying Down Basic Safety Standards for Protection Against the Dangers Arising from Exposure to Ionising Radiation. EC, Brussels. COM(2012) 242 final. Available at: http://ec.europa.eu/energy/nuclear/radiation_protection/doc/2012_com_242.pdf (last accessed 7 December 2013).

IAEA, 2014. Radiation Protection and Safety of Radiation Sources. International Basic Safety Standards (BSS). IAEA Safety Standards Series GSR Part 3. International Atomic Energy Agency, Vienna.

ICRP, 2000. Avoidance of radiation injuries from medical interventional procedures. ICRP Publication 85. Ann. ICRP 30(2).

ICRP, 2005. Low-dose extrapolation of radiation-related cancer risk. ICRP Publication 99. Ann. ICRP 35(4).

ICRP, 2007a. Managing patient dose in multi-detector computed tomography (MDCT). ICRP Publication 102. Ann. ICRP 37(1).

ICRP, 2007b. The 2007 Recommendations of the International Commission on Radiological Protection. ICRP Publication 103. Ann. ICRP 37(2–4).

ICRP, 2007c. Radiological protection in medicine. ICRP Publication 105. Ann. ICRP 37(6).

ICRP, 2009. Education and training in radiological protection for diagnostic and interventional procedures. ICRP Publication 113. Ann. ICRP 39(5).

ICRP, 2010. Radiological protection in fluoroscopically guided procedures performed outside the imaging department. ICRP Publication 117. Ann. ICRP 40(6).

ICRP, 2012. ICRP statement on tissue reactions/early and late effects of radiation in normal tissues and organs – threshold doses for tissue reactions in a radiation protection context. ICRP Publication 118. Ann. ICRP 41(1/2).

ICRP, 2013a. Radiological protection in cardiology. ICRP Publication 120. Ann. ICRP 42(1).

ICRP, 2013b. Radiological protection in paediatric diagnostic and interventional radiology. ICRP Publication 121. Ann. ICRP 42(2).

IEC, 2010. Medical Electrical Equipment. Part 2-43: Particular Requirements for the Basic Safety and Essential Performance of X-ray Equipment for Interventional Procedures. IEC-60601-2-43. International Electrotechnical Commission, Geneva.

Kim K.P., Miller D.L., Berrington de Gonzalez A., et al., 2012. Occupational radiation doses to operators performing fluoroscopically-guided procedures. Health Phys. 103, 80–99.

Matsubara K., Koshida K., Noto K., et al., 2011. Estimation of organ-absorbed radiation doses during 64-detector CT coronary angiography using different acquisition techniques and heart rates: a phantom study. Act. Radiol. 52, 632–637.

Miller D.L., 2013. Efforts to optimize radiation protection in interventional fluoroscopy. Health Phys. 105, 435–444.

Miller D.L., Balter S., Dixon R.G., et al., 2012a. Quality improvement guidelines for recording patient radiation dose in the medical record for fluoroscopically guided procedures. J. Vasc. Intervent. Radiol. 23, 11–18.

Miller D.L., Hilohi C.M. and Spelic D.C., 2012b. Patient radiation doses in interventional cardiology in the U.S.: advisory data sets and possible initial values for U.S. reference levels. Med. Phys. 39, 6276–6286.

Miller D.L., Vañó E., Bartal G., et al., 2010. Occupational radiation protection in interventional radiology: a joint guideline of the Cardiovascular and Interventional Radiology Society of Europe and the Society of Interventional Radiology. Cardiovascular Interventional Radiology Society of Europe; Society of Interventional Radiology. J. Vasc. Intervent. Radiol. 21, 607–615.

Perisinakis K., Seimenis I., Tzedakis A., et al., 2013. The effect of head size/shape, miscentering, and bowtie filter on peak patient tissue doses from modern brain perfusion 256-slice CT: how can we minimize the risk for deterministic effects? Med. Phys. 40, 011911.

Picano E., Vañó E., Domenici L., et al., 2012. Cancer and non-cancer brain and eye effects of chronic low-dose ionizing radiation exposure. BMC Cancer. 12, 157.

Ryan E.R., Thornton R., Sofocleous C.T., et al., 2013. PET/CT-guided interventions: personnel radiation dose. Cardiovasc. Intervent. Radiol. 36, 1063–1067.

Sánchez R., Vañó E., Fernandez J.M., et al., 2010. Staff radiation doses in a real-time display inside the angiography room. Cardiovasc. Intervent. Radiol. 33, 1210–1214.

Sánchez R.M., Vañó E., Fernández J.M., et al., 2014. Brain radiation doses to patients in an interventional neuroradiology laboratory. AJNR. Am. J. Neuroradiol. 35, 1276–1280.

Tapiovaara, M., Siiskonen, T., 2008. PCXMC: a Monte Carlo program for calculating patient doses in medical x-ray examinations, second ed. STUK – Radiation and Nuclear Safety Authority of Finland, Helsinki.

Thierry-Chef I., Simon S.L., Land C.E., et al., 2008. Radiation dose to the brain and subsequent risk of developing brain tumors in pediatric patients undergoing interventional neuroradiology procedures. Radiat. Res. 170, 553–565.

Vañó E., Kleiman N.J., Durán A., et al., 2010. Radiation cataract risk in interventional cardiology personnel. Radiat. Res. 174, 490–495.

Vañó E., Escaned J., Vano-Galván S., et al., 2013a. Importance of a patient dosimetry and clinical follow-up program in the detection of radiodermatitis after long percutaneous coronary interventions. Cardiovasc. Intervent. Radiol. 36, 330–337.

Vañó E., Fernández JM, Sánchez R.M., et al., 2013b. Realistic approach to estimate lens doses and cataract radiation risk in cardiology when personal dosimeters have not been regularly used. Health Phys. 105, 330–339.

Vañó E., Kleiman N.J., Durán A., et al., 2013c. Radiation-associated lens opacities in catheterization personnel: results of a survey and direct assessments. J. Vasc. Intervent. Radiol. 24, 197–204.

Zhang D., Cagnon C.H., Villablanca J.P., et al., 2012. Peak skin and eye lens radiation dose from brain perfusion CT based on Monte Carlo simulation. AJR. Am. J. Roentgenol. 198, 412–417.

SAGE

Annals of the ICRP

Non-cancer effects: science and values aspects of protection decisions

T. Lazo

Division of Radiation Protection and Radioactive Waste Management, Organisation for Economic Co-operation and Development, Nuclear Energy Agency, 12 Boulevard des Iles, 92130 Issy-les-Moulineaux, France; e-mail:edward.lazo@oecd.org

Abstract–The Nuclear Energy Agency organised its third workshop on 'Science and values in radiological protection' in November 2012 in Tokyo. One of the issues addressed, non-cancer effects, had also been addressed in the first two science and values workshops (Helsinki, Finland, 2008; Vaux-de-Cernay, France, 2009), but presented several new elements of relevance to International Commission on Radiological Protection discussions of the evolution of the system of radiological protection. Radiological protection science, both epidemiological and biological, now suggests that stroke and heart disease may well be caused by radiation exposure at doses of the order of 0.5 Gy or less. Further, it is possible that such detriments may be caused by either chronic or acute exposures. While significant uncertainties remain, the need to consider non-cancer detriment in risk assessment and in the development of protection strategies is now a significant scientific and ethical question. This paper will present the results of the Nuclear Energy Agency science and values workshop discussion of non-cancer risks, and of the questions and possible future directions raised during the workshop.

Keywords: Science and values; Non-cancer effects

This paper does not necessarily reflect the views of the International Commission on Radiological Protection.

1. INTRODUCTION

Radiological protection was founded on, and continues to be, a combination of science and value judgements. As described by Lauriston S. Taylor,[1] radiological protection is '... not only a matter for science. It is a problem of philosophy, and morality, and the utmost wisdom' (Taylor, 1957). In the development of radiological protection policy and its practical application, there is always a need for radiological protection policy makers, practitioners, and other stakeholders to better understand the evolving interactions between science and values. At the same time, there is a need for radiological protection scientists to better understand the broad processes of radiological protection decision-making and to better interact with these processes, particularly by providing input from their research. Rolf M. Sievert, the Swedish physicist who laid the foundations of modern radiation physics, clearly expressed the dynamic link between policy, research, and application: 'The establishment of maximum permissible radiation levels is a non-scientific task, which must be based primarily on scientific knowledge and judgment' (Sievert, 1958).

Achieving a mutual understanding among radiological protection policy makers, practitioners, researchers, industry, and non-governmental organisations is expected to facilitate the prioritisation of research and the framing of decision-making in the future, and to improve the quality and robustness of radiological protection.

The Committee on Radiation Protection and Public Health (CRPPH), a standing technical committee of the Organisation for Economic Co-operation and Development (OECD)/Nuclear Energy Agency (NEA), has long been aware of the need to develop a shared understanding of emerging challenges for radiological protection among scientific and regulatory communities, and other concerned stakeholders. The CRPPH expert group on the implications of radiological protection science published a report in 2007 on 'Scientific issues and emerging challenges for radiological protection' (OECD/NEA, 2007a). This report represents a broad summary of key scientific challenges that could arise from on-going research on radiological protection. Additionally, the CRPPH expert group on the collective opinion published its report on 'Radiation protection in today's world – towards sustainability' (OECD/NEA, 2007b). The report identifies key emerging challenges to the radiological protection system in order to assist decision makers at all levels to better address these within their relevant contexts.

In follow-up to these activities and in response to the existing need to foster mutual understanding among all stakeholders concerned, the CRPPH initiated a longer-term process of reflection on scientific and societal issues that might challenge radiological protection in the coming decade. The first step in this process was a workshop organised in collaboration with the Radiation and Nuclear Safety Authority of Finland to address some of these issues. This workshop, entitled 'Science and values in radiological protection', took place from 15 to 17 January 2008 in Helsinki, Finland (OECD/NEA, 2011). At this workshop, more than 60

[1]Radiation physicist and pioneer in the field of radiation safety, founding member of both the International Commission on Radiation Units and Measurements and the International Commission on Radiological Protection (ICRP).

scientists, researchers, representatives of regulatory authorities, and political decision makers from 22 countries (including countries that are not members of the OECD) met to discuss new trends in radiological protection.

A second workshop in the series was then organised in collaboration with France's IRSN (Institut de radioprotection et de sûreté nucléaire) and sponsored by the French Ministry of Ecology, Energy, Sustainable Development, and Land-use Planning. This second science and values workshop took place from 30 November to 2 December 2009 in Vaux-de-Cernay, France (OECD/NEA, 2011). More than 70 participants from 19 countries (European countries, Argentina, Canada, Chinese Taipei, Japan, Korea, and the USA) attended this workshop.

As a result of these two meetings, the CRPPH recognised that mutual understanding of scientific evidence and radiological protection values and practice is important both for obtaining optimal protection and for identifying the gaps in knowledge that are most relevant for radiological protection. Most of the participants at both science and values workshops agreed that while there is no immediate need to change the current principles, extended dialogue among all concerned stakeholders is necessary in order to facilitate the integration of challenging scientific phenomena into existing regulatory frameworks. This type of exchange forum between regulators and scientists was welcomed, and could serve as a model way of moving forward.

As such, and in view of the significant social concerns following the Fukushima accident, the CRPPH agreed to hold a third science and values workshop in Japan, co-organised by the Japanese Nuclear Regulation Authority, the Japanese Ministry of Education, Culture, Sports, Science, and Technology, and the Japanese National Institute of Radiological Sciences (OECD/NEA, 2013). In this workshop, 134 participants from 16 countries came together to share their experiences, knowledge, and issues, and to provide recommendations to further enhance radiological protection. The workshop benefitted from the active and devoted participation of a spectrum of stakeholders, including local Fukushima Prefecture residents and leaders, journalists, physicians, and radiological protection professionals.

For all three of these workshops, the overall objective was to better understand how science and values aspects may influence the evolution of the system of radiological protection, and to better understand how science and values aspects should be included and transparently articulated in radiological protection decision-making. To achieve this, each workshop selected three relevant topics. For each topic, plenary session presentations were given: one on the science aspects of the topic, and one on the values aspects. Parallel breakout sessions were held, one on each topic, to discuss, in detail, the science and values elements that are taken into account when making radiological protection decisions. Table 1 provides a list of the three topics chosen for discussion at each meeting.

In all three workshops, the topic of non-cancer effects was addressed, although with an evolving name. In the first science and values workshop it was called 'circulatory diseases', in the second science and values workshop, it was called

Table 1. Breakout session topics addressed in each science and values workshop.

Helsinki (2008): the first science and values workshop
- Non-targeted effects
- Individual sensitivity
- Circulatory diseases

Vaux-de-Cernay (2009): the second science and values workshop
- Radon as a public health issue
- Medical exposures in diagnostic and screening procedures
- Radiation-induced vascular effects

Tokyo (2012): the third science and values workshop
- Assessment and management of low dose exposures and public health
- Protection of children and self-help behaviour approaches
- Non-cancer effects

'radiation-induced vascular effects', and in the third science and values workshop, it was called 'non-cancer effects'. In spite of the changing title used for this topic, there is a clear line of evolution of scientific understanding and of values consideration in this topic. These workshops are summarised in two NEA publications (OECD/NEA, 2012, 2013).

2. WORKSHOP RESULTS

The three workshops showed similar results with respect to non-cancer diseases. There was broad agreement on the existence of clear epidemiological evidence of radiation-induced non-cancer effects above 0.5 Gy, and that the evidence is inconclusive at lower doses. It was also recognised that radiation-induced non-cancer effects are not specifically addressed by the system, at present, but may have significant impacts on the assessment of morbidity and mortality. However, from 2008 to 2013, there was significant advancement of scientific understanding in the areas of epidemiology and radiation biology, which was reflected in the evolution of the workshop findings and conclusions over this period.

2.1. The Helsinki workshop

The first science and values workshop concluded that if available Japanese risk estimates and the assumption of a linear non-threshold imply potential changes in radiological protection principles in relation to non-cancer effects, there may be a significant need for revision of protection criteria. The radiological protection system is risk-informed, not risk-based, and the main emphasis is on individual dose restrictions. As an indication, including non-cancer effects in risk estimates may suggest that current dose limits be lowered by 30–50%. It was noted that application of the precautionary principle in judging whether or not to modify existing radiological protection criteria should include not only assessment of the change in detriment, but also assessment of the costs and other consequences that such a change would cause. The breakout session recognised that any potential change should be made in the light of evolving science and rigorous value judgements.

2.2. The Vaux-de-Cernay workshop

The second science and values workshop took these conclusions a step further, proposing that, given the strengthening evidence base from epidemiological studies and research on mechanisms, the policy implications of radiation-induced vascular effects should be given much more serious consideration – or, if previously considered, these policy implications should be fully and appropriately acted upon.

Further studies (on both mechanisms and epidemiology) were being undertaken on the potential link between radiation exposure and vascular disease, particularly for exposure to chronic and low exposures. More information was emerging and circulating, thereby increasing professional awareness of this issue. Critical reviews were being made of the existing data and literature, and investigations were being made of plausible biological mechanisms. Regulators were increasingly becoming aware of the challenges raised by this new knowledge, and were reflecting on its potential implications for the current protection system.

Notwithstanding the above, questions were raised as to whether enough was being done, and sufficiently quickly given the apparent and increasing strength of the epidemiological evidence (particularly for chronic exposures where increased risks have been observed for accumulated exposures in the range of 0.1–0.5 Gy). Some felt that under the circumstances, a more proactive response was needed from the radiation protection community; others argued for a more measured approach, pending the development of a stronger evidence base and, in particular, the establishment of a causal relationship, or plausible mechanism, for the induction of vascular diseases following chronic or low dose exposures. It was widely recognised by participants that value judgements would inevitably have a major role to play in the development of an evidence-based policy.

Pending further developments (particularly clarification of, or progress on, the various issues listed above), a number of actions were identified that could, in the interim, mitigate the impact of radiation-induced vascular disease (assuming its occurrence following exposure to low and/or chronic doses were to be confirmed). These included the following:[2]

- giving even more emphasis to optimisation;
- adopting dose constraints for organs (e.g. in the Radiation Protection and Safety of Radiation Sources: International Basic Safety Standards; General Safety Requirements Part 3); and
- giving greater attention to doses accumulated over a lifetime.

[2]Also see the conclusions of the Article 31 group of experts following a European Union scientific seminar on 'Emerging evidence for radiation induced circulatory diseases' (EC, 2008). The group of experts, referred to under Article 31 of the European Atomic Energy Community treaty, provides the European Commission with an opinion on basic standards for health and safety.

The evidence base for the induction of vascular diseases from exposure to high acute doses of radiation was then well established, and was already influencing clinical practice (e.g. in radiotherapy of the breast). This knowledge and changing clinical practice do, however, need to be disseminated more widely. It is necessary to foster greater awareness in medical personnel of the importance of minimising the exposure of the heart when it is in the vicinity of the organ of tissue being irradiated during radiotherapy.

2.3. The Tokyo workshop

The third science and values workshop proposed that new epidemiological evidence (here, the reference is to new studies of the Mayak populations) suggests that chronic exposure at levels as low as 500 mSv may cause a small excess risk of stroke and heart disease. A cautious attitude is needed in the interpretation of epidemiological studies because of confounding factors (smoking and drinking habits, etc.). There are many models in the literature that have been tested in order to derive a dose incidence function. According to some experts, the most appropriate model may be a logistic curve, but this has not been used thus far for radiation protection purposes.

Current biological understanding of radiation effects provides mechanisms that might apply at different dose ranges. A better understanding of these mechanisms is still needed, and it would be helpful in deriving the most appropriate dose response function.

Based on current knowledge, it seems relevant to address the issue of non-cancer effects within the system of radiation protection. The system, as provided by ICRP, will, of course, not change immediately; time is needed for a shared reflection to clarify the issue. Additionally, it seems that the system is robust enough that major conceptual changes are not expected.

In order to ensure that ethical values are fully applied, the issue of the detriment due to deterministic effects must be clarified. This must be achieved in close co-operation with experts from the radiation protection community and from all fields. Breakout session participants recommended that ICRP should set up a task group devoted to assessment of the detriment associated with tissue reactions (deterministic effects) produced by radiation exposures.

As there are still uncertainties regarding epidemiology as well as radiobiology, it was recommended that research efforts should be maintained in order to improve knowledge and decrease uncertainties.

Based on scientific studies, at the time of the third science and values workshop, there appeared to be more information on non-cancer effects than was available on cancer effects in 1966. A delicate remaining question, if non-cancer effects are to be considered in the radiation protection system, is the issue of imputation. What are we going to say in 10 years to a Fukushima worker who develops a cardiovascular disease? What about people who are chronically exposed in contaminated territories? Therefore, the workshop recommended the re-evaluation of the consequences of the new scientific findings for non-cancer effects in the radiation protection system provided by ICRP.

A final recommendation was made for ICRP to use common language when undertaking this re-evaluation.

3. CONCLUSION AND RECOMMENDATIONS

The three science and values workshops represent a significant evolution of thinking with regard to how the radiological protection framework addresses non-cancer effects.

3.1. Moral challenge

Publication 9 (ICRP, 1966) justified its use of protective measures by stating:

> *The mechanism of the induction by radiation of leukaemia and other types of malignancy is not known. Such induction has so far been clearly established after doses of more than 100 rads [1 gray], but it is unknown whether a threshold dose exists below which no malignancy is produced. [...] As the existence of a threshold dose is unknown, it has been assumed that even the smallest doses involve a proportionately small risk of induction of malignancies. [...] The Commission is aware that the assumptions of no threshold and of complete additivity of all doses may be incorrect, but is satisfied that they are unlikely to lead to the underestimation of risks.*

In fact, scientific papers presented during the three workshops suggest that, today, more is known about the risks and detriments of radiation-induced non-cancer effects than was known in 1966 with regard to radiation-induced cancer effects. This does not imply that it is now necessary to change the international system of radiological protection, because, in addition to the evolution of science since 1966, there has been significant social evolution. It does, however, suggest that there is a moral challenge to fully investigate non-cancer effects and, in a multi-party fashion, decide what should and should not be done.

3.2. What are we doing now?

As part of addressing this moral challenge, there are many on-going aspects of study, including:

- reinforcing scientific studies on non-cancer effects;
- increasing professional awareness of non-cancer effects, particularly in the medical area, where patients can receive significant doses approaching or exceeding 0.5 Gy;
- critically reviewing existing data/literature;
- strengthening the evidence database from epidemiological studies, and continuing research on non-cancer disease mechanisms;

- challenging the features of the current radiological protection system in light of evolving science and value judgements; and
- investigating the policy implications that the inclusion of radiation-induced non-cancer disease in the system of radiological protection may have on regulation and application.

3.3. Remaining value questions

As part of the resolution of the moral dilemma posed by radiation-induced non-cancer effects, there are many value questions that will need to be addressed, and the responses to which will help set the direction for the resolution of the moral dilemma. For example:

- How much additional risk is suggested by new studies?
- What are the implications of the additional detriment due to these effects?
- Is evidence sufficient to require further precaution in the radiological protection system for workers and for the general public, and in this context, how important is consistency given the precedent of cancer risk regulation?
- If radiation-induced non-cancer effects are to be taken into the system, how should this be accomplished in terms of the existing system and overall risk management?

3.4. Recommendations

In partial response to these questions, the science and values workshops have developed a few recommendations to assist in clarifying the situation.

- The 'detriment' of an effect is defined as the total harm to health experienced by an exposed group and its descendants as a result of the group's exposure to a radiation source (ICRP, 2007). In order to ensure that ethical values fully apply, there is a need to clarify the issue of the detriment due to deterministic effects. This must be achieved in close cooperation with experts from the radiation protection community and from all fields.
- As such, it is recommended that ICRP should create a task group on the detriment associated with deterministic effects. The time is approaching to review and report on this issue.
- In support of these efforts, there is a need for a synthesis of available knowledge, particularly radiobiology inputs, and for this, more assistance from the United Nations Scientific Committee on the Effects of Atomic Radiation would be helpful.
- There is a need to maintain research efforts in order to provide more reliable answers (e.g. decrease uncertainties) in terms of improved understanding of mechanisms, and of improved understanding of risks through epidemiological studies.

In terms of strengthening protection efforts now, before decisions on how to account for non-cancer effects are made, it is also recommended that more ways are found to spread the message of ICRP recommendations in the context of the wider range of safety issues in daily life, to appropriately apply the principle of the optimisation of protection.

REFERENCES

EC, 2008. Radiation Protection No. 158. EU Scientific Seminar 2008, Emerging Evidence for Radiation Induced Circulatory Diseases. Proceedings of a scientific seminar, 25 November 2008, Luxembourg. Available at: http://ec.europa.eu/energy/sites/ener/files/documents/programme_0.pdf.

ICRP, 1966. Recommendations of the International Commission on Radiological Protection. ICRP Publication 9. Pergamon Press, Oxford.

ICRP, 2007. The 2007 Recommendations of the International Commission on Radiological Protection. Ann. ICRP 37(2–4).

OECD/NEA, 2007a. Scientific issues and emerging challenges for radiological protection: Report of the Expert Group on Emerging Challenges for Radiological Protection. OECD 2007, NEA No. 6167.

OECD/NEA, 2007b. Radiation protection in today's world – towards sustainability. OECD 2007, NEA No. 6165.

OECD/NEA, 2011. Science and Values in Radiological Protection: Summary of the CRPPH Workshops held in Helsinki (2008) and Vaux-de-Cernay (2009). OECD 2011, NEA No. 6364. Available at: http://www.oecd-nea.org/rp/pubs/2011/6364-sciences-values-in-rp.pdf. OECD, Paris.

OECD/NEA, 2013. 3rd Workshop on Science and Values in Radiological Protection Decision Making, and 6th Asian Regional Conference on the Evolution of the System of Radiological Protection: Workshop 2012 Summary Report, Tokyo, Japan, 6–8 November 2012, © OECD 2013, NEA/CRPPH/R(2013)5.

Taylor L.S. 1957. The philosophy underlying radiation protection. Am. J. Roentgenol. Radium Ther. Nucl. Med. 77, 914–919.

Sievert, R.M., 1958. The Work of the International Commission on Radiological Protection. United Nations International Conference on the Peaceful Uses of Atomic Energy, Geneva, 1–13 September 1958, Geneva, Switzerland, Vol. 21, Session 5a, pp. 3–7.

SAGE

ICRP
Annals of the ICRP

Implications of the implementation of the revised dose limit to the lens of the eye: the view of IRPA professionals

J. Broughton[a], M.C. Cantone[b], M. Ginjaume[c], B. Shah[a], R. Czarwinski[d]

[a]*Society for Radiological Protection, Dartington Hall, Totnes, TQ9 6EN, UK; e-mail: admin@srp-uk.org*

[b]*Department of Physics, University of Milan, Italy*

[c]*Institut de Tècniques Energètiques, Universitat Politècnica de Catalunya, Spain*

[d]*Federal Office for Radiation Protection Berlin, Germany*

Abstract–In April 2011, the International Commission on Radiological Protection issued a statement on reduction of the equivalent dose limits for the lens of the eye, and strongly recommended its consideration in the revision of the International Atomic Energy Agency's International Basic Safety Standards on Radiation Protection. The reduced dose limit was incorporated in the final version of the Basic Safety Standards. As significant concern was expressed by radiation protection professionals worldwide, the International Radiation Protection Association (IRPA) established a task group to assess the impact of implementation of the revised dose limit for the lens of the eye for occupational exposure. IRPA Associate Societies (ASs) were asked for their views using a questionnaire addressing three topics: implications for dosimetry, implications for methods of protection, and wider implications. The responses received indicate various methods of approach and express different points of view, reflecting nuances of particular ASs or specific professional groups. Topic experts nominated by ASs were selected to assist with collation of responses, and a report was produced by the task group. Conclusions were drawn on the three issues, including potential cost implications. A number of recommendations were drawn from the responses received including: the request for more understanding about the relationship between exposure of the lens of the eye and cataract formation, and further guidance to assist implementation; the importance of economic and social considerations when introducing the limits into national regulations; the

This paper does not necessarily reflect the views of the International Commission on Radiological Protection.

need to propose or define procedures related to employment of people with existing or pre-cataract conditions; and the practical aspects relating to dosimetry and protective equipment.

Keywords: Dose limit; Lens of eye; Views of radiation professionals

1. INTRODUCTION

The primary purpose of the International Radiation Protection Association (IRPA) is to provide a medium whereby those engaged in radiation protection activities in all countries may communicate more readily with each other, and thereby advance radiation protection in many parts of the world. IRPA's vision is 'to be recognized as the international voice of the radiation protection profession'.

In April 2011, the International Commission on Radiological Protection (ICRP) revised its eye dose threshold for cataract induction to 0.5 Gy; the previous threshold doses for vision-impairing cataracts were 5 Gy for acute exposures and >8 Gy for highly fractionated exposures. Further, ICRP recommended a reduction in the dose limit for occupational exposure in planned exposure situations (in terms of equivalent dose) for the lens of the eye from 150 mSv to 20 mSv year^{-1}, averaged over defined periods of 5 years, with no dose in a single year to exceed 50 mSv (ICRP, 2012).

The International Atomic Energy Agency (IAEA) swiftly introduced the requirement for the revised dose limit to the lens of the eye, with less than the usual time for consultation; this left many radiation practitioners somewhat puzzled regarding the need, the implications (both practical and financial), and the possible timescale for introduction of any possible change to legislation. It took some time for the concerns to surface, but by October 2012 at a meeting of the European Presidents of IRPA Associate Societies (ASs), sufficient interest had been expressed for the IRPA President, Renate Czarwinski, to commission John Broughton, President of the Society for Radiological Protection (SRP), to form and lead a task group to undertake a survey of all IRPA ASs. The requirement was to return a rapid initial response of the reaction and possible concerns of radiation practitioners worldwide to the implications of the introduction of such a change to the dose limit.

2. THE IRPA TASK GROUP: OBJECTIVES AND METHODOLOGY

The IRPA task group was established in December 2012 under the leadership of the SRP/UK in close cooperation with the Italian Association of Radiation Protection. IRPA ASs were asked to complete a questionnaire to provide views and comments, addressing three different topics: implications for dosimetry, implications for methods of protection, and wider implications of implementing the revised limit.

The questionnaire was circulated to all 48 ASs within IRPA. Answers were obtained from 12 ASs (covering 16 countries), including most of the larger organisations. Input was received from regions including Europe, North and South America, and Asia.

After collation of the responses by nominated experts, a report was produced by the IRPA task group, and general and specific conclusions were drawn[1] (Broughton et al., 2013; Czarwinski, 2013).

3. SUMMARY OF FINDINGS

3.1. General

Various methods of approach and differing points of view were expressed in the responses received. Considerable disparity was found in the following aspects:

- cost implications for the procedures aimed to reduce the dose to the eye;
- implications related to the employment of people with existing cataract or pre-cataract conditions; and
- current perceptions of future compensation claims related to the new dose limit to the lens of the eye.

As choices and decisions will remain largely within each country, it will be useful to have further interchanges aimed at achieving better understanding of the various aspects considered in the report. It is recommendable that a harmonised approach should be sought in implementation of the reduced dose limit.

3.2. Specific

As well as some general conclusions, five specific aspects were considered and addressed, namely: field of impact, eye lens dosimetry, protection of workers, wider implications, and potential cost implications/socio-economic effects.

3.2.1 Field of impact

There is broad consensus that the main impact of the new limit will be in the medical sector (interventional radiology and cardiology). Minor concerns emerged in the fields of veterinary x rays. Some other working areas still need investigation concerning their relevance to higher exposures for the lens of the eye, and therefore for protective measures.

3.2.2 Lens dosimetry

In the field of interventional radiology and cardiology procedures, workers could be exposed to doses close to, or in excess of, the revised limit. It is particularly important, in these cases, to know that the dosimeter will definitely record a dose indicative of the dose received by the lens of the eye.

[1]Thanks are given to all of the ASs that responded to the questionnaire, particularly the Topic Experts: (Topic 1) José Miguel Fernández-Soto, Spain and Mercè Ginjaume, Spain; (Topic 2) Steven King, USA and Denisa Nikodemová, Slovakia; (Topic 3) Keiichi Akahane, Sumi Yokoyama, Japan and Bela Csakany, Hungary.

The relationship between dose and cataract formation is not well understood, and the causality needs better clarification.

There is a need for international recommendations to ensure harmonisation of radiation protection criteria, and for an agreed standardised system of dose recording (double-dosimetry system or single badge) with appropriate empirical formula to record the personal dose equivalent at 10 mm depth $H_p(10)$ and at 3 mm depth $H_p(3)$. Arrangements should be confirmed or put into effect for assessing and recording the total dose to the lens of the eye of itinerant workers.

3.2.3 Protection of workers

By considering the protection of workers, the analysis by the task group indicated that established radiation protection techniques and shielding devices to reduce the dose to the lens of the eye are available to all ASs, but they are used somewhat sporadically. Training for effective and consistent use of these techniques presents a significant hurdle because of limitations of both time and availability of staff to organise effective training. Radiation protection training, especially in areas where the dose could exceed established limits, is a priority; therefore, funding of training is very important.

The application of methods of protection varies considerably from one location to another, even within the same country, and explicit guidelines relating to their use would be very beneficial. Even if decisions regarding which techniques and equipment should be used must rest at local level, it would be useful if international guidelines could be prepared and issued. Mandatory use of eye protection should be considered for all exposed workers.

3.2.4 Wider implications

In terms of wider implications, the majority of the ASs emphasised that the new limit could affect current methods of working, and could cause problems in relation to employment issues, high costs for possible additional medical eye examinations, possible increases in compensation in legal actions, and difficulties in explaining how to answer queries about previously unrecorded doses which are below current dose limits, but possibly above the new limit.

A number of ASs noted that significant concern and confusion exists among radiation practitioners on the rationale of changing the dose limit, since it appears that fatal and non-fatal effects are being considered in a similar way.

The evidence to support this change is not linked to harm, but to potential changes to the eye that are not considered a very significant detriment. Additional concerns are related to the fact that the findings in the literature are not consistent, the results are tenuous, and the changes to radiation protection practice could have evident socio-economic implications while the risk to the eyes may be considered to be small.

The work of the international organisations (ICRP, IAEA) on this topic seemed to be undertaken in a hurry, with an inadequate period of open consultation (which the community has been accustomed to having for many years). It was felt that the case for the revised dose limit should be made more visible to practitioners.

3.2.5 Potential cost implications

For the survey presented in the task group report, it was not possible to quantify the additional costs imposed by the new dose limit. However, most ASs had some concerns over the implied costs for:

- additional training;
- dosimetry and shielding;
- the possible need to formally classify more workers;
- systematic investigations of workplaces for their relevance;
- the need to recruit extra staff if current specialist staff are likely to approach the dose limit; and
- enhanced medical eye examinations for workers.

4. RECOMMENDATIONS

From the responses received, representing the voice of radiation protection professionals, a series of recommendations has been derived, grouped under three main headings.

4.1. Understanding and guidance

Support should be given to further studies on the relationship between radiation exposure of the lens of the eye and cataract formation. A more rational explanation should be provided for using the same numerical dose limit for the eye as for non-fatal deterministic and fatal stochastic effects. In addition, it will be of help if the relevant international organisations could explain the reason for the revised dose limit more clearly, so that it is more easily understood by the practitioners.

A further study leading to an international protocol for monitoring doses to the lens of the eye is also needed, and this should be promoted by the IRPA Executive Council. The introduction of guidelines to correctly identify groups of workers who could be exposed to lens doses close to the limit is recommended. Such guidance, to be provided by regulators, is necessary to assist in implementation of the changes introduced, including requirements on management, means of protection, and monitoring procedures and systems.

4.2. Practical aspects

There is a need for a new system for detecting and reporting cataract data, or at least the need for clear comparison of existing systems. Further investigation is required on the validity and limitations of a whole-body dosimeter (at the collar above the apron) to assess $H_p(3)$. Further investigation on the effectiveness of the common protection methods (lead glasses and screens) is also recommended, as well as dissemination of the results to users via training sessions.

It will also be necessary to ensure that dosimeters and protective equipment are comfortable to use and do not interfere significantly with the effectiveness of medical procedures. Optimisation of radiation exposure is required for workers who could be exposed to levels approaching the dose limit, as well as radiation protection for the public. Procedures for emergency situations need to be taken into consideration.

4.3. Social, economic, and management considerations

In general, economic and social considerations have to be taken into account according to national characteristics (e.g. national infrastructure) when introducing new dose limits (as in the present case for the lens of the eye) into national regulations. There is also a need to better detail the costs, showing the additional costs incurred, while reducing eye exposure, in comparison with the total costs of the procedure and the overall costs of any required installation. The new dose limit results in even greater emphasis on the training of key groups of workers who are likely to be most highly exposed.

There is a need to address training on the use of protective equipment, increasing awareness about effectiveness, and correct wearing of relevant dosimeters.

There is a need to define procedures relating to the employment of people with existing cataract or pre-cataract conditions, taking into consideration: (1) the risk of discriminating against people seeking employment on the basis of a condition that is quite common; and (2) the risk of inducing additional deterioration of visual acuity for exposed workers.

IRPA will continue its work to ensure that these findings are integrated into the ongoing international discussion of these implementation issues.

REFERENCES

Broughton, J., Cantone, M.C., Ginjaume, M., et al., 2013. Report of Task Group on the implications of the implementation of the ICRP recommendations for a revised dose limit to the lens of the eye. J. Radiol. Prot. 33, 855–868.

Czarwinski, R., 2013. Implications of the implementation of the reduced dose limits for the lens of the eye: an IRPA activity. J. Radiol. Prot. 33, E15–E16.

ICRP, 2012. ICRP statement on tissue reactions/early and late effects of radiation in normal tissue and organs – threshold doses for tissue reactions in a radiation protection context. ICRP Publication 118. Ann. ICRP 41(1/2).

SAGE

ICRP
Annals of the ICRP

Engaging with local stakeholders: some lessons from Fukushima for recovery

R.S. Hayano

Department of Physics, The University of Tokyo, 7-3-1 Hongo, Bunkyo-ku, Tokyo 113-0033, Japan; e-mail: hayano@phys.s.u-tokyo.ac.jp

Abstract–The Fukushima Daiichi nuclear power plant accident contaminated the soil of densely populated regions in Fukushima prefecture with radioactive caesium, which poses significant risks of internal and external exposure to the residents. Applying the knowledge of post-Chernobyl accident studies, internal exposures in excess of a few mSv per year would be expected to be common in Fukushima. However, extensive whole-body-counter surveys have shown that the internal exposure levels of residents are much lower than estimated; in 2012–2013, the Cs-137 detection percentages (the detection limit being $\sim$300 Bq body^{-1}) were approximately 1% for adults and practically 0% for children. These results are consistent with those of many other measurements/studies conducted to date in Fukushima. As a consequence, risks from external exposure assume greater importance for the majority of residents in Fukushima due to the lower contribution from internal exposure. In both cases, average doses remain low, although some residents are exposed to higher-than-average risks; it is these members of the population who need to be identified and followed-up. Consequently, it is essential to re-establish communication at all levels in society.

Keywords: Fukushima Daiichi nuclear power plant accident; Internal exposures; External exposures; ICRP *Publication 111*

1. INTRODUCTION

The author works at CERN in Geneva on 'antimatter', and began generating graphs soon after the Fukushima Daiichi accident (Tanaka, 2012) in order to better understand the situation. These graphs and other information were shared using Twitter (@hayano), and these activities caught the attention of many people;

This paper does not necessarily reflect the views of the International Commission on Radiological Protection.

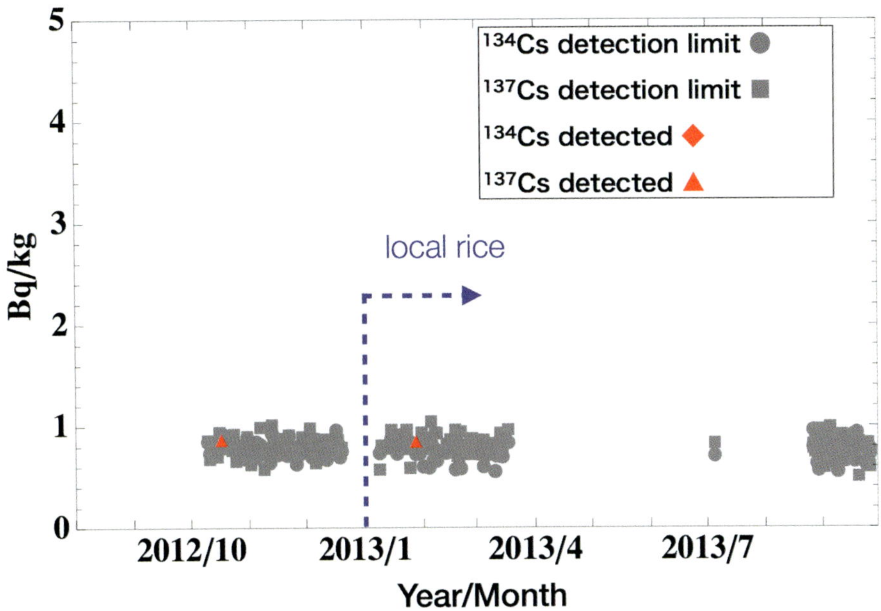

Fig. 1. Results of radioactivity measurements in school lunches in Fukushima city.

as such the number of 'followers' increased from some 3000 to more than 150,000 within a few days. A recent study conducted by Tohoku University showed that the author had the seventh most influential Japanese Twitter account following the Great East Japan Earthquake and the Fukushima Daiichi nuclear power plant accident. Being bidirectional, Twitter is quite different from conventional media such as television and newspapers. Before Summer 2011, there was an increasing trend of tweets on food safety, especially by worried mothers.

At that time, no data were available regarding the level of internal contamination. As such, the author proposed to the Japanese Government that school lunches should be measured systematically for radioactive caesium contamination. This proposal was adopted by the Government, and was included in the 2012 and 2013 Japan Fiscal Year 2012 budget.

Fig. 1 shows the results of the school lunch tests conducted in Fukushima city. The detection limit was approximately $1\,Bq\,kg^{-1}$. Positive results were reported in two cases, but they were at the level of the detection limit and were not considered to be significant. This is despite the fact that Fukushima city started to use local rice in January 2013. Similar tests conducted in all other cities, towns, and villages in Fukushima prefecture have shown that school lunches are practically free of radiocaesium.

These activities related to the measurement of radioactive caesium contamination in school lunches attracted the attention of medical doctors in Fukushima, who were struggling to establish reliable whole-body-counter (WBC) measurements in 2011.

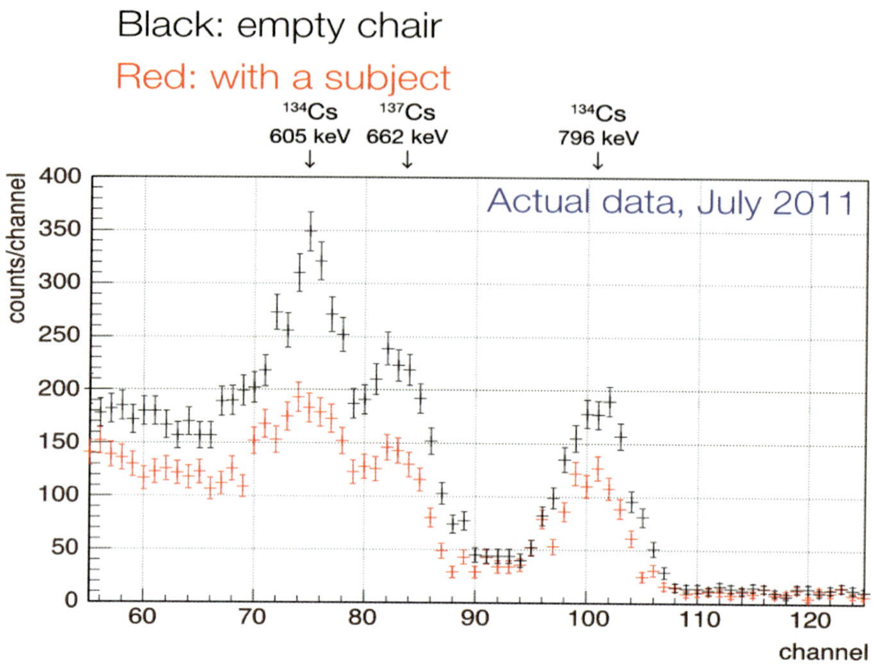

Fig. 2. Spectra measured with a chair-type whole-body counter at Minamisoma General Hospital in July 2011.

Fig. 2 shows an example of the difficulties that doctors were facing. The two spectra were taken with a chair-type WBC (with insufficient shielding), installed in July 2011 at Minamisoma General Hospital, located 23 km north of Fukushima Daiichi nuclear power plant. The spectrum shown in black is a background measurement in July 2011, which clearly shows peaks for Cs-134 and Cs-137. Superimposed in red is the spectrum taken with a subject. As shown, the caesium peak heights are lower with the subject sitting on the chair; the shielding effect by the body was more visible than that of the internal contamination. It was very difficult to interpret such spectra correctly, and this started the author's involvement in WBC measurements in Fukushima.

2. INTERNAL EXPOSURES

The Cs-137 deposition map presented in Fig. 3 for the contaminated area indicates that a typical level of Cs-137 in densely populated cities such as Fukushima (population ~ 290,000) and Koriyama (~340,000) is approximately 100,000 Bq m^{-2}. According to the coefficient (2 μSv 100 Bq^{-1} m^{-2}) published by UNSCEAR (1988), the average committed effective dose of a person living in the Fukushima–Koriyama region would amount to approximately 2 mSv year^{-1} for Cs-137 from ingestion alone, and up to 5 mSv year^{-1} when the contribution from Cs-134 is also considered (important in Fukushima for the first 1–2 years after the accident).

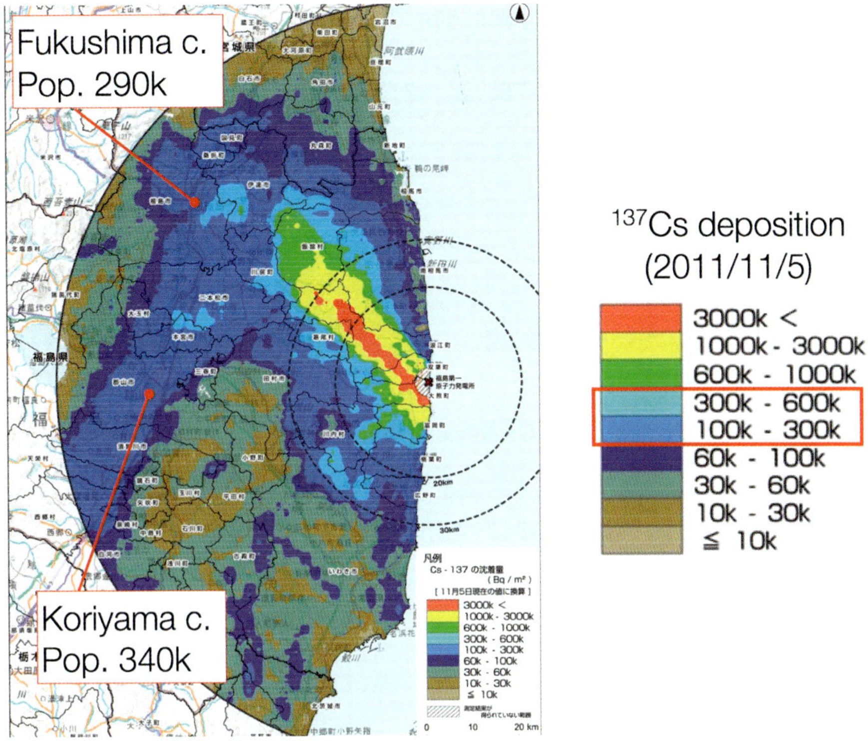

Fig. 3. Results of the airborne monitoring survey by MEXT, Japan (MEXT, 2011).

However, as shown in a recent paper (Hayano et al., 2013), approximately 99% of Fukushima residents do not have detectable levels of radiocaesium in their bodies (detection limit was 300 Bq body^{-1}). For the remaining 1% of residents, the median caesium concentration was approximately 10 Bq kg^{-1} (Fig. 4); a level similar to that received by Japanese adults in the 1960s as a result of global fallout (Uchiyama et al., 1996).

This surprisingly low level of internal exposure is not likely to be due to sampling bias. In Autumn 2012, all school children (1456 enrolled, 1383 measured, coverage 95%) in the town of Miharu, in the suburbs of Koriyama, were measured, and none of them exceeded the detection limit. Questionnaires sent to their parents revealed that 60% of these children were eating local or home-grown rice, and 20% were eating local or home-grown vegetables. These results are consistent with those of many other measurements/studies conducted to date in Fukushima [e.g. rice inspection, foodstuff screening, and duplicate-portion studies (e.g. Harada et al., 2013; Sato et al., 2013)].

Particularly important is the low caesium concentration in rice grown in Fukushima (Nakanishi and Tanoi, 2013). In 2012, Fukushima Prefectural Government tested every bag of brown rice harvested in Fukushima prefecture (more than 10 million 30-kg bags). Only 71 bags exceeded the limit of 100 Bq kg^{-1} set by the Government (Fig. 5).

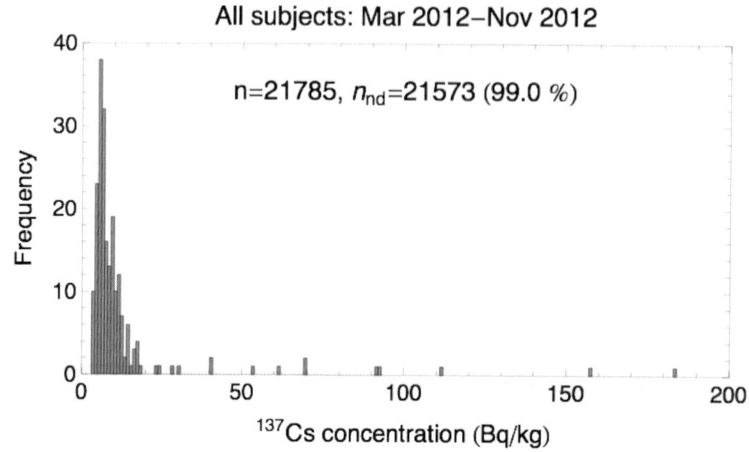

Fig. 4. Cs-137 concentration for 21,785 subjects measured at Hirata Central Hospital in Fukushima prefecture between March 2012 and November 2012 (source: Hayano et al., 2013). Non-exposed (non-detected) subjects have been excluded from the plot.

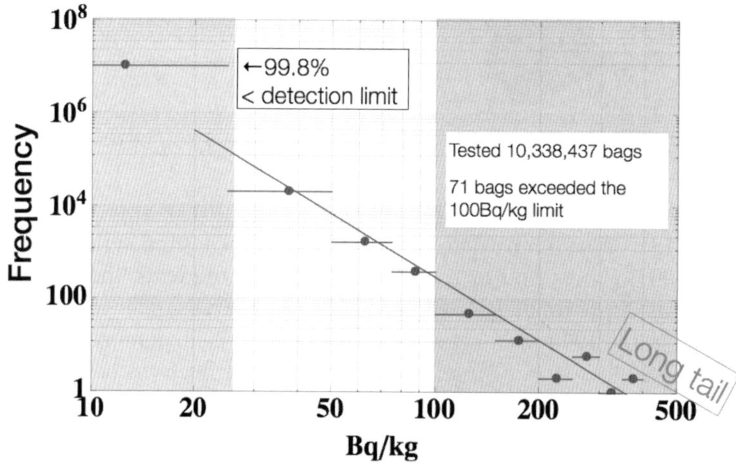

Fig. 5. Results of the brown rice inspection conducted by Fukushima Prefectural Government in 2012 (note: log-log scale).

The contamination level of marketed food is kept low, and therefore ordinary Fukushima residents are not exposed to high risk of internal contamination. Fig. 6 shows the WBC results for Hirata Hospital. They have been re-plotted using a log-log scale, which reveals a characteristic long tail indicative of a small number of residents at higher risk of internal contamination.

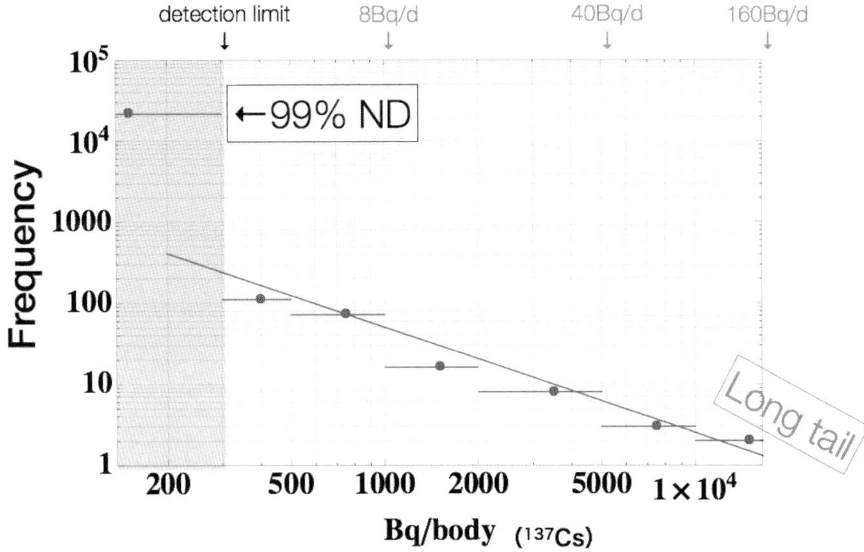

Fig. 6. Whole-body-counter results for Hirata Hospital (including non-exposed subjects) presented using a log-log scale.

On identification of an individual with higher-than-average levels of caesium in their body, the aim is to identify the source of the contaminated food, either by interview or by measurement of the suspected food using a germanium detector. In most cases, it has been possible to identify the source of contaminated food (e.g. wild or home-grown mushrooms, wild boar, etc.). These foodstuffs are known to accumulate high levels of radiocaesium, but as they are not placed on the market, they are not tested for radiocaesium. People who consumed these foods regularly were advised to stop eating them and to come back to the hospital after a few months for whole-body measurements. In individuals who received this advice, body burdens have decreased at a rate consistent with the known biological half-life.

3. EXTERNAL EXPOSURES

In Winter 2011, many municipalities in Fukushima started to distribute 'glass badge' personal dosimeters, the results of which are shown in the top panel of Fig. 7 (results of wearing badges for 2–3 months were converted to annual doses). As shown, most people were subject to external exposures below 2 mSv year^{-1}.

Comparison of 'glass badge' measurements in Fukushima city between Winter 2011 and Winter 2012 (Fig. 8) shows that most external doses were below 1 mSv year^{-1} in Winter 2012. Nevertheless, there is a long tail in the distribution of doses, extending to approximately 5 mSv year^{-1}, where more attention is required. To date, this has not been performed systematically in Fukushima.

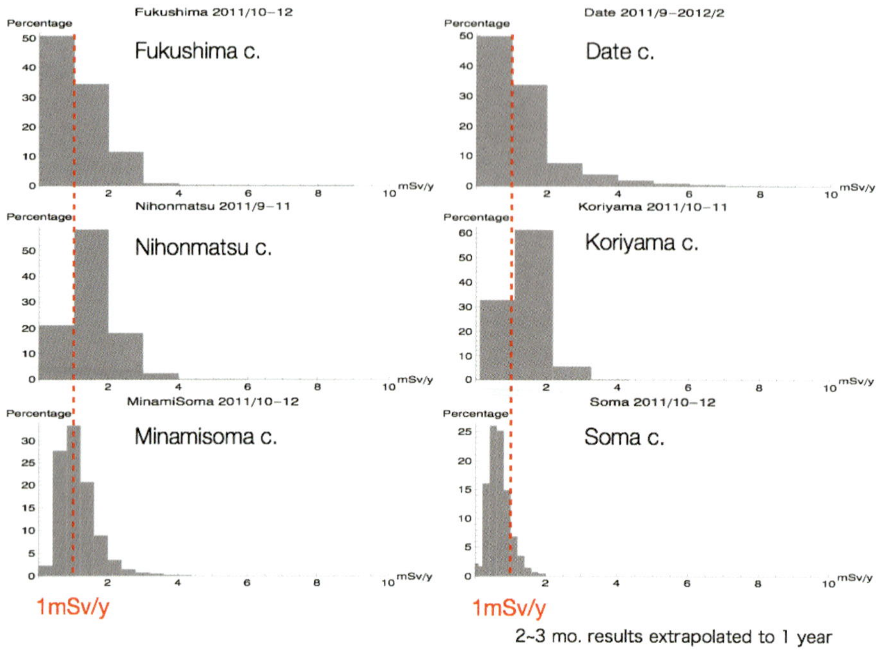

Fig. 7. Results of the 'glass badge' measurements carried out by Fukushima municipalities in Winter 2011.

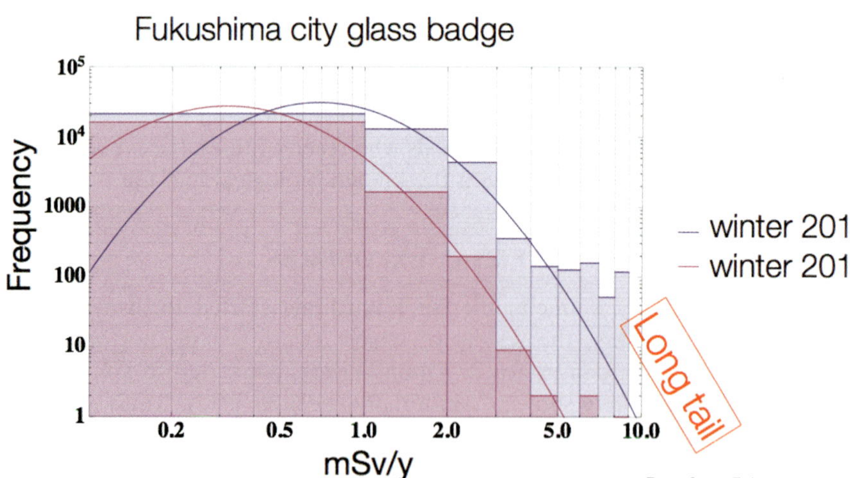

Fig. 8. Comparison of the Fukushima city 'glass badge' results for Winter 2011 (blue) and Winter 2012 (red), presented using a log-log scale.

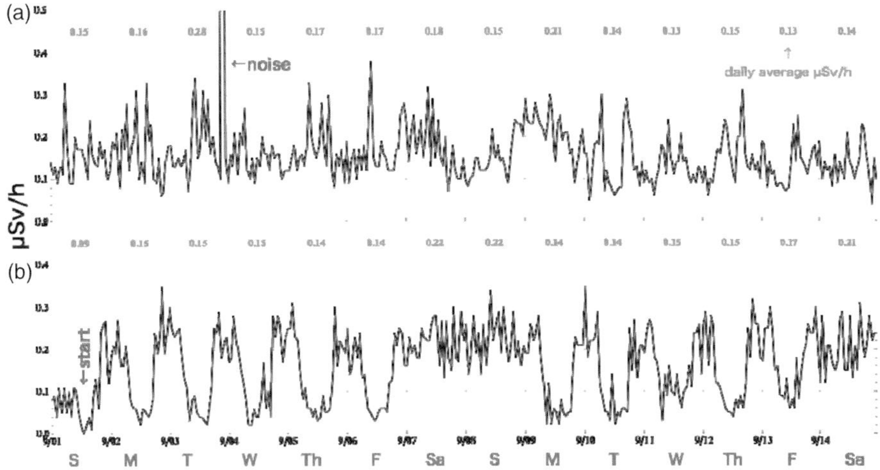

Fig. 9. One-hour integrated dose recorded by personal dosimeters. (A) A farmer who returned recently to the former 20-km-exclusion zone in Tamura city. (B) A school teacher in Fukushima city.

The 'glass badge' dosimeters are useful for identifying people at higher risk, but as they are typically worn for 2–3 months, it is difficult to relate the behaviour of each person to the integrated dose. This is why the use of a solid-state personal dosimeter is currently being promoted, which is capable of recording 1-h integrated dose for up to 1 year (AIST, 2012). Initially, the plan was to use this device together with a GPS; however, it was found to be sufficient to use a graph (Fig. 9), and discuss the result with the resident. If an interview is performed within 2–3 weeks of wearing the dosimeter, residents generally remember their whereabouts, and can correlate their behaviour with the dosimeter readings.

4. CONCLUSIONS

Internal and external doses were considered in Fukushima prefecture following the accident at Fukushima Daiichi nuclear power plant. Internal contamination was much lower than predicted initially, with radiocaesium being undetectable in 99% of residents sampled. Risks from external exposures were higher than risks from internal exposures, although most were less than 1 mSv year^{-1}. Whilst average doses remain low in both cases, it is apparent that some residents are being exposed to higher-than-average risks, and it is these members of the population who need to be identified and followed-up. Establishing lines of communication between residents and local medical staff, between central government and municipalities, and with experts will help to develop awareness among residents about exposures and how to reduce them in the future.

REFERENCES

AIST, 2012. Development of small radiation dosimeter. Long battery life of more than 1 year on a button cell battery. AIST Today 4, 23.

Harada, K., Fujii, Y., Adachi, A., et al., 2013. Dietary intake of radiocesium in adult residents in Fukushima prefecture and neighboring regions after the Fukushima nuclear power plant accident: 24-h food-duplicate survey in December 2011. Environ. Sci. Technol. 47, 2520–2526.

Hayano, R., Tsubokura, M., Miyazaki, M., et al., 2013. Internal radiocesium contamination of adults and children in Fukushima 7 to 20 months after the Fukushima NPP accident as measured by extensive whole-body-counter surveys. Proc. Jpn. Acad. Ser. B 89, 157–163.

MEXT, 2011. Results of the Fourth Airborne Monitoring Survey by MEXT. Ministry of Education, Culture, Sports, Science and Technology – Japan, Tokyo. Available at: http://radioactivity.nsr.go.jp/en/contents/4000/3179/24/1270_1216.pdf (last accessed 18 November 2013).

Nakanishi, T.M., Tanoi, K. (Eds.), 2013. Agricultural Implications of the Fukushima Nuclear Accident. Springer, Tokyo.

Sato, O., Nonaka, S., Tada, J., 2013. Intake of radioactive materials as assessed by the duplicate diet method in Fukushima. J. Radiol. Protect. 33, 823–838.

Tanaka, S., 2012. Accident at the Fukushima Dai-ichi nuclear power stations of TEPCO – outline & lessons learned. Proc. Jpn. Acad. Ser. B 88, 471–484.

UNSCEAR, 1988. Exposures from the Chernobyl Accident. Annex D. United Nations, New York, pp. 309–343.

Uchiyama, M., Nakamura, Y., Kobayashi, S., 1996. Analysis of body-burden measurements of 137Cs and 40K in a Japanese group over a period of 5 years following the Chernobyl accident. Health Phys. 71, 320–325.

SAGE

Annals of the ICRP

Experience and current issues with recovery management from the Fukushima accident

M. Kai

Environmental Health Science Division, Department of Health Sciences, Oita University of Nursing and Health Sciences, Oita City, Japan; e-mail: kai@oita-nhs.ac.jp

Abstract–This paper describes the experiences of, and issues with, recovery management following the accident at Fukushima Daiichi nuclear power plant. The Fukushima accident has brought about socio-economic consequences with inevitable changes to daily life, as well as psychological effects. There is heightened concern amongst the population about the risk and effects of radiation at low doses. Experience has shown that the direct involvement of the affected population and local professionals is a decisive factor for management of the recovery phase. The radiological protection system of the International Commission on Radiological Protection (ICRP) seems to be relevant to the recovery requirements of the Fukushima accident, although some problems remain in implementation. Reference levels could play a role in improving the situation by requiring an iterative optimisation process. The Fukushima experience indicated that a routine, top-down approach using radiological criteria alone was unable to deal with the complexity of the problems, and that stakeholder engagement should be explored. The technical knowledge gap between radiation experts and the public caused a lot of confusion. Experts should understand the ethical values attached to recovery, and ICRP should be more active in promoting trustworthy radiological protection advice.

Keywords: Stakeholder involvement; Reference levels; Existing exposure situation; Situation-based approach

1. INTRODUCTION

The accident at Fukushima Daiichi nuclear power plant has had a large impact, not only in the locally affected area but also at a broader national level in Japan. Basic information on the impact mainly originated from measurements of ambient dose rates and radioactivity in soil, water, air, and foodstuffs. The aerial survey

This paper does not necessarily reflect the views of the International Commission on Radiological Protection.

produced a wide range of dose rate maps. In addition, soil sampling was conducted intensively and extensively by the authorities, and researchers provided detailed information on contamination levels in the affected area. These data were collected immediately after the accident to predict the magnitude of the human and environmental impacts. The unprecedented scale of the accident in Japan brought about a nationwide level of social confusion, which can be summarised in three ways. Firstly, socio-economic consequences have caused changes in daily life of affected communities, as well as psychological effects. For example, the evacuation measures taken during the emergency exposure situation have not been lifted in many places since 2011. Secondly, there are many concerns among the population about the risks of low doses, particularly among children and unborn babies. Thirdly, recovery management problems have arisen from slow step-by-step rehabilitation programmes that have moved forwards in a different way from that planned. Lessons learned from the response to the accident should be used to reflect on how radiological protection practices can be improved in the future. The International Commission on Radiological Protection (ICRP) has recommended fundamental principles for application of the radiological protection system (ICRP, 2007), and specific advice for the protection of people living in long-term contaminated areas after a nuclear accident (ICRP, 2009b). This paper focuses specifically on the key issues raised about the ICRP system of radiological protection on recovery management after an accident.

2. UNDERSTANDING RADIOLOGICAL PROTECTION PHILOSOPHY

2.1. Reference levels

ICRP recommends that the level of protection should be the best possible under the prevailing circumstances, maximising the margin of benefit over harm. The ICRP system of radiological protection is based on three exposure situations: existing exposure situations, planned exposure situations, and emergency exposure situations. The experiences and issues that emerged from the situation-based approach in the Fukushima accident have been discussed by Kai (2012) and ICRP TG (ICRP, 2012).

The main protective actions taken in the event of an emergency are sheltering, evacuation, and foodstuff restrictions. Following the Fukushima accident, the evacuation and foodstuff restrictions adopted by the authorities effectively reduced the doses received by people living in the affected area. However, societal concern about the level of protection intensified with transition from the emergency exposure situation to the existing exposure situation.

On 12 March 2011, an emergency evacuation was implemented in a circular zone within 20 km of the accident site. However, on 22 April 2011, the authorities ordered full evacuation of residents from the village of Iitate, located outside the original 20-km zone, as an exceptional case. This was termed 'deliberate evacuation' because it was not a prompt reactive measure in the emergency phase. Iitate became a symbolic village to the population due to this deliberate evacuation. The residents of Iitate faced the dilemma of leaving the village or staying for 2 months following the authority's decision regarding full evacuation. Management of the Iitate situation has led to a lack of understanding regarding the use of radiation protection criteria. It has also had a major effect on the credibility of the authorities.

The Japanese guidelines for nuclear disaster prevention recommended an evacuation criterion of 50 mSv before the Fukushima accident. However, following the accident, the Japanese Government adopted the criterion of 20 mSv year^{-1} based on the lower value of the reference levels proposed in the ICRP system for emergency exposure situations. Protective action was requested to protect people living in a long-term contaminated area with Cs-137 and Cs-134. The new adopted criterion, 20 mSv year^{-1}, created considerable confusion, as the meaning and rationale behind it were not clear for the authorities nor the affected residents.

The concept of intervention levels has long been used in radiological protection in the case of an emergency. In their 2007 Recommendations, ICRP introduced the situation-based approach to replace the process-based approach based on practices and interventions (ICRP, 2007). Reference levels became key drivers in the situation-based approach, and were a key feature of the 2007 Recommendations. They are guides for implementation of the optimisation process, particularly for identifying priorities in management of the situation. However, following the Fukushima accident, 20 mSv year^{-1} was used as an intervention level in the contaminated area (i.e. as a cut-off to identify contaminated areas where protective actions should be taken). The meaning and rationale behind the use of 20 mSv year^{-1} were not understood and explained properly by the authorities, and consequently have been communicated poorly to the public.

A similar problem occurred when the authorities decided to use the same tentative reference level for school playgrounds within the contaminated area. The dose of 20 mSv year^{-1} was understood as the cut-off between safety and danger by most people. Nobody understood why the dose level in these circumstances was much higher than the public dose limit of 1 mSv year^{-1} in the case of a planned exposure situation. Even some radiation experts were unable to understand the difference between contamination from accidental releases and contamination from fully controlled sources, such as in a planned exposure situation. Furthermore, members of the public misunderstood the decision of the authorities, and believed that the annual dose of 20 mSv year^{-1} was to be maintained over a long period of time.

One of the marked differences between reference levels and intervention levels is the inclusion of a time factor for improving the prevailing circumstances. The 2007 Recommendations of ICRP emphasise that a reference level is a key driver to optimise the residual level of dose remaining after implementation of protection strategies (ICRP, 2007). An intervention level is applied to an averted dose for taking a certain protective action, and is also understood as a level above which an action is justified and below which no action is needed (ICRP, 2009a). It is easier for a regulatory body to apply intervention levels for an emergency situation because a single number can be used to distinguish between circumstances where action is required and circumstances where action is not required. On the contrary, reference levels are instruments to identify where protection actions have to be implemented as a priority, and to push improvement in the situation by undertaking an iterative optimisation process. Reference levels are likely to be misunderstood if a single number is used. The dose of 20 mSv year^{-1} may be used as a reference level for an emergency exposure situation, but in this case, a lower reference level should be chosen for managing the recovery phase. After the accident, the authorities did not choose a reference dose between 1 and 20 mSv year^{-1}. An intermediate dose (e.g. 5 mSv year^{-1}) was not chosen because the authorities wanted a single intervention level above which actions were justified and below which no actions were needed.

2.2. Stakeholder involvement

Once an emergency exposure situation begins, it is unlikely that stakeholders will be involved in discussing protective measures. However, ICRP recommends that stakeholders should be involved in the planning for an emergency response, and also as soon as possible in management of the post-accident phase (ICRP, 2009b). When the authorities ordered full evacuation of the residents from Iitate, the residents asked the authorities to keep eight factories open, as well as a nursing home for the elderly. Implementation of the deliberate evacuation was deeply disturbing for the residents of Iitate; nevertheless, the authorities drew a sharp line based on 20mSv year^{-1}. Stakeholder involvement was only possible because the authorities introduced a moratorium for the evacuation. However, the moratorium highlighted an emerging issue on radiological protection, because the preparedness programme did not include the notion of deliberate evacuation.

Stakeholder engagement plays a much more important role once the situation develops into the recovery phase. Previously, radiological protection was expected to be conducted in a top-down fashion. More recently, radiological protection philosophy has evolved through the 2007 Recommendations of ICRP by emphasising the role of relevant stakeholders in the optimisation process (ICRP, 2007). Furthermore, in *Publication 111*, ICRP provided guidance about how to involve stakeholders in radiological protection for recovery management after a nuclear accident (ICRP, 2009b).

After the Fukushima accident, the affected people were drawn to *Publication 111* (ICRP, 2009b) where stakeholder involvement is emphasised for recovery management. Unfortunately, the authorities and some radiation specialists believe that radiological protection should only be considered in terms of applying specific dose criteria to decide on the implementation of protective actions. Currently, ICRP advocates a situation and risk-based approach emphasising the need for an optimisation process and the importance of involving stakeholders. The recovery phase is of sufficient duration to involve stakeholders in the decision-making process. This is quite different from the response during the acute phase of the emergency. To promote stakeholder engagement, ICRP (2009b) recommends that authorities should facilitate setting up local fora involving representatives of the affected populations and relevant experts. Accordingly, a similar forum was set up by local government after the accident at Fukushima, and this played an important role in effective recovery management. One of the difficulties was the selection of representatives from the affected population.

Facilitators may be required to establish effective communication in the forum. Following the Chernobyl accident, a French group appeared to play the role of facilitator. Even then, the uncertainty of future health effects brought about by radioactive contamination can lead residents to lose trust in the authorities. The experience of the ETHOS Project, implemented in Belarus during the 1990s (Lochard, 2013), showed that a key point in the recovery process is to respect the fundamental values of prudence in management of radiological risk, and of dignity of the affected people.

The Fukushima experience indicated that a routine top-down approach based solely on radiological protection criteria was unable to deal with the complexity of the problems resulting from the nuclear accident, and that stakeholder engagement needed to be explored.

2.3. Protective actions for contaminated foodstuffs

Contamination of foodstuffs was of concern not only to the affected population, but also elsewhere in Japan and even worldwide. When the accident occurred, there was no food regulatory law about radioactive materials. Derived intervention levels in the preparedness programme for a nuclear disaster had been developed according to the international standards such as *Publication 63* (ICRP, 1992). These intervention levels had no power of regulatory control. Responding to the accident, the Ministry of Health, Labour, and Welfare (MHLW) immediately gave notification of provisional regulation values, based on intervention levels. Foodstuff regulation started using the provisional regulation values on 17 March 2011.

In July 2011, some contaminated beef exceeding the provisional regulation values were delivered to consumers outside the contaminated area in Fukushima, and trust in the regulatory control was completely lost. Although management of contaminated foodstuffs was reinforced, some consumers were reluctant to buy agricultural products from Fukushima, even though they were controlled below the provisional regulation values. A new foodstuff regulation for the management of recovery was required. The MHLW made a request to the Food Safety Commission under the Cabinet Office. This situation led to a heated debate at national level about what should be considered as a safe level in the future. The media made frequent reports about contaminated foods, and public concern centred on consumer safety, with many consumers considering that the provisional regulatory values were too high.

After public consultation, the Food Safety Commission reported that there was no evidence that lifetime doses below 100 mSv led to detectable incidence of radiation-induced cancer. Finally, the MHLW decided to select 100 Bq kg^{-1} based on 1 mSv $year^{-1}$, and the new foodstuff criteria were introduced in April 2012. The authorities explained that 1 mSv $year^{-1}$ was the intervention exemption level used in the Codex guideline. This argument accelerated the idea that 1 mSv $year^{-1}$ was the threshold for safe annual dose. Most members of the public, other than Fukushima's farmers and their supporters, welcomed the new criteria. However, some voices were raised to draw attention to the need to find a balance between the interests of producers, distributors, and consumers. The Radiation Council, an advisory board to the authorities on radiological protection, emphasised that the new criteria may not contribute to decrease the contamination levels, as most foodstuffs already met the new criteria. There was, however, a strong view that there was no need to protect farmers to the detriment of consumers' safety. After introducing the new foodstuff criteria, there was also concern that the public chose foodstuffs with much lower contamination levels, even below the new criteria. This illustrates the complexities of foodstuff regulation as a protective action.

2.4. Decontamination of the land

The greatest programme for managing recovery in Fukushima is decontamination of the land. The Ministry of the Environment set some criteria for decontamination. Contaminated areas were targeted if ambient dose rates were more than 0.23 $\mu Sv\,h^{-1}$, taking into account a background dose rate of 0.04 $\mu Sv\,h^{-1}$. It was assumed that people stay indoors for 8 h day^{-1} with a reduction factor of 0.4 applying to the excess dose rate, and that annual doses reach 1 mSv. Originally, it was intended that the criterion was not a designated goal for decontamination but an investigation level. However, before deciding on the criterion, it was recognised that areas with higher dose rates would not be given priority for decontamination. The cost and effectiveness of decontamination strategies in the evacuated areas had been analysed, and these results could help to prioritise the areas suitable for decontamination from the viewpoint of cost and effectiveness of reducing the dose (Yasutaka et al., 2013). Most affected residents considered values of less than 1 mSv $year^{-1}$ to be a safe dose. Decontamination has been performed using an implicit criterion of 1 mSv $year^{-1}$, as no legally binding target dose exists. This situation has brought about various difficulties for returning people to the evacuated areas. In order to resolve these issues, dialogue with the relevant stakeholders needs to be considered, as social and ethical aspects are key factors of the decision-making process.

3. CREDIBILITY OF RADIOLOGICAL PROTECTION

3.1. Inconsistencies between experts and ICRP's message

The credibility of the experts and authorities was lost after the accident, and this was reported frequently by the media. The unprecedented scale of the accident in Japan led to misleading communication between experts and residents. Some radiation experts emphasised that no effects had been shown for radiation doses below 100 mSv. This statement was probably intended to mitigate unnecessary fear among residents about the low dose exposure under the particular circumstances after the accident. However, in the 2007 Recommendations, ICRP states that epidemiological and experimental studies provide evidence of radiation risk, albeit with uncertainties, at doses of approximately 100 mSv or less in the case of cancer, and there is increased likelihood of deterministic effects and a significant risk of cancer at doses above 100 mSv (ICRP, 2007). Most members of the public could not understand how both statements could be true, and they gradually stopped trusting the experts.

There is a gap in understanding of cancer risk at low doses. ICRP recommends that reasonable protection should be used no matter how low the radiation dose. Even before the accident, this message sometimes heightened concerns about medical exposure. When experts faced the problem, they tried to explain that the radiation either had no effect or that a larger benefit was derived compared with the risk. The risk–benefit approach is generally an effective method of communicating with patients. Seemingly, the public believe that the radiation risk after the Fukushima accident should be avoided at much lower doses. In this situation, experts often tried to communicate to the public that there is no significant radiation risk below 100 mSv. In reality, many people felt that this statement would stop the decontamination of contaminated areas. Finally, it seems that the public has gradually understood ICRP's message about the linear, non-threshold model for radiation protection.

The difference between the public and experts regarding the risks at low doses comes not only from scientific knowledge but also from an ethical aspect. People recognised that there is scientific controversy (Yasui, 2013) about cancer risk at low doses. The controversy was, for example, illustrated through media reports on the contradictory statements concerning health effects of radiation on children in Ukraine following the Chernobyl disaster. These scientific health issues on low-dose exposure could not convince the public. Some experts felt that the public did not understand that the risks from low-dose radiation would be quite low. Consequently, the experts tried to emphasise that there were no radiation effects below 100 mSv. This message partly caused the loss of credibility of the experts and authorities after the accident.

After the Fukushima accident, the authorities tried to apply ICRP's recommendations for recovery. However, the public felt that the authorities made a makeshift law because there were no rules about recovery from a nuclear disaster until 2011. Moreover, they felt that the public dose limit of 1 mSv year^{-1} that applies for planned exposure situations was relaxed to adjust to the accident. The loss of trust in the authorities and radiation experts was inevitable.

3.2. Issues to be challenged

Currently, there is a marked gap between the public's and ICRP's positions. Following the Fukushima accident, it has become apparent that differences in perception have brought about much confusion in society. These have caused socio-economic consequences with inevitable changes to daily life, as well as psychological effects. Radiological protection considers the control of health effects using the risk-based approach. Since the accident, most radiation experts have acknowledged the difficulties of the risk-based approach for radiological protection. The experts felt that even a low probability of cancer induction could not convince the public, as they were unable to understand the probability that cancer might occur at low doses. The concept of risk should be used to reduce doses on a prospective basis, but is actually used to calculate the probability of cancer no matter how low the dose and where epidemiological limitations exist.

Contrary to the concept, risk itself cornered the affected people into a dead-end. The experts failed to find a scientific way to discuss tolerability, and compare radiation risk with other risks. When the experts tried to explain that no effects have been observed for exposure levels below 100 mSv, they lost the public's trust.

So, where should we go? Risk quantification is needed to promote radiological protection in a scientific way, but currently lacks what the risk means. Cancer risk should be addressed from the viewpoint of cancer risk management to an individual. Talking about risk comparison, it may be more important to estimate individual risk, rather than average risk among a population. Age and lifestyle behaviours, such as smoking, can be taken into account. Obviously, risk comparison should be used for risk management. To facilitate risk management, it will be necessary for communication between experts and affected individuals to be based on trust.

ICRP, like the United Nations Scientific Committee on the Effects of Atomic Radiation and the World Health Organization, plays a key role in disseminating good science about radiation risk. However, ICRP is in a unique position with respect to maintaining a trustworthy radiological protection system. In the personal opinion of the author, if scientific controversy occurs in relation to an unexpected situation, ICRP should provide timely scientific information and radiation protection advice.

4. CONCLUSIONS

In the case of the Fukushima accident, the ICRP system of radiological protection seems to be relevant, although there have been problems in its implementation. Reference levels could improve the situation by requiring an iterative optimising process. The Fukushima experience indicated that a 'classical' top-down approach relying on radiological criteria was unable to deal with the complexity of the problems, and that stakeholder engagement should be explored. The gap in understanding of radiation risks from low doses between radiation experts and the public has caused much confusion. Experts should recognise that ethical issues, as well as scientific uncertainty, exist in radiological protection. ICRP should aim to be more proactive in order to promote trustworthy radiological protection.

REFERENCES

ICRP, 1992. Principles for intervention for protection of the public in a radiological emergency. ICRP Publication 63. Ann. ICRP 22(4).

ICRP, 2007. The 2007 Recommendations of the International Commission on Radiological Protection. ICRP Publication 103. Ann. ICRP 37(2–4).

ICRP, 2009a. Application of the Commission's Recommendations for the protection of people in emergency exposure situations. ICRP Publication 109. Ann. ICRP 39(1).

ICRP, 2009b. Application of the Commission's Recommendations for the protection of people living in long-term contaminated areas after a nuclear accident or a radiological emergency. ICRP Publication 111. Ann. ICRP 39(3).

ICRP, 2012. Report of ICRP Task Group 84 on Initial Lessons Learned from the Nuclear Power Plant Accident in Japan vis-à-vis the ICRP System of Radiological Protection.

International Commission on Radiological Protection, Ottawa. Available at: http://www.icrp.org/docs/ICRP%20TG84%20Summary%20Report.pdf (last accessed 9 February 2015).

Kai, M., 2012. Some lessons on radiological protection learned from the accident at the Fukushima Daiichi nuclear power plant. J. Radiol. Prot. 32, N101–N105.

Lochard, J., 2013. Stakeholder engagement in regaining decent living conditions after Chernobyl. In: Oughton, D., Hansson, S.O., Baxter, M. (Eds.), Social and Ethical Aspects of Radiation Risk Management. Elsevier Science, Oxford, pp. 311–332.

Yasui, S., 2013. An analysis of the argument over the health effects of low-dose radiation exposure caused by the accident at the Fukushima Daiichi APP in Japan. J. Risk Res. 16, 937–944.

Yasutaka, T., Naito, W., Nakanishi, J., 2013. Cost and effectiveness of decontamination strategies in radiation contaminated areas in Fukushima in regard to external radiation dose. PLOS ONE 8, e75308.

SAGE

ICRP
Annals of the ICRP

Decision making for late-phase recovery from nuclear or radiological incidents: new guidance from NCRP

A.F. Nisbet[a], S.Y. Chen[b]

[a]*Public Health England, Chilton, Didcot, Oxon OX11 0RQ, UK; e-mail: anne.nisbet@phe.gov.uk*

[b]*Illinois Institute of Technology, USA*

Abstract–In 2010, the US National Council on Radiation Protection and Measurements (NCRP) established a scientific committee (SC5-1) to prepare a comprehensive report on the framework and approach for optimising decision making in late-phase recovery from nuclear or radiological incidents that lead to wide-area contamination. The NCRP report builds on recommendations from the International Commission on Radiological Protection's (ICRP) *Publication 111* which specifically addresses the protection of people living in long-term contaminated areas. Based on this approach, the report addresses all relevant dimensions: health, environment, economic, psychological, cultural, ethical, and political. NCRP, like ICRP, considers optimisation to be the best approach to decision making for balancing these multiple risk factors in situations involving wide-area contamination where the conventional clean-up approach may encounter some serious constraints. The NCRP report describes optimisation as an iterative process that can be broken down into a series of steps, all of which involve deliberations with stakeholders as a necessary element for a community-focused recovery effort. The steps, elaborated on in the report, range from defining the situation to a series of actions involving assessing impacts, evaluating options, developing a strategy, and demonstrating its successful implementation. In conclusion, the NCRP report makes a series of recommendations aimed at enhancing and strengthening late-phase recovery following a major nuclear or radiological incident.

Keywords: Nuclear and radiological incidents; Optimisation; Late-phase recovery; Wide-area contamination; Stakeholders

This paper does not necessarily reflect the views of the International Commission on Radiological Protection.

1. INTRODUCTION

In the USA, efforts on emergency preparedness have focused primarily on triaging the initial response to the event; the more complex, long-term issues relating to recovery have not been fully explored. It is clear from the accidents at the nuclear power plants in Chernobyl and Fukushima that the radiological impact can affect wide areas and last for years, making planning for recovery particularly important. Similar challenges would have to be faced following the use of radioactive or nuclear material in terrorist events such as those resulting from a radiological dispersal device (RDD) or improvised nuclear device (IND). Some guidance has been published (Federal Emergency Management Agency, 2008) for the protection of public health in the response to an RDD or IND, but no specific radiological criteria have been prescribed for long-term recovery; instead a site-specific optimisation process has been recommended, the details of which require further development.

1.1. Widespread radiological contamination and related issues

Late-phase recovery is concerned with a range of issues arising from the spread of long-lived radionuclides over a wide area. The scale of the recovery effort required can provide a significant challenge to society as witnessed in the aftermath of the accident at Fukushima Daiichi nuclear power station in 2011. An important consideration for restoring affected communities is mitigating the impact of the radiological contamination. In doing so, however, decisions have to be made regarding issues such as radiation monitoring, health surveillance, future land uses, priorities for remediation and time scales for implementation, clean-up criteria, decontamination technologies, human and financial resources, waste management, and appropriate and effective communication strategies, taking into account sociopolitical factors, cultural perspectives, human health and public welfare needs, and ecological risks.

1.2. Background

The US Department of Homeland Security has supported NCRP (2010–2013) in establishing a scientific committee (SC5-1) to prepare a comprehensive report that defines, in more detail, the process and procedures for optimising late-phase recovery from a wide range of radiological incidents that lead to wide-area contamination. The principle of optimisation has been embedded in radiation protection for many decades, and remains a tenet to support remediation actions to address late-phase recovery issues (ICRP, 2009). It has been the task of SC5-1 to elaborate guidance on optimisation that would be applicable to late-phase recovery for wide-area contamination resulting from either a nuclear emergency or terrorist attack. It requires a very different strategy from conventional clean-up, where the primary focus is on reducing radiation risk. In particular, due to resource limitations, far broader considerations apply and decisions will inevitably involve trade-offs and entail complex deliberations with stakeholders. As major incidents involving nuclear or radiological sources have been rare, one important aspect of SC5-1's work has been to evaluate past incidents (e.g. those involving nuclear facilities, atomic testing or military-related activities, clandestine acts, and planning exercises) for their relevance with respect to the role of optimisation. The NCRP report (NCRP, 2014) has undergone an extensive peer review process.

2. KEY ELEMENTS OF LATE-PHASE RECOVERY

Whilst it is not possible to be too prescriptive about the time frame for phases of response and recovery, the National Disaster Recovery Framework (NDRF) (Federal Emergency Management Agency, 2011a) identifies four periods: preparedness (ongoing); short term (days); intermediate (weeks to months); and long term (months to years). Whilst late-phase recovery takes place predominantly in the long term, it has its roots in the short term and continues into the intermediate phase. For example, some prompt actions taken to decontaminate buildings or land during the intermediate phase may influence waste management strategies in the late phase. Late-phase recovery is primarily concerned with rehabilitation, remediation, and recovery (i.e. permanent housing solutions, decontamination strategies, rebuilding infrastructure and businesses, and following up health care). Key elements facilitating the success of late-phase recovery are resilience, involvement of the whole community, stakeholder engagement, and effective risk communication.

2.1. Resilience

For a severely impacted community, the objective of the long-term recovery is to create normal and acceptable living conditions in the most expedient manner, with a goal to re-establish and sustain the local economic viability. Resilience requires adequate resources to sustain the functionality and vitality of the community. It also requires a community that is readily adaptable under adverse situations. With proper preparedness, such as by strengthening the support infrastructure and better public awareness and training, a community can substantially improve its overall resilience prior to an incident.

2.2. Whole community

In recent years, there has been growing concern in the USA about the adequacy of community preparedness for major disruptions caused by disasters or terrorist acts. Recognising the limitations of the Government's role and its effectiveness in responding to such an event, the Federal Emergency Management Agency began to develop a concept that involves the 'whole community' (Federal Emergency Management Agency, 2011b) in preparedness for such situations. The concept applies throughout all phases of an incident, although it is particularly important in the late-phase recovery effort as there will be considerably diverse and complex issues that will involve a broad base of stakeholders. Advancing such a concept also necessitates related actions by the Government and responsible authorities to develop and adapt appropriate policies to facilitate and implement the approach. Such policies may include the creation and support of a self-help programme as a means to engage and empower citizens in recovering from an incident.

2.3. Stakeholders

Large accidents, such as those witnessed in Chernobyl and Fukushima, have shown that recovery from a severe event is not likely to return the affected communities to pre-accident conditions. Rather, a 'new normality' would be created in the longer term in which stakeholders may find it acceptable to continue to live and/or work in contaminated areas. Thus, the important issues including choice of decontamination methods, clean-up criteria, waste disposal sites, and future land use would require extensive interactions with the local community and other stakeholders to reach a consensus. NDRF provides guidance to promote effective recovery by clarifying the roles and responsibilities for stakeholders both pre- and post-emergency. The International Radiation Protection Association (IRPA, 2008) has also published guiding principles for radiation protection professionals in stakeholder involvement. These aim to promote the participation of all relevant parties in the process of reaching decisions involving radiological protection, which may impact on the well-being and quality of life of workers and members of the public, and on the environment. These principles can be applied at each step of the recovery process.

2.4. Risk communication

Numerous studies have highlighted the importance of effective risk communication in enabling people to make informed choices following disasters, including nuclear and radiological incidents (Becker, 2007; Covello, 2011). Effective communication requires accurate information that can be disseminated in a timely manner in order to enhance the response effort and mitigate potential psychological and social impacts, including discrimination. It is thus important to address such issues in the pre-event planning stage, recognising that the later phases of recovery will necessitate a more sophisticated approach towards communication to address the complex decisions that have to be made and the uncertainties involved. The information needs of stakeholders will be great, and it is therefore important that all available communication methods are used to disseminate and share information. There will be a need to use traditional media outlets (television, radio, online news), supplemented by full use of other delivery channels such as social media. Effective risk communication can help people to find peace and be connected, hopeful, adaptable, and cooperative, instead of feeling unsafe, anxious, isolated, pessimistic, inflexible, uncooperative, helpless, dependent, fatalistic, and victimised.

2.5. Latest guidance from ICRP

ICRP (2009) provides guidance for the protection of people living in long-term contaminated areas resulting from either a nuclear accident or a radiation emergency. This elaborates on the 2007 Recommendations (ICRP, 2007) which advocate a system of protection based on the type of exposure situation (planned, emergency, or existing) to which the fundamental principles of justification and optimisation apply. Late-phase recovery is considered by the Commission as an existing exposure situation to which reference levels of dose can be applied to assist the optimisation process for both planning and response. ICRP (2009) suggests that the reference level should be selected in the lower part of the 1–20 mSv $year^{-1}$ band, recognising that national authorities may take into account the prevailing circumstances and the timing of the overall rehabilitation programme. ICRP (2009) acknowledges that living in long-term contaminated areas is a complex situation that cannot be managed by radiological protection considerations alone, and must address all relevant dimensions such as health, environment, economic, social, psychological, cultural, ethical, and political. Furthermore, ICRP highlights practical aspects for implementing protection strategies, and emphasises the need to involve the affected population and local professionals directly in the management of the situation. The work carried out by NCRP SC5-1 further elaborates on the latest guidance from ICRP by providing a step-by-step framework for optimisation.

3. PROCESS OF OPTIMISATION

Optimisation is a multifaceted approach that is the best method for balancing multiple risk factors. It can be depicted as an iterative seven-step process that involves stakeholders at each step (Fig. 1). The multitude of issues cannot be resolved by a top-down government-driven approach; instead, the recovery effort must be community based, with the government playing a supporting role to support initiatives and facilitate actions.

3.1. Define situation

Establishing an accurate and detailed characterisation of the contamination and presenting it in an understandable manner is an important element to defining the situation. This includes determining the radionuclide composition of the deposit, its mobility, spatial variability, and location of hotspots. This process relies on monitoring and surveillance of buildings, pavements, infrastructure, parks, surface waters, soils, produce, livestock, and commodities. Other important aspects of defining the situation include establishing land use, population size, its distribution and composition, habits, and activities.

3.2. Assess impacts

Environmental monitoring data coupled with assessment models may be used to calculate projected doses to adults and children living in the affected area, taking their habits into account. The situation can be complex due to the involvement of multiple radionuclides, multiple surfaces and media, and multiple exposure pathways. When assessing impacts, focus should be on doses from the various exposure scenarios, not activity concentrations on (or in) various media. This is because the time and effort required for removing contamination beyond certain levels from everywhere does not automatically lead to a reduction in doses, and can generate unnecessarily large amounts of waste. The assessments must be realistic and take into account prevailing environmental conditions and the potential for elevated background radiation coming, for example, from direct shine from adjacent sites or

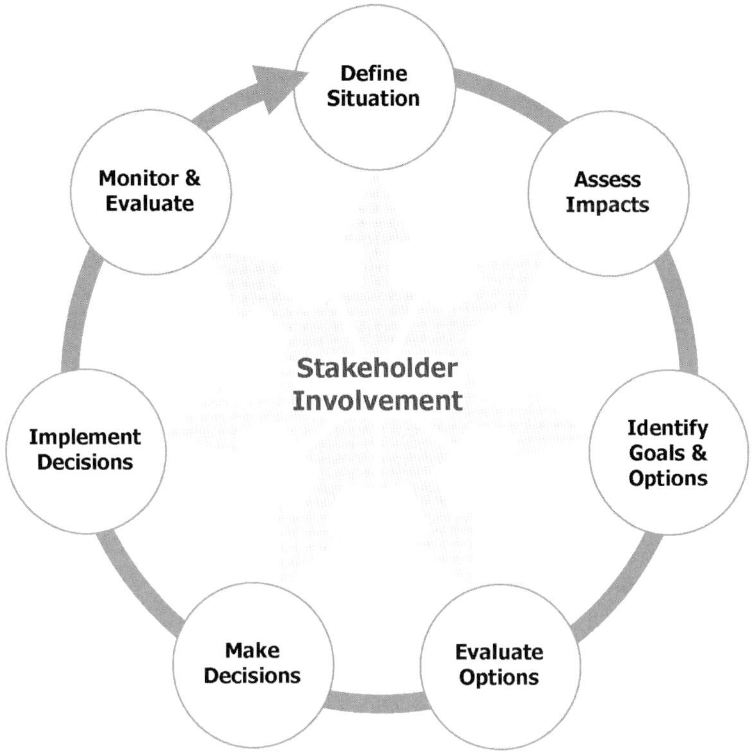

Fig. 1. Optimisation process for late-phase recovery.

contaminated objects such as trees. Local knowledge can play a critical role in the impact assessment process.

3.3. Identify goals and options

For a radiation emergency, the primary goal of the entire recovery process will be to develop an agreed strategy for returning areas affected by the emergency to a state as close as possible to that existing before the release of radioactivity, and the population to a lifestyle where the accident is no longer a dominant influence. It is important that the public participate fully in establishing the goals for recovery, be they based on radiological, economic, environmental, or other criteria. When setting radiological goals, it is important to establish how the level of radiological risk (dose) will be equated with measurable levels of radioactivity in the environment. Other goals of recovery may include targets for restoring businesses or for minimising waste generation.

There are many options available for managing recovery. Options may include controlling access, modifying individual behaviours, intervening in food production systems and drinking water supplies, or by decontaminating inhabited areas.

Identification and selection of these options will depend on the goals of recovery; some options will be very effective at reducing doses but generate large volumes of waste for which no disposal route is available, and other options may be less effective but provide reassurance to the population. In meeting different recovery goals, it may be necessary to reconcile options to optimise the overall recovery strategy.

3.4. Evaluate options

Evaluation of options involves scrutinising their key attributes to decide whether the agreed goals for recovery can be met. This should be carried out at the local level and in conjunction with stakeholders. Key attributes include: effectiveness, feasibility, capacity, time scales of implementation, constraints (legal, societal, and environmental), waste generation, doses to implementers, costs, societal impact, and acceptability to stakeholders. To assist in comparison between options and for selecting and combining options, datasheets such as those published in the UK Recovery Handbooks for Radiation Incidents (HPA, 2009) can be used to record information systematically on key attributes of each recovery option. Various techniques, such as cost benefit and multi-attribute analysis, are available to assist the evaluation of options that should take place in conjunction with stakeholders.

3.5. Decision making

Unlike emergency situations, where prompt response towards triaging the incident is an over-riding consideration, more time can be expended in the late phase to develop comprehensive and effective schemes for involving stakeholders in decision making. No matter how valid the recovery strategy, or how well the science has been conducted, if it is not accepted by the stakeholders, it will fail. Considerations should include issues that are specific to local and regional needs, including cultural and ethnic sensitivities, applicability of existing policies and legislation, and requirements for the development and trialling of appropriate decontamination technologies. The discussions should involve the citizens of the affected communities.

3.6. Implementation

Once decisions have been reached regarding the recovery strategy, implementation must be accompanied by documentation on the basis and rationale for the decisions (including prioritisation for recovery options), and there must be communication of the decision to stakeholders, including the programme of implementation, the technologies that will be used, criteria by which their success will be evaluated, and the relevant time scales. The entire decision process and resulting recovery plan must maintain transparency throughout. It is important that the recovery plan is sufficiently flexible to allow adjustments and improvements to be made during implementation. Sometimes, technologies are new or under development and have to be trialled on a small scale before consideration and approval is given for their wider application.

3.7. Monitor and evaluate

A long-term monitoring programme is a key element to evaluating the success of the recovery strategy. It is recommended that various measurable milestones for recovery are established and agreed with input from the community; these may include short- to medium-term targets for projected dose; restoration of utilities, transport infrastructure, local businesses, agricultural production, and tourism; and the transfer of waste to safe storage for managed disposal. These targets provide a means of monitoring and evaluating progress, and may assist in deciding when specific recovery activities can be scaled down. In addition to long-term monitoring of residual contamination in the environment, other public health objectives (e.g. referrals), economic indicators (e.g. employment statistics, numbers of hotel rooms filled), or environmental targets (volumes of waste) may be evaluated.

3.8. Challenges to optimisation

ICRP, NCRP, and other international organisations suggest that late-phase recovery from wide-area contamination is best managed through an iterative optimisation process whereby radiological criteria represent one of many factors that have to be taken into account when making decisions. Consequently, there are no pre-set clean-up goals, and the return to a new normality is guided by the priorities set by local communities who are actively involved in the entire process. This is in marked contrast to the current clean-up approach carried out under statutory regulatory provisions that focuses on radiological risk, precautionary decision making, and clean-up goals close to background. Although thorough and generally effective in many remediation situations, it requires a lengthy process and may take several decades to decontaminate a site (GAO, 2012). Such a protracted time frame is not conducive to a community's expedient recovery to decent living conditions. Whilst some may oppose the concept of optimisation, it should be seen as a logical and practicable way forward for managing wide-area contamination, which will ensure balanced and timely recovery in a region affected by a serious nuclear or radiological event.

4. RECOMMENDATIONS

Eight recommendations to enhance and strengthen late-phase recovery from radiological or nuclear incidents for wide-area contamination have been proposed by NCRP; they suggest well-balanced consideration of factors for circumstances that go well beyond those experienced in conventional clean-ups. The recommendations are summarised below.

- Develop a national strategy promoting community resilience.
- Integrate late-phase recovery into planning and ensure it is exercised.
- Embrace the optimisation paradigm for managing non-conventional wide-area contamination.
- Ensure that stakeholder engagement and empowerment underpin the optimisation process.

- Develop a communication plan as an integral part of the preparedness strategy.
- Develop adaptive and responsive policies including those for waste management.
- Conduct research and development to address the impact of wide-area contamination specifically.
- Establish a mechanism to integrate lessons learned from past incidents.

The recommendations are considered to be applicable to all types of nuclear or radiological incidents, be they associated with nuclear accidents or malevolent acts. Central to the recommendations is the need to further develop approaches and methods that can be implemented on an incident and site-specific basis. Such activities are vital to the preparedness for any future events.

5. CONCLUSIONS

Given the potentially large magnitude of impacts caused by a nuclear or radiological incident, it has been recognised that the traditional, statutory approach to long-term recovery, and specifically to the remediation and clean-up of contaminated areas, may not be feasible due to the magnitude of issues involved. NCRP has proposed a framework for optimisation that provides a balanced approach to addressing complex issues typical of late-phase recovery in a nuclear or radiological incident that may involve wide-area contamination. The framework, which builds on the guidance published by ICRP, is underpinned by stakeholder involvement, and balances the protection of human health against the available resources, business and economic targets, waste management, and societal needs.

REFERENCES

Becker, S.M., 2007. Communicating risk to the public after radiological incidents. BMJ. 335, 1106–1107.

Covello, V.T., 2011. Risk communication, radiation, and radiological emergencies: strategies, tools, and techniques. Health Phys. 101, 511–530.

Federal Emergency Management Agency, 2008. U.S. Department of Homeland Security, planning guidance for protection and recovery following radiological dispersal device (RDD) and improvised nuclear device (IND) incidents. Fed. Regist. 73, 45029–45049.

Federal Emergency Management Agency, 2011a. National Disaster Recovery Framework. September 2011, Federal Emergency Management Agency, Washington. Available at: http://www.fema.gov/pdf/recoveryframework/ndrf.pdf (accessed February 2015).

Federal Emergency Management Agency, 2011b. *A Whole Community Approach to Emergency Management: Principles, Theme, and Pathways for Action. FDOC 104-008-1*, December 2011, Washington: U.S. Government Printing Office.

GAO, 2012. Superfund: Status of EPA's Efforts to Improve its Management and Oversight of Special Accounts, GAO-12-109. US Government Accountability Office, Washington, DC.

HPA, 2009. UK Recovery Handbooks for Radiation Incidents. HPA-RPD-064. Health Protection Agency, Chilton. Available at: https://www.gov.uk/government/publications/uk-recovery-handbook-for-radiation-incidents-and-associated-publications (accessed February 2015).

ICRP, 2007. The 2007 Recommendations of the International Commission on Radiological Protection. ICRP Publication 103. Ann. ICRP 37(2–4).

ICRP, 2009. Application of the Commission's recommendations to the protection of people living in long-term contaminated areas after a nuclear accident or a radiation emergency. ICRP Publication 111. Ann. ICRP 39(3).

IRPA, 2008. Guiding Principles for Radiation Protection Professionals on Stakeholder Engagement (IRPA 08/08). International Radiation Protection Association. Available at: http://www.irpa.net/members/54494/%7B86D953FC-5B32-4BF9-91CE-739C8F615F4B%7D/Stakeholder-Engagement-Guiding-Principles.pdf (accessed February 2015).

National Council on Radiation Protection and Measurements, 2014. Decision Making for Late-Phase Recovery from Major Nuclear or Radiological Incidents. NCRP Report No. 175. NCRP, Bethesda, Maryland.

SAGE

ICRP
Annals of the ICRP

French policy for managing the post-accident phase of a nuclear accident

F. Gallay, J.L. Godet and J.C. Niel

Autorité de Sûreté Nucléaire, 15 rue Louis Lejeune, CS 7001392541, Montrouge cedex, France; PH: +33 (0)1 46 16 40 00; FAX +33 (0)1 46 16 41 55; e-mail: Florence.gallay@asn.fr

Abstract–In 2005, at the request of the French Government, the Nuclear Safety Authority (ASN) established a Steering Committee for the Management of the Post-Accident Phase of a Nuclear Accident or a Radiological Emergency, with the objective of establishing a policy framework. Under the supervision of ASN, this Committee, involving several tens of experts from different backgrounds (e.g. relevant ministerial offices, expert agencies, local information commissions around nuclear installations, non-governmental organisations, elected officials, licensees, and international experts), developed a number of recommendations over a 7-year period. First published in November 2012, these recommendations cover the immediate post-emergency situation, and the transition and longer-term periods of the post-accident phase in the case of medium-scale nuclear accidents causing short-term radioactive release (less than 24 h) that might occur at French nuclear facilities. They also apply to actions to be undertaken in the event of accidents during the transportation of radioactive materials. These recommendations are an important first step in preparation for the management of a post-accident situation in France in the case of a nuclear accident.

Keywords: Nuclear accident; Emergency response; Protective measures

This paper does not necessarily reflect the views of the International Commission on Radiological Protection.

1. INTRODUCTION

Following the French Government Directive of 7 April 2005, the French Government tasked the Nuclear Safety Authority (ASN), together with relevant ministerial offices, to establish a policy framework to prepare and implement the necessary steps to address post-accident situations following a nuclear accident or a radiological emergency. In June 2005, the ASN formed a Steering Committee for the Management of the Post-Accident Phase in the Event of Nuclear Accident or a Radiological Emergency Situation (CODIRPA), in charge of elaborating the policy elements. The scenarios selected to develop this policy were medium-scale nuclear accidents causing short-term radioactive release (less than 24 h) that might occur at French nuclear facilities. Over nearly 7 years, 14 working groups, coordinated by ASN and involving representatives from ministerial offices, expert agencies, local information commissions around nuclear installations, non-governmental organisations, elected officials, licensees, international experts etc., developed the policy elements that were finally brought together in a summary report published in November 2012 entitled 'Policy elements for post-accident management in the event of nuclear accident' (ASN, 2012). This report begins by specifying the definitions of each stage of a nuclear accident. It then lists the principles adopted by CODIRPA to support management efforts subsequent to a nuclear accident.

2. DEFINITIONS OF THE EMERGENCY PHASE AND THE POST-ACCIDENT PHASE

2.1. The emergency phase

The emergency phase generally consists of:

- a period of threat resulting from failures in the facility, during which time the operator implements actions to return the facility to an adequate level of safety, thereby preventing potential releases;
- a period of radioactive release into the environment, should the operator have been unable to restore safe operating conditions in the facility; and
- an exit period from the emergency phase, during which time the facility returns to safe operating conditions, significant radioactive release ceases, and any further threat of release disappears.

The emergency phase is characterised by the need to take action very quickly in order to cope with the actual or potential release of radioactive substances into the environment that are likely to lead to substantial exposure of the population.

2.2. The post-accident phase

The post-accident phase generally consists of:

- a transition period (which can last up to a few weeks or a few months after the accident), during which time understanding of the actual state of contamination of the various components of the environment is still unclear and the risks of exposure to affected individuals can still be high; and
- a long-term period (which can last up to several years or even several decades after the accident), during which time long-term protective actions to control exposures, and support and recovery measures are implemented.

3. OBJECTIVES, PRINCIPLES, AND KEY POINTS OF POST-ACCIDENT MANAGEMENT

3.1. Three basic objectives

Three basic objectives have been set out in terms of managing the post-accident phase of a nuclear accident.

- Protection of the population from the dangers of ionising radiation.
- Provision of support to the populations affected by the consequences of the accident.
- Recovery of the affected territories, from social and economic perspectives.

3.2. Four management principles

Four principles have been set out in order to determine the actions to be undertaken in preparing for post-accident management and in the event of an accident.

- Anticipation – the issues at stake in post-nuclear accident management must be taken into account from as early as the end of the emergency phase.
- Justification – the measures adopted, particularly those aiming to protect the affected populations, must be justified, meaning that the expected benefits, particularly in terms of radiological harm prevented, must exceed the risk and drawbacks inherent in their implementation.
- Optimisation – population exposure to ionising radiation must be kept to a level as low as reasonably achievable, taking into account economic and societal factors.
- Shared construction and transparency – post-accident management must involve the population, elected officials, business community, and other stakeholders. The transparency of the information provided is one of the prerequisites for this cooperative spirit to be effective.

3.3. Six key points in post-accident management

There are six key points to keep in mind when implementing the objectives and principles listed above.

- Immediate delimitation of the contaminated territories (to be adjusted over the course of the transition period and beyond). This serves as the structural framework by which actions designed to protect the populations will be managed. This zoning makes it possible to implement restrictions on the consumption and release-to-market of locally-produced foods (i.e. the main source of population exposure).
- The population affected by the consequences of the accident must be given the benefit of medical and psychological care, dosimetric and epidemiological monitoring, financial support, and compensation for the damages incurred.
- The radiological situation, particularly the levels of contamination of foodstuffs and water, must be characterised as quickly as possible in order to understand the extent of the contaminated territories and the potential impact of the contamination. Subsequently, a long-term radiological surveillance relying on standard practices must be implemented and maintained throughout the post-accident phase.
- A water management plan, specific to tap water, should be considered, taking into account the exposure due to contamination of the resource, while adapting the measures to be implemented and possible restrictions on water supply or distribution in accordance with the potential risk.
- A new governance based on the awareness and active participation of those affected is needed, especially where the radiological conditions allow revival of business activity and revitalisation of the affected territories.
- Action to mitigate contamination and to manage the contaminated products may generate large amounts of waste from varying sources and of different types. This makes it necessary to gradually replace the temporary solutions adopted during the emergency phase with lasting solutions.

4. MEASURES TO BE ENACTED OR INITIATED AT THE END OF THE EMERGENCY PHASE

These measures, to be implemented across contaminated territories, must be prepared and, if possible, planned upstream from the emergency phase.

4.1. Post-accident zoning and monitoring of the deposited radioactivity

The first post-accident zoning is established on the basis of predictive modelling of future population exposure to radioactivity in the inhabited zones, and to food chain contamination due to radioactive deposits. Distinction is to be made between two zones, each of which has a distinctive purpose:

- a 'population protection zone' (ZPP), inside which action is needed in order to lower population exposure to atmospheric radioactivity and ingestion of contaminated foods as much as possible; and
- a 'heightened territorial monitoring zone' (ZST), which is broader and more focused on economic management, within which specific monitoring of foodstuffs and farmed crops is to be initiated.

In particular situations of high external exposure, it may be necessary to relocate a fraction of the population residing within the ZPP. In such cases, residents will be temporarily relocated for a duration that will vary according to the level of exposure in their living environment.

4.2. Early protection and care measures for the affected population

Once the radioactive release has ended and safe operating conditions have been restored at the facility, sheltering can be lifted. A decision then has to be made, taking into account foreseeable concentrations, regarding whether the people living in the ZPP should remain there or be relocated. The return of individuals evacuated during the emergency phase will also have to be considered at this stage.

In situations where the population's radiological exposure could come from the ingestion of contaminated foodstuffs, actions must be adopted to prohibit the consumption and release-to-market of foodstuffs following the end of the emergency phase.

As soon as feasible, a radiological control appropriate to each farming production sector should be implemented in the ZST in order to allow compliant products to be released to the market. As a precautionary measure, movement should be restricted in certain ZPPs, or even ZSTs, within which radioactive substances tend to accumulate (e.g. forests, green areas, etc.).

Another important measure to consider in organising care for the population is the establishment of public reception and information centres at the end of the emergency phase. These structures should be organised by coordination between the reception and grouping centres called for by the French Organization of Civil Security Response. They must be operational once early protective actions are lifted, and must provide response to the following important needs: reception; listening; census taking; provision of medical and psychological support; information; accommodation; emergency grants; and financial assistance to the population.

5. PLANNING POST-ACCIDENT MANAGEMENT DURING THE TRANSITION PERIOD

During the transition period, management efforts will have to shift into an increasingly participatory mode, featuring, in particular, greater involvement of the local authorities in the decision-making processes, thereby paving the way for the long-term perspective. This will necessarily bring about regular adjustments to the post-accident management programme. The post-accident management programme should be developed along 10 lines of action:

- receiving the population;
- lowering population exposure to deposited radioactivity;
- addressing public health issues;
- refining the characterisation of the radiological situation in the environment;
- improving the radiological quality of the environment;
- managing waste;
- improving stakeholder involvement through a well-designed governance model;
- supporting and re-deploying business activity;
- providing grants and compensation; and
- informing about the situation and its evolution.

6. A TERRITORIAL PROJECT FOR LONG-TERM MANAGEMENT

The long-term post-accident period begins when the radiological consequences of the release have been characterised to a sufficient degree of precision. Hence, those active in the affected territories (i.e. elected officials, business community, professionals, and populations) can take action to manage their exposure and to set out to address the future of their territories, alongside the public authorities.

This process requires:

- a sufficiently precise understanding of the radiological conditions in the environment, foodstuffs, and people so that those affected can protect themselves adequately (this knowledge may be improved with time);
- public authorities to be re-organised appropriately but remain subject to change (depending on the circumstances);
- stakeholders to be involved in decision-making and measures taken to rehabilitate living conditions, which will be all the more effective in that they will have been planned in advance; and
- the conditions necessary to allow re-establishment of social and business activity, and the development of a shared project within the territory.

7. CONCLUSIONS

The first elements of a national policy on post-accident management were drawn up with respect to medium-scale nuclear accidents triggering short-term (less than 24 h) radioactive release that might occur at French nuclear facilities. The territories affected by radioactive deposits resulting from accidents of this type may extend across significantly broader distances than those of the off-site emergency plans (i.e. from 10 to several dozen kilometres).

The extent of the territories actually affected by the accident cannot be identified during the preparation stage, as it depends on environmental and meteorological factors. The extent can only be outlined within the specific context of a confirmed accident situation, once the radioactive release has ended, on the basis of the expert advice provided.

The CODIRPA report should be taken into account during the preparation stage, at least where it concerns early measures to be enacted upon at the end of the emergency phase and, at the local level, in operational documents specific to each nuclear site.

Once the first phase of investigation into post-accident management principles is completed, it will be necessary to consider certain topics in greater depth during the preparation stage (defining levels for the release-to-market of foodstuffs and other goods, levels for the clearance of waste, etc.).

A long-term release (over several days) involving large-scale accident management will be addressed in the next stage of the CODIRPA. While the guiding principles are generally the same, the timeline and feasibility of these measures may differ. Indeed, it may become necessary to undertake actions concurrently, in different territories, to protect the population directly exposed to the release of radioactive substances into the atmosphere and the population exposed to deposits of these substances in the environment, as demonstrated for the Fukushima Daiichi accident in Japan.

REFERENCE

ASN, 2012. Policy Elements for Post-accident Management in the Event of Nuclear Accident. Document drawn up by the Steering Committee for the Management of the Post-Accident Phase of a Nuclear Accident (CODIRPA). Autorité de sûreté Nucléaire: Paris (France), Available at: http://www.french-nuclear-safety.fr.

SAGE

Annals of the ICRP

Review of the ICRP system of protection: the approach to existing exposure situations

D.A. Cool

US Nuclear Regulatory Commission, Washington DC, VA 20555, USA; e-mail: Donald.Cool@nrc.gov

Abstract–The International Commission on Radiological Protection (ICRP) system of protection consists of existing, planned, and emergency exposure situations. With the 2007 Recommendations in ICRP *Publication 103*, a coherent approach has been established that emphasises the optimisation of protection with appropriate constraints or reference levels in each exposure situation. Existing exposure situations pose unique challenges because the source of exposure already exists, and it may not always be possible to control the source directly. This is the case for naturally occurring sources, which are ubiquitous in the environment and vary widely in the magnitude of exposures that may be received by individuals. Decisions on protection strategies must consider a graded, pragmatic, and flexible approach for dealing with exposure of members of the public, and those that may be occupationally exposed while working with naturally occurring sources. Although limits are not applicable, aspects of the management approach for planned exposure situations may be appropriate, depending upon the magnitude of exposures.

Keywords: NORM; Existing exposure situations; System of protection

1. INTRODUCTION

The International Commission on Radiological Protection (ICRP) system of protection was updated and elaborated in *Publication 103* (ICRP, 2007). The system of protection aims to contribute an appropriate level of protection for people and the environment against the detrimental effects of exposure to ionising radiation, without unduly limiting the desirable human actions that may be associated with

This paper does not necessarily reflect the views of the International Commission on Radiological Protection.

radiation exposure. Application of the framework is intended to manage and control exposures to ionising radiation so that tissue reactions (deterministic effects) are prevented, and the risks from stochastic effects are reduced as much as reasonably achievable.

2. SYSTEM OF PROTECTION

The system of protection is organised with three fundamental principles, three exposure situations, and three categories of exposure.

2.1. Principles

The first principle is justification. Decisions that alter the radiation exposure situation should do more good than harm. In the context of existing exposure situations, this justification may be more complicated than usual when introducing a source. The decisions that alter the radiation exposure are usually associated with altering the pathways, and may be intrusive or conflict with previously well-established behaviours. The second principle is optimisation. Exposures should be kept as low as reasonably achievable, taking into account the economic and societal factors that may be present. Optimisation is conducted within restrictions on individual exposure, referred to as 'reference levels' for existing exposure situations, in order to limit possible inequities in dose distribution.

The third principle is the application of dose limits. In planned exposure situations, the total dose to any individual should not exceed the appropriate limits as recommended by ICRP. This is a particular case of the more general concept of restricting individual exposures, and is important in planned exposure situations to ensure that multiple sources do not result in cumulative exposures that are inappropriate. The dose limits are not applicable in existing exposure situations. This has been one of the difficulties that many have seen with application of the system, particularly when there is a desire to control the situation in the same manner as a planned exposure situation.

2.2. Exposure situations

The principles of protection are applied in exposure situations. In its simplest terms, an exposure situation exists when there is a source, pathways of exposure from that source to a person, and exposure of individuals. Fig. 1 illustrates the components of any exposure situation.

According to *Publication 103*, 'Protection can be achieved by taking action at the source, or at points in the exposure pathways, and occasionally by modifying the location or characteristics of the exposed individuals' (ICRP, 2007, Para. 169). In practical terms, the system of protection can only be applied when there are reasonable mechanisms for control. Preferences are always given first to actions on the source itself, such as shielding. The source can be secured to prevent access. Steps can be taken to avoid radioactive materials from migrating to the environment, be that the laboratory bench or the environment around a facility. In other cases, and

Fig. 1. Elements defining an exposure situation.

particularly for most existing exposure situations, there is little that can be done about the source itself, and actions on the pathways or on the individuals must be considered.

After considering the source, protection control actions may be applied to the pathways from the source to an individual. This may be the amount of time spent in the area of the source, or the processing of effluents to remove radioactive materials before release. The possibilities are dependent upon the source being considered, and the circumstances, location, and amount of time that an individual may be present. The combination of the source and the pathways leads to the exposure of the individual. As can be seen, the categorisation of the exposure situation is clearly dependent upon the characterisation of the source and the characterisation of the pathways.

Using these elements, ICRP has defined three exposure situations that cover all of the situations in which individuals may be exposed. Existing exposure situations are situations where the source or sources already exist when decisions to control them are taken. Quite simply, there is a source present in the environment, and one must decide what to do about it. The source was already there when it was recognised or characterised, and a conclusion was made that controls need to be applied. Thus, characterisation of exposures is a prerequisite to control. At this point, it is important to note that the Commission considers long-term exposures resulting from a nuclear accident as an existing exposure situation.

Planned exposure situations result when exposures are the result of a deliberate introduction and operation of sources. Exposures can be anticipated and fully controlled. This is the case considered most often, and was the situation within which the protection framework was introduced and initially elaborated. In planned exposure situations, the decision is made to use a source for a particular purpose. By making that decision, it is possible to consider the characteristics of the source and the pathways, and provide appropriate controls. Emergency exposures result (almost always) from the loss of control of a planned source. These situations require urgent and timely actions in order to mitigate exposure. The controls that have been applied to the source are no longer present because something has happened that resulted in their unexpected loss. There is urgency in the actions, and the timing, to regain control and prevent exposures. The Commission considers exposures resulting from malicious acts as an emergency exposure situation.

As noted above, an important consideration in exposure situations is the time frame over which the exposure may occur, and the time available to take protective actions. In an emergency, there is urgency to take actions, to prevent deterministic tissue reactions, and to regain control of the source. In contrast, in existing exposure situations, there is no urgency requiring rapid actions to ensure protection. Time is available to consider the source, the pathways, and the possible protective actions more carefully. As the source already exists, protective actions can be implemented, and should be effective immediately. However, this does not mean that full control and an optimised situation is accomplished immediately. In many cases, a progressive approach of improving protection may occur in order to review the exposures continually, and consider how to further improve protection.

2.3. Exposure categories

Within the exposure situations, the exposed individuals may be categorised as medical, occupational, or public. From the Commission's standpoint, occupational exposure is exposure incurred at work as a result of the exposure situation, that can reasonably be regarded as being the responsibility of the operating management. Medical exposure is the exposure of patients in diagnostic, interventional, and therapeutic procedures. This category has also included those individuals who may be specifically providing comfort and care, or voluntarily participating in research activities. Public exposure encompasses all other exposures, thus becoming anything that is not occupational or medical. This definition, perhaps oversimplified as 'anything else', completes the theoretical structure, although regulatory organisations often find it necessary to be more prescriptive in some circumstances. Individuals may fall into the three categories, respectively, as workers, patients, or members of the public. In fact, any particular individual may be in all three categories at various points during their daily activities.

3. EXISTING EXPOSURE SITUATIONS

There are many types of existing exposure situation. For example, aircraft crew exposure to cosmic radiation is an existing exposure situation. The cosmic ray dose rate exists, dependent upon altitude, geographical location, etc. Likewise, exposure to radon in dwellings and workplaces is, for the most part, an existing exposure situation, because the naturally occurring decay chain of uranium has not been introduced deliberately.

Naturally occurring radioactive material (NORM) presents another existing exposure situation. This includes consideration of materials that are already present on the earth's surface, and the transport of naturally occurring materials from deep within the ground to the surface as a result of activity such as drilling for oil or gas. Two other situations also fall within the general category of existing exposure situations. These are sites that may have contamination from past activities, perhaps from long ago, which may not, in light of today's understanding and measurements, be considered acceptable. Likewise, contamination from a nuclear accident or radiation emergency eventually falls within this category; as it now exists in the environment, little can be done to modify the source, and actions have to be considered over the long term. The transition from an emergency exposure situation to an existing exposure situation continues to be an issue of considerable discussion, and is currently being examined by ICRP Committee 4 during updates to *Publications 109* and *111* (ICRP, 2009a,b) in light of the experience from Fukushima Daiichi.

Existing exposure situations have a number of unique features, some of which have already been hinted at in the above discussion. First, and perhaps obvious, existing exposure situations are ubiquitous, and there can be wide differences between different types of situation. As such, each situation needs to be characterised and considered carefully before starting to take actions for control. The time frame for taking protective actions is not urgent, and characterisation is important to the process of justifying and optimising the recommended actions.

Control of the exposure to individuals is mainly through the pathways, as the source itself will not, in many cases, be amenable to direct control and modification. This does not mean that actions on the source are excluded, and consideration can be given to such actions when it is reasonable. Likewise, because of the wide variability, there may be a large distribution in individual exposures. This has been seen many times, such as in the contaminated zones in Chernobyl where individual habitats and locations, even within a single village, resulted in widely variable individual exposures.

The ubiquity of existing exposure situations means that workers may be receiving exposures that are adventitious, and not part of the specific work that is being done. Further, the sources may not be under the control of the operating management or employer because the source exists within the environment. In many situations with NORM, there may be a lack of experience with radiation protection, or even a lack of awareness that exposures are occurring.

The lack of existing radiation protection culture can make the introduction of control measures more difficult to accept by the individuals who are being exposed. However, it is these individuals, be they workers or simply members of the public, who are living in the existing exposure situation who can have the greatest influence on their own exposure. Informed personal behaviours, based on information, support, and knowledge, can be one of the most significant mechanisms to reducing the risk to these individuals. This individualised action has been referred to as 'self-help' protective actions.

Finally, existing exposure situations are unique because they require a long-term perspective. The sources exist, are characterised, and actions are taken. This does not mean that the situation goes away. Instead, this is an ongoing process, and it is important to continue to consider the source, the protective actions, and what might be reasonable and possible to improve the situation further over time.

The introduction of the three exposure situations in *Publication 103* (ICRP, 2007) was designed to emphasise the commonality of the approach to radiation protection, irrespective of the circumstances. Most are familiar with the classic approach when the introduction of the source is planned. A dose limit controls the sum of all the exposures of an individual. Further, a restriction of the exposure from a particular source to a particular individual serves as a boundary of what is considered as acceptable in planning the exposure and optimising the protection.

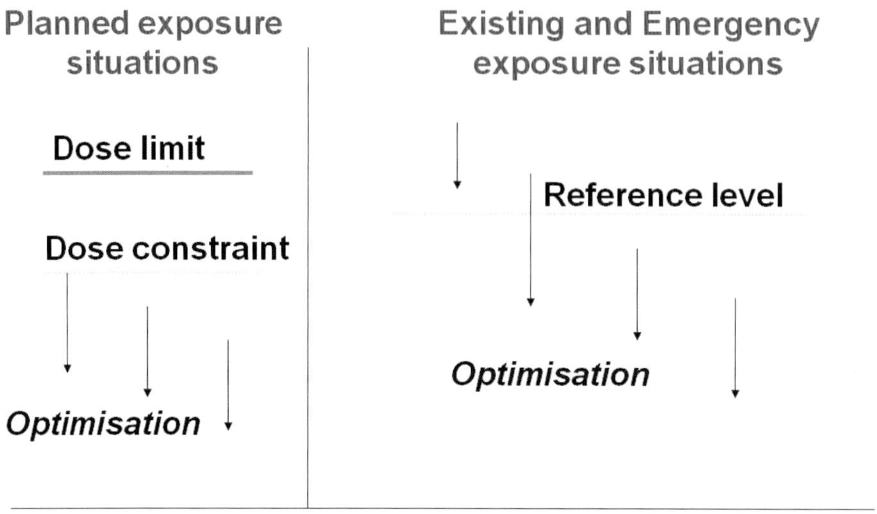

Fig. 2. Commonality of the approach for planned, existing, and emergency exposure situations.

What was not at all obvious previously was that the same approach was equally applicable in existing and emergency exposure situations. In this case, a restriction of what would be considered a priori as acceptable for planning can be established as a reference level. Optimisation acts to improve protection, irrespective of whether the initial exposure levels are greater than, or less than, the reference level. Optimisation is the unifying approach to radiation protection, and always challenges one to consider if there are reasonable and appropriate measures to further reduce exposures to levels that are as low as reasonably achievable, economic and social factors being taken into account. The commonality of the approach is illustrated in Fig. 2.

4. APPLICATION TO NORM

NORM is one particular example of existing exposure situations that presents a number of challenges and areas where there is confusion. ICRP has classified NORM as an existing exposure situation. Primarily, the source is the various concentrations of the natural radionuclides in the earth's crust. Human activities may create new pathways of exposure or alter the pathways and/or concentrations that may be present. In many cases, a combination of preventative and corrective actions can be taken.

As shown above, the protective approach, namely optimisation of exposure below a dose restriction, is fundamentally the same as for other exposure situations. In fact, NORM exposure can be managed in the same manner as a planned exposure situation. Regulatory organisations may choose to use the common tools applied to other planned situations in the case of NORM when it is reasonable to do so, and the characteristics of the source and pathways can be handled as such. There are also situations where the source can be removed, or where the natural material is, in fact, extracted or modified to make a more traditional radiation source.

One consideration for NORM is whether there is some modification to the concentrations and locations of the material. Without modification, NORM exposure may be unamenable to controls. However, humans tend to modify their environment, and thus the source and pathways become different from the naturally existing environment.

Modifications of the source may be adventitious or may be deliberate. In this case, 'deliberate' refers to specific activities that use the radioactive materials for some purpose. These could become planned exposure situations, because there is the opportunity to plan what is being done, consider the source, provide controls over the source such as containment, ventilation, etc., and control exposures adequately. Adventitious modification is more complicated because there may, or may not, be an awareness of the source, either historically or in the initial activities. Nevertheless, in all cases, a modified source is present, there are pathways, and there will, as a consequence, be exposed individuals.

For an existing exposure situation, it is possible, under the ICRP exposure categories described above, to organise exposures as occupational or public on the basis of whether the situation can reasonably be regarded as being the responsibility of the employer. In either case, control actions have to be justified, and protection is provided by application of optimisation within restrictions on what is considered as appropriate restrictions on individual exposure. It should be noted that this differentiation into occupational vs public is not on the basis of the magnitude of the exposures. The exposures may reasonably be regarded as being within the responsibility of the employer, and not be a challenge to an appropriate reference level. Likewise, this differentiation is not based on the choices that might be made by an employer, or regulatory organisation, regarding the types of requirements that may be employed in a regulatory context.

ICRP Committee 4 has been carefully examining the practical approach to radon, NORM, and other situations. One key to this is the use of a graded approach. Pragmatically, protection should be commensurate with the risks that are presented by the specific situation, and the opportunities to provide protection. The strategies used must be appropriate and adapted for the circumstances. Existing exposure situations, and NORM in particular, need to be characterised so that the level of ambition for protection can be established, there can be realism in what is considered, and the effectiveness of the actions can be assessed and modified as necessary.

Regulatory organisations often wish to have specifications for all the requirements so that compliance can be assessed, and enforcement taken when necessary. While appropriate, it must be remembered that this must also be influenced by the risks that are presented by the situation. In certain circumstances, it is reasonable and appropriate to specify legal responsibilities including specific mandatory provisions, again in a graded approach. Of course, there may be circumstances where there are no reasonable mechanisms for control, or where the levels of exposure are such that the most appropriate oversight is an exemption.

ICRP has stated many times that constraints and reference levels are not 'limits'. This is particularly important in existing exposure situations. The reference level indicates the boundary that, from a planning standpoint, one should not plan to exceed. This differs from 'compliance'. It may not be possible to achieve all exposures below the desirable value immediately, and it would not be appropriate for a regulatory organisation to translate the reference level into an enforceable limit. On the other hand, if the situation is well characterised and controls are well established, it may be reasonable to apply the types of requirements and structures that are used in planned situations.

There are a number of complications that need to be taken into account when considering existing exposure situations. First, the principle of application of dose limits is actually a specific example in planned exposure situations of the broader concept for restricting individual doses, which applies in all exposure situations. Dose limits are explicit in planned exposure situations, where there is clearly the ability, and the duty, to manage an individual's exposure to within the recommended limits, and, in fact, further restrict individual doses through the use of constraints in the process of optimisation. In an existing exposure situation, dose limits do not formally apply. However, the underlying concept of restricting individual doses, through the use of an appropriate reference level in the optimisation process, still functions to ensure that there is no inequity in exposures to certain individuals.

When regulatory requirements are established, there is a tendency to specify requirements clearly. Regulators often use the concept 'limits' because they are clear, precise, and unambiguous. While the ICRP system of protection does not specify dose limitation as one of the principles for existing exposure situations, the use of such a regulatory construct could be appropriate, depending on the specifics of the exposures, and the degree to which the activities are well characterised and can be well controlled.

Another complication is what might be called 'back-fitting'. When a situation is characterised and a conclusion is reached, there need to be some controls to provide radiation protection properly. The imposition of those requirements may be difficult and may be expensive. No one particularly likes someone coming into their activities and demanding that they expend resources to make changes, particularly if they cannot fully appreciate the risks that may have been unrecognised previously. In this case, there must be careful consideration of the justification for establishing requirements, and a balance drawn between the radiation risks, other risks that may be present or introduced by the introduction of controls, and the costs and benefits of the control measures.

5. CONCLUSIONS

The overall objective of the ICRP system is protection of people and the environment, and this is applicable irrespective of the exposure situation. In order to justify and optimise protection effectively, it is necessary to know what is being dealt with. Characterisation of the exposure, and the available options, is critical to obtaining an optimised protection strategy. That strategy and approach must match the circumstances if it is to be effective and sustainable. Further, responsibilities must be clear and stakeholders must be involved. Unlike in an emergency exposure situation, where the urgency of action requires that some preplanned protection strategies are implemented with limited information, existing exposure situations can be well characterised and the affected stakeholders can be involved in the decisions. It is clear that if these individuals are not involved in their own protection decisions, they will be much more difficult to implement. It is very easy to become alienated when there is a perception that you are simply a victim, or have something imposed upon you.

ICRP's system of protection consistently establishes that exposures are to be optimised. While there may be a variety of expectations and regulatory tools in a graded approach based on the level of risk, the over-riding approach is optimisation, and looking to see if there are reasonable steps that can improve protection.

Individual exposures are considered in the restriction of individual dose by constraints or reference levels. These restrictions are the boundary for the optimisation process, and help to ensure that there is equity for all individuals. In some cases, it is reasonable and possible to apply the regulatory structures of planned exposure situations, including limits, for a particular case of existing exposure based on ethics and safety.

Finally, it must always be remembered that existing exposure situations are not a 'one size fits all'. Just as each case is unique, so too are the decisions to be made, and the control tools that may be used have to be commensurate with the risks.

REFERENCES

ICRP, 2007. The 2007 Recommendations of the International Commission on Radiological Protection. ICRP Publication 103. Ann. ICRP 37(2–4).

ICRP, 2009a. Application of the Commission's recommendations for the protection of people in emergency exposure situations. ICRP Publication 109. Ann. ICRP 39(1).

ICRP, 2009b. Application of the Commission's recommendations to the protection of people living in long-term contaminated areas after a nuclear accident or a radiation emergency. ICRP Publication 111. Ann. ICRP 39(3).

SAGE

ICRP
Annals of the ICRP

Application of the Commission's recommendations to naturally occurring radioactive material

J-F. Lecomte

Institute for Radiological Protection and Nuclear Safety, F-92260 Fontenay-aux-Roses, France; e-mail: jean-francois.lecomte@irsn.fr

Abstract–Since publication of the 2007 Recommendations (ICRP *Publication 103*), the International Commission on Radiological Protection has focused on preparing a series of publications dedicated to different types of existing exposure situations, such as radon exposure, cosmic exposure in aviation, and exposure to naturally occurring radioactive material (NORM). The publication related to NORM will present the main types of corresponding activities, and describe the characteristics of NORM exposure. It will also develop a conceptual framework for the practical application of the Commission's system to NORM exposure. In particular, the publication will explain why NORM activities are generally considered to be existing exposure situations, and when some of them should be managed as planned exposure situations. It will indicate when the workers should be considered as occupationally exposed. It will also provide recommendations regarding application of the three principles of radiological protection. The need to consider the justification of the re-use or recycling of residues carefully will be highlighted. Guidance will be provided for selection of the reference level, and for implementation of the optimisation process through a graded approach including both prevention and mitigation of exposures. Flexibility will be recommended for the application of dose limits, notably when the situation is managed as a planned exposure situation.

Keywords: Natural exposure; Protection strategy

This paper does not necessarily reflect the views of the International Commission on Radiological Protection.

1. INTRODUCTION

At its meeting in Berlin, Germany in October 2007, the Main Commission of the International Commission on Radiological Protection (ICRP) approved the development by Committee 4 of a new report related to application of the Commission's recommendations on radiological protection of workers, the public, and the environment to exposures resulting from industrial processes using naturally occurring radioactive material (NORM). Task Group 76 was created, the aim of which is to develop recommendations to cover the broad range of activities associated with the processing, production or use, and disposal of materials with enhanced levels of NORM. The report will also clarify the issues concerning the type of exposure situations, the categories of exposure, and the basic principles to be applied for the management of NORM. Task Group 76 is continuing its work, and its membership was revised in 2013.

2. CHARACTERISTICS OF NORM EXPOSURE

NORM is ubiquitous and present in almost all materials on earth. Human activities have the potential to either increase or decrease radiation exposures to NORM. The primary source is not controllable or is only partially controllable. However, the source may be modified unintentionally or intentionally.

NORM exposure occurs as a result of industrial processes or work activities. These activities may increase both occupational and public exposures. A wide range of processes or activities may be concerned. The main types of NORM industries are:

- mining and mineral processing industries;
- coal, oil, and gas production;
- some metal production industries (thorium, niobium, zircon, titanium, etc.);
- phosphate industry and production of some building materials; and
- water treatment.

This list is not intended to be exhaustive or exclusive, but rather representative of the situations that pose an exposure situation of concern.

Typical NORM industries process a wide range of raw materials and activity concentrations, producing a variety of products and wastes. These industries may or may not be of concern, depending on the activity concentrations in the raw materials handled, the processes adopted, the uses to which final products are put, the re-use and recycling of residues, and the disposal of wastes. Exposures may occur during various stages of production or from the use of products, residues, and waste. In many cases, there is a lack of awareness of the presence of radioactive material in these products.

Concentrations may be enhanced in products, by-products, or waste generated by some mineral processing industries. In general, exposures from these industries involve small doses to large numbers of people, and the resulting population doses can be greater than those from artificially produced radionuclides.

Exposure of workers is often adventitious as it is not normally considered to be part of the job in which they are engaged. There is often a lack of awareness about radiation protection culture in the industry. In these industries, risk reduction strategies depend on changing personal behaviour, and need a long-term perspective when implemented.

Due to the large variations in activity concentration in NORM industries, and the uses, as well as a wide range of industrial practices, to which these materials are put, it is not possible to adopt a simple generic approach for their safe management.

3. EXPOSURE SITUATIONS AND CATEGORIES OF EXPOSURE

The Commission's system of radiological protection is described in *Publication 103* (ICRP, 2007). According to Paragraph 44, it 'applies to all radiation exposures from any source, regardless of its size and origin'. In particular, according to Paragraph 45, 'the Commission's Recommendations cover exposures to both natural and man-made sources. The Recommendations can apply in their entirety only to situations in which either the source of exposure or the pathways leading to the doses received by individuals can be controlled by some reasonable means. Sources in such situations are called controllable sources.'

3.1. Exposure situations

An exposure situation is the process that includes a natural or man-made radiation source and the transfer of radiation through various pathways leading to the exposure of individuals. The recommendations in *Publication 103* (ICRP, 2007) organise radiological protection according to three types of exposure situation: planned exposure situations, emergency exposure situations, and existing exposure situations. Planned exposure situations are situations resulting from the deliberate introduction and operation of sources. Exposures can be anticipated and fully controlled. Emergency exposure situations result from the loss of control of the source, and urgent action is necessary in order to avoid or reduce undesirable consequences. They also include exposures resulting from a malicious act or from any other unexpected situation. Existing exposure situations are situations where the source already exists when a decision to control the related exposure is taken. They include naturally occurring exposures as well as exposures from past events and accidents, in addition to practices. Characterisation of exposure is a prerequisite to control.

NORM exposure situations have the characteristics of existing exposure situations as NORM are ubiquitous and present in almost all materials on earth. NORM is mentioned as an example of existing exposure situations in Paragraph 284 of *Publication 103* (ICRP, 2007). The source is natural and already exists when a decision on control has to be taken. NORM industries create or alter pathways with the potential to either increase or decrease radiation exposures from these materials. When such alterations arise without modifying the source, or when the source is

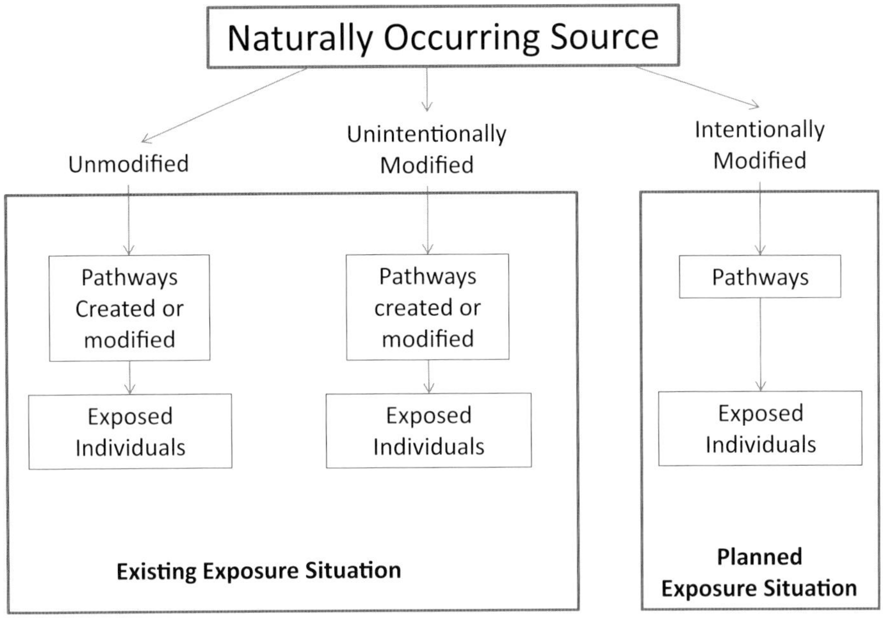

Fig. 1. Types of naturally occurring radioactive material exposure situations according to the state of the source.

modified unintentionally, the exposure situation is an existing exposure situation. However, when the source is removed and modified intentionally, the exposure situation becomes a planned exposure situation (see Fig. 1). When the primary source is not (or only partially) controllable, the modification of the source is controllable. NORM industries are not likely to give rise to an emergency exposure situation, although high concentrations of NORM in a place may require the prompt implementation of protective actions.

3.2. Categories of exposure

The Commission distinguishes between three categories of exposure: medical exposures of patients, occupational exposures, and public exposures. Medical exposures of patients occur in diagnostic, interventional, and therapeutic procedures. NORM industries do not lead to medical exposures.

Occupational exposure is radiation exposure of workers incurred as a result of their work. However, because of the ubiquity of radiation, the direct application of this definition to radiation would mean that all workers should be subject to a regime of radiological protection. The Commission therefore limits its use of 'occupational exposures' to radiation exposures incurred at work as a result of situations that can reasonably be regarded as being the responsibility of the operating management (ICRP, 2007). The Commission also considers that workers who are not regarded as being occupationally exposed to radiation are usually treated in the same way as members of the public.

In the case of NORM industries, if the source has been modified intentionally and introduced deliberately, the exposure at work is considered to be occupational exposure, because it is clear that the exposure can be regarded as the responsibility of the operating management. The exposure situation is then managed by controlling the source, the pathways, and individual exposures. If the source has not been modified, the exposure at work is considered to be adventitious, the workers are treated in the same way as members of the public, and the exposure situation is managed by controlling the pathways. In a case where the source has been modified unintentionally, the exposure at work may also be seen as adventitious, and is consequently not usually regarded as the responsibility of the operating management. However, if the protection of a worker needs the application of individual requirements, such as training, individual monitoring, and health surveillance, the exposure should be considered as occupational exposure, and the exposure situation should be managed by controlling both the pathways and individual exposures. Whatever the case, the health and safety of workers continue to be the responsibility of their employer.

Public exposure is exposure that is neither medical nor occupational.

4. APPLICATION OF THE PRINCIPLES

4.1. Justification

In the ICRP system of protection, the principle of justification is one of the two source-related fundamental principles (ICRP, 2007, Para. 203). In application of this principle, any decision that alters the radiation exposure situation should do more good than harm. This means that by introducing a new radiation source, reducing existing exposure, or reducing the risk of potential exposure, one should achieve sufficient individual or societal benefit to offset the detriment it causes.

In the case of NORM exposure, the intentional modification of a source should be justified like the deliberate introduction of a source. When the source has not been modified, or has been modified unintentionally, the strategies of protection should be justified. In all cases, particular attention should be paid to justification of the re-use or recycling of residues. Among the considerations to take into account are the level of exposure, the fact that the re-use or recycling should not give rise to exposure of concern (e.g. some building materials which may become the main source of radon exposure in some countries), and ethical considerations such as the dissemination of radiation in consumer products.

4.2. Optimisation of protection

The Commission considers that the principle of optimisation has a key role in NORM industries. The first step is to characterise the relevant exposure situation in terms of the nature of the exposure, the benefits from the exposure situation to individuals and society, as well as other societal criteria, and the practicability of preventing or reducing the exposures. Assessment of these attributes should enable the selection of an appropriate range for the constraint or the reference level.

The management of NORM should be subjected to a graded approach consistent with the optimisation principle. Optimisation requires that the likelihood of incurring exposures, the number of people exposed, and the magnitude of the exposures should all be kept as low as reasonably achievable, taking into account economic and societal factors. The level of protection should maximise the margin of benefit over harm, and should be the best possible under the prevailing circumstances.

The optimisation principle is particularly important for the management of NORM residues as doses to the public can be comparable to those from background radiation, and may exceed the public dose limit for planned exposure situations.

An endpoint for optimisation must not be set in advance, as the optimised level of protection will depend on the situation and past experience with the management of similar situations. The result of an assessment may be that action to reduce exposures is not warranted, depending on the controllability of the source of exposure, and the prevailing economic, societal, and cultural circumstances. However, the concept of exemption, which is a regulatory concept, should not drive the protection strategy.

The Commission recommends that all industries involved in the production and handling of NORM should develop a protection strategy through a NORM management plan. The essential elements of a management plan to be considered in any operation involving materials containing NORM include:

- identification of potential sources of exposure to workers and members of the public;
- identification of potential impacts on the environment;
- identification of exposures to workers and members of the public (determine who is exposed, where, when, and how);
- selection of the relevant dose restrictions (reference levels or dose constraints);
- selection of the best protective option in the prevailing circumstances, including prevention and mitigation measures; and
- management of environmental impacts of wastes, residues, and discharges.

The NORM management plan should cover all aspects of the management of radiation exposures and radioactive waste, and should include:

- a description of all processes where doses to workers or the public may arise;
- a plan to manage radiation exposures;
- a radioactive waste management plan;
- an assessment of the current or projected use of material that may be recycled;
- an assessment of the potential impact of manufactured items containing NORM;
- responsibilities of the operator/employer and of the employees;
- any relevant occupational health and safety issues;
- any relevant environmental protection issues;
- ways to involve stakeholders;

- appropriate monitoring and review programmes;
- demonstration of compliance with relevant radiological protection criteria; and
- periodic review of the status of the operation in relation to continuing controls.

Appropriate prevention and mitigation procedures should be chosen for each situation, as appropriate. A monitoring programme should be implemented to check the adequacy and efficiency of control measures, and data collected should be used to justify continuation of the present control measures, or to indicate a need to change them.

4.3. Dose restrictions

4.3.1. Reference levels and dose constraints

Reference levels and dose constraints are very similar concepts for the planning of optimisation. The reference level represents, in emergency and existing controllable exposure situations, the level below which the ambition is to reduce all doses from a source to a level that is as low as reasonably achievable, economic and societal factors being taken into account. The reference level is also the level above which it is judged, from a planning standpoint, to be inappropriate to allow exposures to occur. A dose constraint is a prospective and source-related restriction on the individual dose from a source in a planned exposure situation, which serves as an upper bound on the predicted dose in the optimisation of protection for that source. It is a level of dose above which it is unlikely that protection is optimised for a given source of exposure, and therefore action must almost always be taken.

According to *Publication 103* (ICRP, 2007), the chosen value for a reference level or a dose constraint depends upon the prevailing circumstances of the exposure situation under consideration. In order to provide guidance for selecting appropriate values, the Commission has defined a dose scale (ICRP, 2007, Table 5) reflecting the fact that, within a continuum of risk (linear non-threshold assumption), the risk that may be acceptable depends on the exposure circumstances. This scale is divided into three bands reflecting the more or less important need for action, which is dependent on the characteristics of the exposure situation: controllability of the source; individual or societal benefit from the situation; and requirements with regard to information, training, and dosimetric or medical surveillance. Numerically speaking, the three bands are <1 mSv, 1–20 mSv, and 20–100 mSv (in acute or annual doses).

As far as NORM exposure is concerned, reference levels for public exposure, including the exposure of workers treated as members of the public, should be typically selected, within a graded approach, in the lower range of the second band, between 1 and 20 mSv, set on an annual basis. A specific graded approach may be implemented in workplaces in order to allow the management of workers as occupationally exposed in case, despite all reasonable efforts, the doses cannot be maintained consistently below the selected reference level. For workers considered as occupationally exposed in an existing exposure situation, the relevant dose restriction should be selected in the band between 1 and 20 mSv $year^{-1}$. In case of a planned

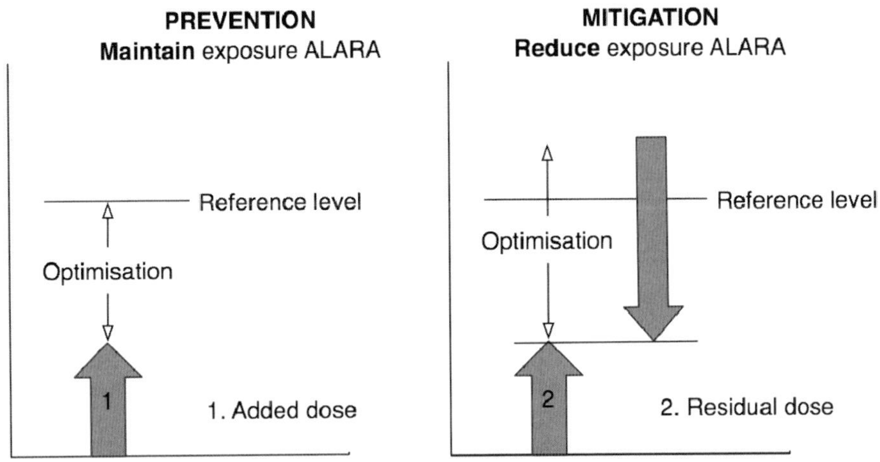

Fig. 2. Implementation of the optimisation principle. ALARA, as low as reasonably achievable.

exposure situation, the dose constraint should be selected in the first band, <1 mSv year^{-1}, for public exposure, and in the second band, between 1 and 20 mSv year^{-1}, for occupational exposure.

In practice, derived reference levels should be determined in measurable quantities (dose rate, activity, or activity concentration) for the operational implementation of the optimisation process.

The optimisation process is implemented for NORM exposure through the management plan. When implementing optimisation, the distinction should be made between prevention aiming to maintain the expected exposure (added dose) to a level that is as low as reasonably achievable under the prevailing circumstances, and mitigation aiming to reduce the residual dose to a level that is as low as reasonably achievable (see Fig. 2).

4.3.2. Application of dose limits

According to *Publication 103* (ICRP, 2007, Para. 203), the principle of application of dose limits is the third fundamental principle of the ICRP system. It is individual related and only applies for planned exposure situations. The total dose to any individual from regulated sources in planned exposure situations other than medical exposure of patients should not exceed the appropriate limits recommended by the Commission. Dose limits apply to NORM exposure in planned exposure situations. The dose limit recommended by the Commission for occupational exposure is expressed as an effective dose of 20 mSv year^{-1}, averaged over defined 5-year periods (100 mSv in 5 years), with the further provision that the effective dose should not exceed 50 mSv in any single year (ICRP, 2007, Para. 244).

5. CONCLUSIONS

The purpose of the report of Task Group 76 is to provide guidance on application of the Commission's system of radiological protection to NORM exposure. As such exposure occurs as a result of industrial processes, one could assume that it can be managed in the same way as exposure occurring in the nuclear industry. This is true in some cases, when the source has been removed and modified intentionally. In this case, the exposure situation is a planned exposure situation and can be managed as such. However, in most cases, the source, whether unmodified or modified unintentionally, is not controllable or is only partially controllable. The corresponding exposure situations are existing exposure situations and should be managed as such. As for radon exposure, NORM exposure of workers may be adventitious, in which case the workers should be treated as members of the public. The wide range of industrial processes, activity concentrations in materials, and distribution of individual exposures is another challenge to address. Regardless of the NORM exposure situation, it is managed mainly on the basis of the optimisation process. The corresponding protection strategy, implemented through a management plan, should be based on an ambitious, realistic, and graded approach. Further reflection is needed by Task Group 76 to determine appropriate recommendations for NORM exposure.

REFERENCE

ICRP, 2007. The 2007 Recommendations of the International Commission on Radiological Protection. ICRP Publication 103. Ann. ICRP 37(2–4).

SAGE

ICRP
Annals of the ICRP

What should a radiation regulator do about naturally occurring radioactive material?

J. Loy
Federal Authority for Nuclear Regulation, P.O. Box 112021, Abu Dhabi, United Arab Emirates; e-mail: john.loy@fanr.gov.ae

Abstract–The standard regulatory framework of authorisation, review and assessment, inspection and enforcement, and regulation making is directed principally towards ensuring the regulatory control of planned exposure situations. Some mining and industrial activities involving exposures to naturally occurring radioactive material (NORM), such as uranium mining or the treatment and conditioning of NORM residues, may fit readily within this standard framework. In other cases, such as oil and gas exploration and production, the standard regulatory framework needs to be adjusted. For example, it is not sensible to require that an oil company seek a licence from the radiation regulator before drilling a well. The paper discusses other approaches that a regulator might take to assure protection and safety in such activities involving exposures to NORM, including the use of conditional exemptions from regulatory controls. It also suggests some areas where further guidance from the International Commission on Radiological Protection on application of the system of radiological protection to NORM would assist both regulators and operators.

Keywords: NORM; Exposure situations; Regulation

1. NORM, OIL AND GAS AND THE SYSTEM OF RADIOLOGICAL PROTECTION

Early in the life of the United Arab Emirates (UAE) Federal Authority for Nuclear Regulation (FANR), the Abu Dhabi National Oil Company (ADNOC) asked if FANR needed or proposed to regulate exposures to naturally occurring

This paper is the view of the author and cannot be taken as the position of the United Arab Emirates Federal Authority for Nuclear Regulation. In addition, this paper does not necessarily reflect the views of the International Commission on Radiological Protection.

radioactive material (NORM) in ADNOC's operations. The answer given was – no, or at least, not yet.

Sometime later, ADNOC approached FANR with a proposal to construct a treatment and disposal facility to deal with the NORM residues arising from the activities of its operating companies engaged in oil and gas production in the Emirate of Abu Dhabi and offshore. Here, FANR's response was clear that both the system of radiological protection and the UAE's nuclear law required that the NORM facility be licensed by FANR in order to operate.

Are these two answers consistent with the system of radiological protection and the UAE's regulation of exposures to radiation?

The production of oil and gas in general creates an exposure situation. Radioactive material, principally Ra-226, is brought to the surface and the consequent exposure of workers in the production facilities may arise. Should this exposure be characterised as an existing exposure situation or a planned exposure situation?

There are arguments for either characterisation. Certainly the radioactive material is existing, and exposure is incidental to another purpose and, in this sense, is not at all planned. On the other hand, the likelihood of exposures arising from the activities involved in oil and gas production is known in advance and can be planned for.

The answer does matter for a regulator. A planned exposure situation should be regulated. The scope for regulation of a planned exposure situation includes authorisation or licensing, which follows from review and assessment of an application undertaken before the activity leading to the exposure can begin. One can expect to see application of the principle of optimisation of protection and safety through the use of dose constraints, and application of dose limits for individuals.

An existing exposure situation is one that does not usually benefit from full regulation because the limited controllability of the source and exposure does not lend itself to meeting firm regulations. An existing exposure situation may be better dealt with through regulatory guidance. In particular, the guidance should offer a graded approach in the context of some accepted and acceptable reference level of dose arising from the exposure situation.

2. THE UAE SYSTEM FOR REGULATION OF RADIOACTIVE MATERIAL

The UAE Federal Law by Decree No. 6 of 2009 concerning the Peaceful Uses of Nuclear Energy (UAE Nuclear Law; available at www.fanr.gov.ae) establishes the regulatory framework for radioactive material. Any person is prohibited from undertaking 'regulated activity' with 'regulated material' unless they are licensed to do so by FANR. Regulated activities with regulated material include possession, use, manufacture or handling, import or export, transportation, storage, and disposal.

Regulated material is defined in the UAE Nuclear Law as including radioactive material. In turn, radioactive material is defined as 'material designated by the Authority (FANR) as being subject to regulatory control because of its radioactivity'.

Is NORM captured by this definition? In its 'Basic Safety Standards for Facilities and Activities involving Ionizing Radiation other than in Nuclear Facilities' (FANR-REG-24; available at www.fanr.gov.ae), FANR adopted the international exemption levels as being the dividing line above which material requires regulatory control because of its radioactivity. Effectively, this means that with regard to NORM, any material with an activity concentration of any radionuclide in the uranium and thorium decay series exceeding 1 Bq g^{-1} is defined as 'regulated material'.

The UAE Nuclear Law allows FANR, consistent with international commitments and agreements, to exempt certain facilities and activities from licence requirements, provided that they do not represent a substantial threat to achieving the priorities and objectives of the UAE Nuclear Law.

It seems clear that the activity of gathering together material containing NORM residues (at least those above 1 Bq g^{-1}), processing it, and disposing of conditioned NORM residues is an exposure situation that is planned in advance and, in terms of the UAE Nuclear Law, involves several regulated activities with regulated material (possession, handling, transport, storage, and disposal). The proposals for a NORM facility thus require review and assessment, licensing, and inspection of a programme of radiation protection involving the optimisation of protection and safety, and the limitations of individual dose. The degree of regulatory control imposed should, of course, reflect the assessed hazard.

Dealing with exposures to NORM arising from oil and gas production activities seems much less clear in regulatory terms. The exposures arising from these activities are probably better characterised as existing exposure situations. It can certainly be argued that oil and gas production does not involve any regulated activity with regulated material, in that the activities have an entirely different purpose and the exposure is incidental to that other purpose, rather than the intended purpose of the activity being to apply the regulated material to achieve a certain end (such as a medical diagnosis).

3. EXPOSURE OF AIR CREW – IS IT LIKE EXPOSURE OF OIL AND GAS PRODUCTION WORKERS?

The exposure of air crew to cosmic radiation is the most studied occupational existing exposure situation, apart from occupational exposure to radon. The International Atomic Energy Agency in its recently revised Basic Safety Standards (IAEA GSR Part 3; available at www.iaea.org/standards) lays out an approach for the protection and safety of air crew, which is treated as an existing exposure situation, as follows:

> *5.30 The regulatory body or other relevant authority shall determine whether assessment of the exposure of air crew due to cosmic radiation is warranted.*
>
> *5.31. Where such assessment is deemed to be warranted, the regulatory body or other relevant authority shall establish a framework which shall include a reference level of dose*

and a methodology for the assessment and recording of doses received by air crew from occupational exposure to cosmic radiation.

5.32. In accordance with Para. 5.31:

(a) *Where the dose of aircrew members is likely to exceed the reference level, employers of air crew:*
   (i) *shall assess doses and keep records;*
   (ii) *shall make records available to aircrew members;*

(b) *Employers:*
   (i) *shall inform female aircrew members of the risk to the embryo or fetus due to exposure to cosmic radiation and of the need for early notification of pregnancy;*
   (ii) *shall apply the requirements of Para. 3.114 in respect of notification of pregnancy.*

The four elements of this regulatory approach are:

- a decision as to whether assessment of exposure is warranted;
- where warranted, the establishment of a framework by the regulatory body to include a reference level of dose, a dose assessment, and recording methodology;
- employers to assess doses, keep dose records, and make them available to workers where doses are likely to exceed the reference levels; and
- a general duty of care of employers towards pregnant workers.

The major difference between the exposures of air crew and the exposures of oil and gas production workers is that the radiation dose expected to arise from any given flight can be found through readily available software. The quantities of NORM, and hence the resulting exposures to oil and gas production workers, differ widely from field to field. However, it is hoped that measurement of activity concentrations of NORM in any field, and generic modelling of the behaviour of workers, will allow average exposures to be assessed.

Following this approach, a possible mode for the control of exposures in the oil and gas industry is that the regulatory body, after consultation with the industry, establishes (through a regulatory guidance document) a reference level of dose and a methodology for dose assessment.

The consequent obligations on employers can be achieved through a regulation or a conditional exemption from regulatory control requiring:

- the operator to make an exposure assessment using the determined methodology and data derived from measurements of activity concentrations of NORM in local conditions;

- if the assessed dose from the exposure assessed is below the reference level, there is no need for any further action, other than the duty of care towards pregnant workers; and
- if the assessed dose is above the reference level, the employer shall implement a scheme for assessment and recording of the doses assessed for individual workers, with the information to be made available to workers.

Otherwise, prudent occupational health and safety measures can be assumed to assure the optimisation of protection and safety.

4. WHAT SHOULD ICRP DO?

As a general comment, there is a need for further analysis and exploration by the International Commission on Radiological Protection (ICRP) of existing exposure situations. Currently, the concept is dealt with briefly, with the notable exception of exposures to radon in workplaces and dwellings.

A vital piece of work to be undertaken is exploration of the concept of reference levels, from which further guidance may come on the basis for choosing a certain reference level for a regulatory scheme for NORM, such as proposed in this paper. Is there a one-size-fits-all generic reference level that should be applied to all industrial activities that result in exposure to radiation from NORM? Or should a reference level be set at a level that represents a low/average/higher level of exposure occurring in a particular industry? Or is there some other way?

Finally, ICRP needs to advocate pragmatism and flexibility, supported by radiation protection principles, when it comes to NORM.

Annals of the ICRP

Radiological protection in North American naturally occurring radioactive material industries

D.B. Chambers

SENES Consultants (an ARCADIS Company), 121 Granton Drive, Suite 12, Richmond Hill, Ontario, Canada, L4B 3N4; e-mail: dchambers@senes.ca

Abstract–All soils and rocks contain naturally occurring radioactive material (NORM). Many ores and raw materials contain relatively high levels of natural radionuclides, and processing such materials can further increase the concentrations of natural radionuclides, sometimes referred to as 'technologically enhanced naturally occurring radioactive material' (TENORM). Examples of NORM minerals include uranium ores, monazite (a source of rare earth minerals), and phosphate rock used to produce phosphate fertiliser. Such activities have the potential to result in above background radiation exposure to workers and the public. The objective of this paper is to review the sources and exposure from NORM in North American industries, and provide a perspective on the potential radiological hazards to workers and the environment. Proper consideration of NORM issues is important and needs to be integrated in the assessment of these projects. Concerns over radioactivity and radiation amongst non-governmental organisations and the local public have resulted in the cancellation of NORM mining and mineral extraction projects, as well as inhibition of the safe use of by-product materials from various NORM industries. This paper also briefly comments on the current regulatory framework for NORM (TENORM) in Canada and the USA, as well as the potential implications of the recent activities of the International Commission on Radiological Protection for NORM industries.

Keywords: NORM; Mining; Rare earths; Phosphate; Oil and gas; Regulation

---

This paper does not necessarily reflect the views of the International Commission on Radiological Protection.

1. INTRODUCTION

People have always been exposed to natural background radiation; radioactivity from radionuclides found naturally in soils and rocks; radioactivity in food, water, and air; and radiation from space. While the levels of radiation and radioactivity vary widely from place to place, by perhaps a factor of 10 or so, there is nowhere on earth that is not radioactive. According to the United Nations Scientific Committee on the Effects of Atomic Radiation (UNSCEAR), the nominal per caput annual dose from all natural sources of radiation is approximately 2.4 mSv with a range of (approximately) 1–13 mSv year^{-1} (UNSCEAR, 2000).

1.1. What is NORM?

Most naturally occurring radioactive material (NORM) contains radionuclides from the so-called long-lived 'primordial' decay chains, resulting from the decay of U-238, U-235, Th-232, and other long-lived radionuclides, such as K-40. All soils and rocks are naturally radioactive, and hence all ores are also radioactive. This is illustrated in Fig. 1 which shows the nominal range of radioactivity in soils and a variety of ores.

Elevated levels of natural background radiation are seen in many occupational settings, especially in the mining and processing of ores, the extraction of oil and gas, and the production of phosphate fertilisers. A few examples of radioactivity levels in a variety of NORM are summarised in Table 1, which illustrates the wide range of 'typical' concentrations of uranium and thorium in NORM.

Fortunately, in most cases, radionuclide concentrations in NORM do not pose any material risk to the environment or humans. However, when NORM is concentrated due to human activities such as oil and gas, water treatment waste, and coal combustion, it is important to consider such enhancement of the radioactivity in the products, by-products, residues, or wastes arising from these industrial processes (IAEA, 2006).

The radionuclides of interest for NORM are from the uranium and thorium natural decay series. As suggested above, NORM is found in mining and mineral processing, as well as in many industrial waste streams including overburden, mine spoils and tailings from uranium mining, metal mining and processing wastes, rare earth mining and extraction, oil and gas production (including fracking), and the production of phosphate fertilisers. Each of these NORM sources is discussed briefly below.

2. EXAMPLES OF NORM INDUSTRIES

2.1. Uranium mining

According to the World Nuclear Association (WNA, 2013), uranium was produced in some 20 countries in 2012 through one or more of the three main methods, i.e. underground mining, open pit mining, and in-situ leaching (ISL) (sometimes referred to as 'in-situ recovery'). Conventional mines, either underground or open

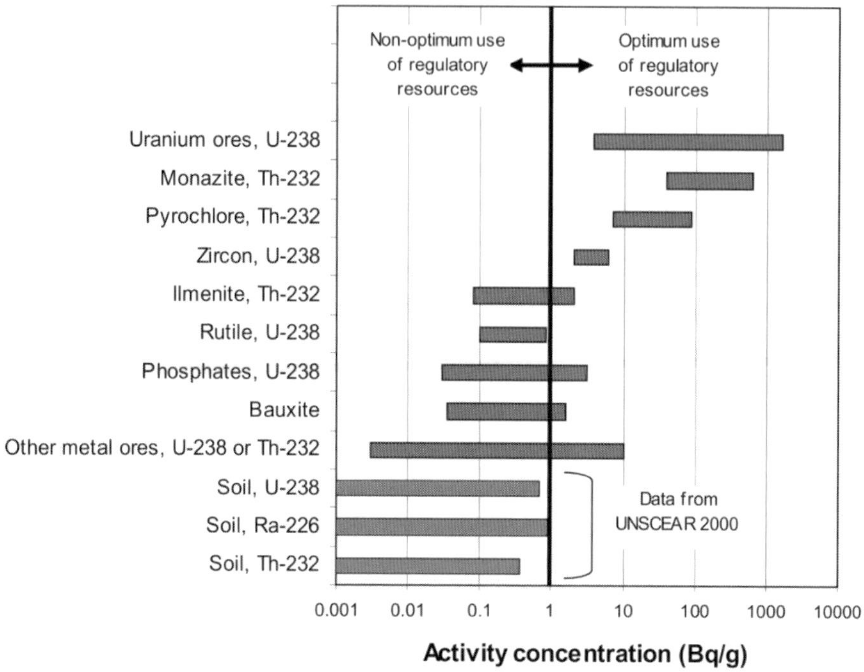

Fig. 1. Uranium and thorium in soils and ores (Wymer, 2008). UNSCEAR, United Nations Scientific Committee on the Effects of Atomic Radiation.

pit, have a mill where the ore is crushed, ground up, and then leached[1] to dissolve the uranium and separate it from the host ore. At the mill of a conventional mine or the treatment plant of an ISL operation, the uranium which is now in solution is separated by ion exchange before being precipitated, dried, and packed. This product, uranium oxide concentrate, is also referred to as 'yellowcake' and 'mixed uranium oxides' (U_3O_8). In addition, uranium can be recovered as a by-product from phosphate fertiliser production, and from mining of other minerals including copper and gold where the ores contain economic quantities of uranium. In such situations, the treatment process to recover uranium may be more complex.

During uranium mining and processing, workers may be exposed externally to gamma rays emitted from the ores, process materials, products, and tailings, and internally through inhalation of aerosol particles (inhalable dust) containing long-lived alpha activity, including radon progeny. Environmental exposures vary by region, type of mine, and processing, but include releases of dust and radon to the air; releases of uranium, radium, and other radionuclides to the surface water; and potential effects on groundwater. For modern mines and processing plants, most

[1]Depending on the mineralogy of the ore, various processes including either sulphuric acid or alkaline (carbonate) leach are employed to liberate the uranium from the host ore. The method used to recover uranium has implications for protection of both workers and the environment.

Table 1. Examples of radioactivity levels in naturally occurring radioactive material.

Source	Radionuclide(s) with highest activity concentration	Typical activity concentration (Bq g^{-1})
Monazite sand	Th-232 series	40–600
Metal ores, e.g. Nb/Ta, Cu, Au	U-238 and Th-232 series	<10
Zircon sand	U-238 series	2–4
Phosphate rock	U-238 series	0.03–3
TiO_2 feedstocks	Th-232	0.001–2
Bauxite	Th-232 series	0.035–1.4
Red mud (alumina production)	U-238, Th-232	0.1–3
Phosphogypsum (H_2SO_4 process)	Ra-226	0.015–3
Niobium extraction slag	Th-232	20–120
Tin melting slag	Th-232	0.07–15
Scale (oil and gas production)	Ra-226	0.1–15,000
Residue (rare earth extraction)	Ra-228	20–3000
Scale (TiO_2 pigment production)	Ra-228, Ra-226	<1–1600
Scale (rare earth extraction)	Ra-226, Th-228	1000
Sludge (oil and gas production)	Ra-226	0.05–800
Residue (niobium extraction)	Ra-228	200–500
Coal	U-238 and Th-232 series	0.01–0.025
Scale (coal mines with Ra-rich inflow water)	Ra-226, Ra-228	<200

Source: adapted from Table 1 of IAEA (2006).

workers receive radiation doses well below the radiation protection guidance developed by the International Commission on Radiological Protection (ICRP). Annex B of UNSCEAR (2008) provides considerable information on current levels of radiation seen by workers and the public arising from uranium recovery operations. Available information indicates that average doses to workers involved with mining and processing uranium in Canada and worldwide are less than 2–3 mSv $year^{-1}$ (Health Canada, 2008; UNSCEAR, 2008).

2.2. Metal mining and processing industries

All natural materials contain some level of uranium and thorium. Thus, it is plausible that the mining and processing of ores other than uranium may also need to consider NORM, the potential concentration of uranium or thorium during processing, and the associated radiological exposures. Examples of metal mining and processing activities involving NORM include the mining and processing of gold, aluminium, iron, zinc, copper, nickel, silver, and more (e.g. NCRP, 1993; NRC, 1999; IAEA, 2003a, 2006; EPA, 2008; UNSCEAR, 2008).

The mining and mineral processing industry is the largest producer of NORM wastes, and legacy issues associated with tailings and mine spoils arising from past uranium and other mining is a significant issue in Canada and the USA. To illustrate, according to the US Environmental Protection Agency (EPA), there are a few billion tonnes of waste, with varying concentrations of NORM, that potentially require consideration during reclamation with requirements for long-term stewardship (EPA, 2008).

2.3. Rare earth elements

All rare earth element (REE) ore deposits contain minor levels of thorium and uranium, and hence are considered as NORM. The name suggests that these elements are rare, but in fact they are relatively abundant in the earth's crust. The reason why they are considered rare is because of their geochemical properties. REEs are not found in a concentrated form, which makes them economically difficult to obtain. As the concentrations of REEs are low, greater amounts have to be mined and then purified, which increases the potential risk to workers, the general public, and the environment (IAEA, 2011; EPA, 2012).

The mining and processing strategies for REE concentrate production are resource and location specific. It is important to understand that the hydrometallurgic REE concentrate specifications typically stipulate very low levels of radioactivity, and essentially all of the radioactivity will remain in the mineral concentration rejects (tailings) and the metallurgic residues. In some cases, the radionuclide concentrations of REE can be much higher than the normal range of background levels and radiological issues. External gamma radiation exposure, for example, will be proportionately high with concentration, and may require consideration in the workplace and for managing tailings and wastes. To illustrate, Pillai (2007) reported measured gamma fields ranging from a few $\mu Gy\ h^{-1}$ to a few 100 $\mu Gy\ h^{-1}$ in a REE processing plant in India. Similar levels would be expected in North American facilities.

In some cases, it may be possible that the concentration of uranium or thorium in process intermediates, tailings, or process wastes could exceed 0.05% by weight, and hence would (in the USA) require a licence from the US Nuclear Regulatory Commission (NRC).

2.4. Phosphate

Phosphate fertiliser is used throughout the world in agriculture to help sustain food production. Phosphate rock, which contains low concentrations of NORM (primarily uranium series radionuclides), is mined and then processed to produce phosphate fertilisers.

Phosphate rock is mined using various mining techniques, and phosphate ore is separated from the rest of the ore. This separation process produces two types of waste: phosphatic clay tailings, which contain approximately 50% of the radioactivity that was present originally; and sand tailings, which contain approximately 10%

of the radioactivity that was present originally. The rest of the radioactivity is carried by the phosphate rock. During the production of phosphoric acid (wet process) and the production of elemental phosphorus (thermal process), the main by-products are phosphogypsum and phosphate slag, respectively. During the wet process, phosphogypsum retains approximately 80% of Ra-226, which is the primary radionuclide of interest (IAEA, 2013).

With the majority of the phosphogypsum in the USA centralised in Florida, the state established the Florida Institute of Phosphate Research (FIPR) to study phosphate issues that affect Florida (Birky, 2002). Studies by FIPR and others have confirmed the low risk arising from the use of by-product phosphogypsum as an agricultural soil amendment and fertiliser, as a road base, and as daily landfill cover (Birky, 2002).

There is increasing interest in using the by-product as a resource. According to Hilton (2006), phosphogypsum is growing at more than 100 million tonnes year^{-1}, and the total worldwide phosphogypsum will double within 30 or so years. The land-based stacks, such as those in central Florida, are often in prime, highly sensitive, increasingly populated areas. FIPR also initiated a global effort, working with the International Atomic Energy Agency (IAEA), referred to as 'Stack Free' to achieve eventual equilibrium between production and consumption of phosphogypsum across the full life cycle of a producing facility, with the aim of considering phosphogypsum as a resource rather than a waste (Birky, 2002; Hilton, 2006).

2.5. Oil and gas

Approximately 64% of gas-producing equipment and 57% of oil-producing equipment contains levels of radioactivity at or near background levels (EPA, 2010). Uranium and thorium isotopes are largely immobile during oil and gas extractions, but Ra-226, Ra-224, Ra-228, and Pb-210 are slightly more soluble, and may become mobilised in the fluid phases of the formation of oil and gas reservoirs that co-exist with saline water (water that has salts dissolved in it). The concentration of salts in the saline water varies depending on the length of time that the water is in contact with the rocks. Brines, which are a solution of salt in water, are brought to the surface along with the oil and gas, where they are separated and disposed. The brines, typically referred to as 'produced water', potentially contain radium from both the uranium and thorium decay chains (EPA, 1999; Zielinski and Otton, 1999; IAEA, 2003b; Gazineu et al., 2004; EPA, 2010).

A separator is used to separate streams of oil, gas, and water, dependent on their densities. The produced water is stored in water tanks and usually deposited back into disposal or recovery wells. When the water and oil mixture is brought to the surface, a radioactive scale containing radium can form inside pipes and vessels. A radioactive sludge accumulates at the bottom of oil stock and water storage tanks (WNA, 2013). Radon gas (Rn-222) is also present, both dissolved in the hydrocarbons and in an aqueous phase, and brought to the surface with the gas. Ra-222 gas can follow the processing and distribution systems, and result in elevations of Pb-210 in thin films on downstream equipment. The management of wastes

Table 2. External gamma dose rates observed near oil-producing and -processing equipment (IAEA, 2013).

Location	Dose rate (μSv h^{-1})
Down hole tubing, safety valves (internal)	<300
Wellheads, production manifold	0.1–22.5
Production lines	0.3–4
Separator (scale, measured internally)	<200
Separator (scale, measured externally)	<15
Water outlets	0.2–0.5

arising from oil and gas production can cause challenges in managing the contaminated equipment, and the associated radioactive scales and sludges. Accumulated radioactivity may also present a hazard to workers from external gamma radiation when working near the contaminated equipment, and from inhalation when opening contaminated equipment. Examples for external exposure rates can be fairly high, as illustrated from measurements made near various types of equipment (Table 2).

2.6. Fracking

Fracking is a technique to introduce fluids under high pressure into a natural-gas-containing rock formation to fracture the rock and prop open the fractures. This allows natural gas to flow more freely from the formation into wells where it is collected.

Fracking produces very large quantities of wastewater, which contain some level of NORM. Wastewater from fracking can be categorised as flowback or produced water. Flowback is essentially fracturing fluid injected into a gas well that returns to the surface when the drilling pressure is released. Water returning to the surface before the gas extraction process begins is considered under this category. Produced water refers to the wastewater emerging after the production has begun, much of which is salty water contained within the shale formation. The characteristics of the produced water depend on the geological conditions (Hammer and Van Briesen, 2012; EPA, 2014a, 2014b). Although the shales being fracked may only contain a small amount of radioactivity, radium is more soluble in the drilling fluid, which mainly contains water, in conditions found underground. As such, when wastewater is brought back up to the surface, it will potentially contain elevated levels of radium. Moreover, drilling fluids can be partially recycled, and with each re-use, radium continues to concentrate. Concentrations of NORM radionuclides in produced water can be fairly high, with concentrations reported from a few Bq l^{-1} to more than 1000 Bq l^{-1} (Warner et al., 2013; EPA, 2014).

Flowback water can mainly be managed by minimising use of produced water, recycling and re-using the water, or transporting the water off site for treatment and eventual release into ground or surface water. Land farming of produced water has the potential to create future liability, as any radionuclides in the produced water would accumulate in the soils where land farming is employed.

EPA began research under its 'Plan to Study the Potential Impacts of Hydraulic Fracturing on Drinking Water Resources' to ensure that natural gas extraction does not present undue hazards to the health of the public. The environmental scope of the project includes the full lifespan of water in fracturing. A progress report was released in 2012 (EPA, 2012).

3. REGULATION

3.1. Canada

The Canadian Nuclear Safety Commission (CNSC) is the federal agency that licenses and regulates nuclear facilities and materials. However, NORM is exempt from the CNSC's jurisdiction (except for import, export, and transport).

Policies surrounding uranium mining/milling and nuclear power generation are varied across Canada, and each province has the authority to prohibit these operations through legislation. The presence of NORM in the process materials and wastes requires that potential radiation exposures and material management be addressed. Such requirements are noted specifically within the provincial or territorial mines acts, and/or applicable provincial workplace safety regulations.

To support provincial agencies, the Federal Provincial Territorial Radiation Protection Committee introduced guidelines that apply to situations where NORM is in its natural state, and to cases in which the concentration of NORM has been increased by processing. The Canadian Guidelines for the Management of NORM state:

> *The basic principle of the Guidelines is that persons exposed to NORM should be subject to the same radiation exposure standards that apply to persons exposed to CNSC-regulated radioactive materials. No distinction is made regarding the origin of the radiation, or whether it is NORM in its natural state or NORM whose concentration of radioactive material has been increased by processing (technologically enhanced NORM or TENORM).* (Health Canada, 2000)

The incorporation of NORM guidelines to provincial mining and occupational acts is at the discretion of the provinces, although the NORM guidelines serve as the default for most provinces.

3.2. USA

Simmons (2009) provided a comprehensive review of the legal aspects of NORM and TENORM ('NORM') in the USA. Much of the following discussion is adapted from Simmons, who noted that NORM in the form of natural uranium and/or thorium is licensable source material when the concentration of uranium or thorium exceeds 0.05% by weight, and a specific licence is needed to mine or process such materials. He also commented that TENORM is currently not defined in any federal regulations of the USA, but that some states have promulgated TENORM regulations with TENORM defined broadly as 'naturally occurring radioactive materials whose radionuclide concentration has become increased by human activity'.

According to Simmons (2009), the application of regulations for NORM in the USA is not well understood. The ultimate regulation for radioactive source materials is based on the Atomic Energy Act (AEA), which has pre-emption on all other regulations that deal with health and safety for uranium and thorium source materials. Under the AEA, NRC has responsibility for 'source material', and the NRC's Part 40 regulations excluded all forms of uranium and thorium below a concentration of 0.05%. In 1961, the definition of source material was revised and only ore below 0.05% concentration was excluded from the source material definition. Chemical mixtures, compounds, solutions, or alloys below this concentration level are exempted as 'unimportant quantities' of source material. Some mineral extraction processes are licensed specifically by NRC because the concentration of materials exceeds 0.05% by weight in some part of the process.

Apart from the licence requirement noted above, and the transport of NORM that is subject to US Department of Transportation regulations (essentially the IAEA regulations), NORM is, in practice, subject primarily to state regulation control regulations. State NORM regulations are typically based on the guidelines suggested by the Conference of Radiation Control Program Directors (CRCPD). CRCPD is a US organisation of state radiation regulators formed to assist in the formulation of consistent radiation protection regulations. As part of their Suggested State Regulations for the Control of Radiation, CRCPD issued Part N on the regulation and control of TENORM (CRPCD, 2004). CRCPD does not have any legal authority over the regulation of radioactive material, including NORM; therefore, application of the CRCPD guidance is discretionary.

3.3. International

Both Canada and the USA follow the guidance developed by IAEA for the safe transport of radioactive material (IAEA, 1996, 2000, 2005).

One further aspect requires a comment. To implement its system of radiological protection, ICRP has established a dose conversion convention for radon to allow doses from exposure to radon to be added to doses from other sources of radiation exposure (IAEA, 2005). ICRP now recommends a doubling of its nominal risk coefficient for radon-induced lung cancer, based on its review of uranium miner epidemiology[2] (ICRP, 2010).

In addition, ICRP proposed that radon should be treated in the same way as other radionuclides in its system of radiological protection (i.e. using ICRP's biokinetic dose models), and provided 'nominal' dose coefficients calculated using its biokinetic dose models and 'nominal' parameter values (ICRP, 2010). The use of dose conversion

[2]ICRP (2010) also noted that the risk of lung cancer from radon for smokers is substantially greater than that for non-smokers. This is an important observation, as current radon risk projection models are all relative risk models, and risk estimation is strongly influenced by the prevalence of smoking in the reference population.

factors based on biokinetic models is very sensitive to model inputs, notably for radon decay product activity, size, and distribution of alpha particles. Unfortunately, mine environments vary widely from place to place and with time, and dosimetrically relevant parameters, notably activity size distribution, are currently very limited. The effect of this change on mine design and operations is not fully understood at present, but will certainly present challenges for the foreseeable future.

4. CONCLUSIONS

Uranium and thorium are ubiquitous in soil, rocks, and ores. Thus, mining and many industrial processes will have NORM as an integral part of their activities. NORM industries considered in this paper are uranium mining, metal mining and processing industries, REE mining and processing, phosphate fertiliser, oil and gas production, and fracking. Exposures and types of radionuclides will vary between industries, but the most prominent NORM radionuclides arise from the uranium decay series and thorium decay series. The levels of uranium and thorium series radionuclides associated with NORM industries vary quite widely; even for industries where the initial concentrations of radionuclides in the ores or feeds are fairly low, there is potential for concentration of the radionuclides at various locations in process streams or wastes. Thus, the presence of NORM and consideration of the potential types and levels of radiation exposure should be an integral part of health and safety planning.

The national nuclear regulators in Canada and the USA do not regulate activities with uranium and thorium below 0.05% by weight. However, in both Canada and the USA, guidance for radiation protection from NORM is available and, broadly speaking, is followed by the provinces and states. Both countries follow the guidance for transport of radioactive materials developed by IAEA.

The work of ICRP is important as radiation protection guidance is typically based on their generic guidance. Thus, it is important to follow the work of ICRP and provide comments to ICRP on proposed new guidance as it is developed. IAEA also provides guidance for many NORM industries on the practical application of radiation protection principles.

REFERENCES

Birky, B., 2002. Radiation Issues Associated with Phosphogypsum Use. Global Gypsum Conference, 4–5 March 2002, Miami, FL, USA.

CRPCD, 2004. Guidelines for NORM CRCPD. Conference of Radiation Control Program Directors, Part N, Regulation and Licensing of Technologically Enhanced Naturally Occurring Radioactive Material (TENORM). Suggested State Radiation Control Regulations, Volume I. April 2004, Frankfort, KY, USA.

EPA, 1999. Technologically Enhanced Naturally Occurring Radioactive Materials in the Southwestern Copper Belt of Arizona. Washington, DC: US Environmental Protection Agency.

EPA, 2008. Technical Report on Technologically Enhanced Naturally Occurring Radioactive Materials from Uranium Mining, 2 Volumes. US Environmental Protection Agency.

EPA, 2010. Oil and Gas Production Wastes. Washington, DC: US Environmental Protection Agency. Available at: http://www.epa.gov/radiation/tenorm/oilandgas.html (accessed 5 March 2015).

EPA, 2012. Rare Earth Elements: a Review of Production, Processing, Recycling, and Associated Environmental Issues. EPA 600/R-12/572. US Environmental Protection Agency.

EPA, 2014a. Study of Hydraulic Fracturing and its Potential Impact on Drinking Water Resources. Washington, DC: US Environmental Protection Agency. Available at: http://www2.epa.gov/hfstudy (last accessed 4 November 2014).

EPA, 2014b. The Hydraulic Fracturing Water Cycle. Washington, DC: US Environmental Protection Agency. Available at: http://www2.epa.gov/hfstudy/hydraulic-fracturing-water-cycle (last accessed 11 April 2014).

FIPR. http://www.fipr.state.fl.us/about-fipr-general.htm.

Gazineu, M.H.P., de Araújo, A.A., Brandão, Y.B., Hazin, C.A., 2004. Radium-226 and radium-228 in scale and sludge generated in the petroleum industry.

Hammer, R., Van Briesen, J., 2012. In Fracking's Wake: New Rules are Needed to Protect our Health and Environment from Contaminated Wastewater. Natural Resources Defense Council, New York.

Health Canada, 2000. Canadian Guidelines for the Management of Naturally Occurring Radioactive Material (NORM). Health Canada, Ottawa.

Health Canada, 2008. 2008 Report on Occupational Radiation Exposures in Canada. Health Canada, Ottawa. Available at: http://publications.gc.ca/ collections/collection_2012/sc-hc/H126-1-2008-eng. pdf (last accessed 4 November 2014).

Hilton, J., 2006. Phosphogypsum Management and Opportunities for Use. Paper Presented at the International Fertiliser Society Conference, 14 December 2006, Cambridge, UK.

IAEA, 1996. Regulations for the Safe Transport of Radioactive Material (amended in 2003). International Atomic Energy Agency, Vienna.

IAEA, 2000. Regulations for the Safe Transport of Radioactive Material – 1996 Edition (Revised). IAEA Safety Standards Series No. TSR- 1 (ST-1, Revised). International Atomic Energy Agency, Vienna.

IAEA, 2003a. Extent of Environmental Contamination by Naturally Occurring Radioactive Material (NORM) and Technological Options for Mitigation. Technical Report Series No. 419. International Atomic Energy Agency, Vienna.

IAEA, 2003b. Radiation Protection and Management of Radioactive Waste in Oil and Gas Industry. Safety Reports Series No. 34. International Atomic Energy Agency, Vienna.

IAEA, 2005. Regulations for the Safe Transport of Radioactive Material – 2005 Edition. IAEA Safety Standards Series No. TS-R-1. International Atomic Energy Agency, Vienna.

IAEA, 2006. Assessing the Need for Radiation Protection Measures in Work Involving Minerals and Raw Materials. Safety Report Series 49. International Atomic Energy Agency, Vienna.

IAEA, 2011. Radiation Protection and NORM Residue Management in the Production of Rare Earths from Thorium Containing Minerals. Safety Report Series 68. International Atomic Energy Agency, Vienna.

IAEA, 2013. Radiation Protection and Management of NORM Residues in the Phosphate Industry. Safety Reports Series No. 78. International Atomic Energy Agency, Vienna.

ICRP, 2010. Lung cancer risk from radon and progeny and statement on radon. ICRP Publication 115. Ann. ICRP 40(1).

NRC, 1999. Committee on Evaluation of EPA Guidelines for Exposure to Naturally Occurring Radioactive Materials. Evaluation of Guidelines for Exposures to Technologically Enhanced Naturally Occurring Radioactive Materials. National Academies Press, Washington, DC.

NCRP, 1993. Radiation Protection in the Mineral Extraction Industry. NCRP Report No. 118. National Council on Radiation Protection and Measurements,Washington, DC.

Pillai, P.M.B., 2007. Naturally Occurring Radioactive Materials (NORM) in the Extraction and Processing of Rare Earths. Proceedings of the International Symposium on NORM, 19–22 March 2007, Seville, Spain.

Simmons, C., 2009. Legal aspects of TENORM regulations. In: Karam, P.A., Vetter, B.J. (Eds.), Naturally Occurring Radioactive Materials (NORM) and Technologically Enhanced NORM (TENORM). Medical Physics Publishing, Madison, WI.

UNSCEAR, 2000. Sources and Effects of Ionizing Radiation 2000. UNSCEAR 2000 Report to the General Assembly, with Scientific Annexes, Vol. 1. Sources. United Nations, New York.

UNSCEAR, 2008. Annex B Exposures of the Public and Workers from Various Sources of Radiation. United Nations, New York.

Warner, N.R., Christie, C.A., Jackson, R.B., Vengosh, A., 2013. Impacts of shale gas wastewater disposal on water quality in Western Pennsylvania. Environ. Sci. Technol 47, 11849–11857.

WNA, 2013. Naturally Occurring Radioactive Material. World Nuclear Association, London. Available at: http://world-nuclear.org/info/Safety-and-Security/Radiation-and-Health/Naturally-Occurring-Radioactive-Materials-NORM/#.UjByNn-0RSM (last accessed 4 November 2014).

Wymer, D.G., 2008. Managing Exposure to NORM – Consensus or Chaos. Keynote Address in Proceedings of an International Symposium on Naturally Occurring Radioactive Material (NORM V), 19–22 March 2007, Seville, Spain.

Zielinski, R.A., Otton, J.K., 1999. Naturally Occurring Radioactive Materials (NORM) in Produced Water and Oil-field Equipment: an Issue for Energy Industry. US Department of the Interior, US Geological Survey.

SAGE

ICRP
Annals of the ICRP

Construction of a naturally occurring radioactive material project in the BeAAT hazardous waste facilities

H. Abuahmad

Abu Dhabi National Oil Company, HSE Division, P.O. Box 898, Abu Dhabi, United Arab Emirates; e-mail: habuahmad@adnoc.ae

Abstract–Naturally occurring radioactive material (NORM) is produced during exploration and production operations of subsidiaries of the Abu Dhabi National Oil Company (ADNOC) in the United Arab Emirates, and accumulates in drilling tubulars, plant equipment, and components. These NORM hazardous wastes need to be managed in such a way that they do not damage human health and the environment. The primary radionuclides of concern in the oil and gas industries are radium-226 and radium-228. These radioisotopes are the decay products of uranium and thorium isotopes that are present in subsurface formations from which hydrocarbons are produced. While uranium and thorium are largely immobile, radium is slightly more soluble and may become mobilised in the fluid phases of the formation (International Association of Oil & Gas Producers, 2008). In order to treat and dispose of NORM waste products safely, ADNOC's subsidiary 'TAKREER' is developing a new facility, on behalf of all ADNOC subsidiaries, within the existing Central Environmental Protection Facilities (BeAAT) in Ruwais city. The NORM plant is envisaged to treat, handle, and dispose of NORM waste in the forms of scale, sludge, and contaminated equipment. The NORM treatment facility will cover activities such as decontamination, volume reduction, NORM handling, and concrete immobilisation of NORM waste into packages for designated landfilling.

Keywords: NORM; Uranium; Thorium; BeAAT

This paper does not necessarily reflect the views of the International Commission on Radiological Protection.

1. UNITED ARAB EMIRATES REGULATIONS

1.1. The Federal Authority for Nuclear Regulation Established by Law 6 of 2009

Regulation No. 24 within Law 6 addresses basic safety standards for facilities and activities involving ionising radiation other than in nuclear facilities (it defines thresholds, licensing, etc.), and applies to the following:

- Ra-226 and Ra-228 activity threshold $>10\,\text{Bq}\,\text{g}^{-1}$ and activity of $>1\times10^5$ Bq for radioactive material of <1 ton; and
- radioactive material in a bulk amount for which the activity concentration of each radionuclide in the U and Th decay series $>1\,\text{Bq}\,\text{g}^{-1}$, or $>10\,\text{Bq}\,\text{g}^{-1}$ for K-40.

There is exemption from licensing when:

- the effective dose does not exceed 1 mSV year^{-1}; or
- regulatory control of the planned exposure situation would provide no net benefit.

2. SELECTION OF ALTERNATIVES FOR TREATMENT AND DISPOSAL OF NORM

2.1. Disposal objectives

The guiding disposal objectives for selecting the best treatment and disposal (T&D) processes for use at BeAAT were determined to be (URS Corp, 2010):

- containment (from erosion, intrusion, dispersal);
- isolation ($<$0.3–1 mSv year^{-1} doses by blocking gamma, radon, and groundwater pathways);
- permanence (minimise maintenance and future liability);
- longevity ($>$200 years); and
- cost-effectiveness.

2.2. Method screening

T&D of NORM-impacted wastes includes segregation/volume reduction methods, treatment/solidification methods, and encapsulation/disposal methods. A variety of methods were identified and screened in terms of relative cost, effectiveness, and compatibility with each other, with the waste streams, and with the BeAAT environment. Three tiers (low, medium, and high) of each of the three parameters (cost, effectiveness, and compatibility) were used to rank each method. Scores from 1 to 3 were assigned for each parameter (with reverse order for costs), and added to estimate a ranking for each method.

2.2.1. Segregation/volume reduction

Twelve segregation and volume reduction methods used or proposed for use with oilfield NORM wastes were identified and ranked. Five of the methods separate NORM from equipment: hydrolysing, high-pressure washing, rattling, reaming, and chemical dissolution. Six other methods separate water and oils from solid NORM wastes: centrifuging, flocculating, solvent extraction, incineration, cyclone separation, and evaporation pond. One method is for volume reduction (compaction) and one method is for liquid phase separation (oil–water separation). The most useful methods for the waste streams at BeAAT were determined to be hydrolysing, high-pressure washing, incinerating, cyclonic separation, evaporation pond, and oil–water separation.

2.2.2. Treatment/solidification methods

Three solidification methods were assessed with the same ranking method. These included concrete cube solidification, grout solidification, and vitrification. The concrete cube and grout methods are similar, involving use of NORM-contaminated water to make concrete or grout, and incorporating oil-free NORM solids (from sludge and scale incineration) into the concrete or grout mix. Large (1 m^3) cubes currently made at BeAAT for hazardous wastes will use up to 40% high-NORM scale in the mix, along with gravel and sand aggregates to make low-permeability monoliths. Grout will use a similar low-permeability mix with lower-NORM sludge solids, targeting flow ability for use as backfill in and around equipment, and in making low-NORM monoliths that could be disposed of directly. Vitrification scored lowest because of its potentially lower suitability for NORM wastes and high energy, as well as handling costs for large waste volumes.

2.2.3. Encapsulation/disposal methods

Nine encapsulation or disposal methods were evaluated. These included concrete vaults, high-integrity containers, bitumen membranes, bitumen volumetric fill, polymerised plastics, grouting (tubing and equipment), containment in equipment, landfill cover/liner systems, and well injection with hydrofracture. The most useful in this category were identified as concrete vaults, routing equipment cavities, and landfill burial. Bitumen surface coatings were also considered for deflecting any infiltrating water from concrete/grout monoliths, primarily for their low costs. Injection well disposal also ranked highly, but this was not considered to be feasible because there are no known opportunities to use it in the BeAAT area.

2.3. Optimum treatment and disposal processes for BeAAT

The combinations of the most promising processes were examined to develop a T&D strategy best suited to the ADNOC wastes and the BeAAT environment. The prime disposal option for meeting the disposal objectives and using the BeAAT site was determined to be landfill disposal. While the dry Ruwais environment is ideal for landfill disposal, the relatively shallow water table could be problematic. Secondary containment in low-permeability concrete canisters was therefore considered to be important in meeting the longevity and isolation objectives. Conversion of the wastes into low-permeability waste forms, such as concrete cubes and grout ingots, was chosen as an additional measure to reduce risks from leaching of radium into the groundwater. Incineration of both sludge and scale wastes appears to be the simplest and least-expensive means of removing hydrocarbons sufficiently from the raw waste streams to render the solid residuals suitable as ingredients for concrete and grout. For NORM-contaminated equipment, a combination of hydrolasing for pipes and tubing, and high-pressure washing for tanks and other vessels offers the most efficient moisture removal with good containment of the removed NORM. Water associated with equipment and facility cleanout, and the incinerator scrubber will be filtered and recycled as much as possible, but will use an evaporation pond to accommodate excess volumes that cannot be used in concrete and grout mixes.

2.3.1. High-NORM treatment and disposal

The combination of processes for the highest-NORM scale wastes include hydrolasing to remove them from pipes and tubing, supplemented by high-pressure washing of scale in separators and other non-tubular equipment. A hydrocyclone separator and filter will remove most scale from the wash water, allowing much of it to be re-used for equipment cleaning. Excess water will be drained to an evaporation pond after removal of solids, and an incinerator will be used to burn off residual hydrocarbons from the drained scale solids. The incinerator scrubber will use NORM-contaminated water from the evaporation pond. Dried scale solids will be moistened with evaporation pond water for safe transfer from the incinerator to the concrete batch plant, where they will be mixed with clean aggregate, evaporation pond water, cement, and pozzolans to form low-permeability concrete. The concrete will be cast and cured in 1-m^3 cubes, similar to those used at present for hazardous wastes at BeAAT. The cubes will be loaded into concrete canisters, grouted to fill residual void spaces. Closed and sealed canisters will be placed in the Class 1 landfill, where their top and side surfaces will be coated with a bitumen moisture barrier to deflect any infiltrating water. Canisters will be covered periodically by flowable fill. The covered canisters will finally be isolated by an engineered cover system for closure of the landfill.

2.3.2. Low-NORM treatment and disposal

A similar combination of processes will be used for low-NORM sludge wastes. These are generally removed from equipment, where necessary, by high-pressure washing. An oil–water separator will be used to separate excess oil from the wash water before it is drained to the evaporation pond. Sludge solids will be burned in the incinerator and treated in the same way as the scale solids, except that the sludge solids will primarily be used to prepare low-permeability, flowable grout. The grout will be used to fill voids in equipment being disposed, concrete canisters, and to seal canister lids. Excess grout material will be cast into cubic or ingot monoliths for disposal in the landfill outside the concrete canisters. The ingots will be coated with a bitumen moisture barrier before burial in the landfill.

2.4. Injection well disposal

Based on the US experience, it may be more cost-effective by reducing waste handling and processing it for deep well injection. This technology requires different treatment equipment such as tanks and pumps to support injection, hydro-fracture, and monitoring. However, at present, no available or suitable deep-disposal formations have been identified in the BeAAT vicinity that could potentially have satisfied the containment, isolation, permanence, and longevity criteria for NORM wastes.

2.5. Chemicals

No liquid chemicals are used for NORM removal, and no spent-chemical wastes are generated. Solid products used for concrete and grout include cement powder, pozzolans, and water reducers. Solid bitumen is melted as necessary for coating concrete and grout surfaces, and small quantities of solvent may be required for bitumen equipment clean-up. Rags and rubbish contaminated with bitumen and solvent will be incinerated with the sludge waste stream.

Welding/cutting gases will be required to support equipment cutting operations. Dust-suppressing agglomeration chemicals may be used as necessary to reduce dust resuspension from NORM-contaminated materials.

2.6. Support facilities

The NORM T&D facility at BeAAT will require a variety of supporting equipment. This includes fork-lifts, crane and rigging, front-loaders, and excavation/construction equipment for handling and moving materials, supplies, equipment, and waste cubes/ingots, as well as for excavation and preparation of landfills. Pumps, filters, cyclone separators, and tubing will be required to support the pressure washing equipment, incinerator scrubber, concrete plant, and evaporation pond. Gamma scintillation meters and beta/gamma meters will be needed for equipment surveys and assessing surface contamination, respectively. Air sampling/monitoring equipment will be required in the equipment cleanout, incinerator, and concrete mixing areas, supported by water sprayers and antidusting additives as needed. Cutting/welding equipment will be needed to open equipment such as separators for internal decontamination. Drainable concrete pads will be needed for equipment staging, cleaning, and supply storage.

2.7. Treatment capacity

The NORM T&D operation is sized to accommodate quantities currently stored as historical waste and the wastes produced in the first year. The T&D operation would handle approximately 370 m^3 of contaminated equipment (plus contingency), 450 m^3 of sludge, and 50 m^3 of scale solids. Disposal capacity allows for disposal of 300 m^3 of contaminated equipment, 130 m^3 of high-NORM concrete, and 520 m^3 of low-NORM grout in 50 15-m^3 concrete vaults. An additional 610 m^3 of low-NORM grout ingots would be disposed of in landfill outside of the concrete vaults. The landfill dimensions are estimated at 30 m x 30 m x 3 m, and the evaporation pond dimensions are estimated at 35 m x 35 m x 2 m. The concrete batch plant accommodates 5-m^3 batches, and the incinerator accommodates approximately 10 m^3 per batch.

3. SAFE NORM HANDLING AND TRANSPORTATION

3.1. NORM handling

Safe handling of NORM-impacted materials requires training and awareness, worker protection, and contamination control. Graded training requirements include NORM awareness training for all BeAAT workers, technical training for workers who clean equipment and handle NORM, and supervisory training for NORM T&D managers and radiation safety officers. Worker protection includes continuing training and safety meetings, isolation of operating areas, and restriction of operating areas to essential personnel. Contamination control includes published waste acceptance criteria, advance notification/scheduling of waste shipments, clear labelling/signage for equipment and operating areas, dust control guards and moist-handling of NORM solids, operation only with trained personnel with proper equipment and personal protective equipment, monitoring for landfill leachate and evaporation pond leakage, and secure transportation of NORM wastes to the site and within the site.

3.2. NORM transportation

Shipments of NORM-impacted wastes to the site should be, as per international standards (IAEA, 2005), regarded as low-specific activity (LSA)-1 or surface study contaminated object (SCO)-1 materials, unless they are determined to be below the 100 Bq g^{-1} Ra-226 or Ra-228 threshold, which is the case for ADNOC's waste. These categories allow minimal packaging provided that:

- there is no escape of radioactivity under routine transport conditions;
- material is transported in exclusive-use vehicles; and
- exclusive-use vehicles are not required for SCO-1 materials with minimal surface contamination.

Other important transportation considerations include advance notification and planning; provision of emergency contingency plans to the vehicle operator; use of appropriate placards and signage; records of shipment description, contents, waste form, volume, and quantity of NORM; and use of strong, tight transportation drums, boxes, and bulk containers. Wastes may not contain free water or liquids that could drain, should the containment leak or become damaged. Contaminated equipment may be shipped without packaging provided that all openings to the contamination are secured by threaded caps or other durable tight closures to prevent leakage.

4. CONCLUSIONS

By developing, designing, and implementing this NORM T&D project, ADNOC was able to solve a historical dilemma in dealing with NORM waste. This solution is a transferable example for adaptation by oil and gas operating companies in the Middle East.

REFERENCES

International Association of Oil & Gas Producers, 2008. Guidelines for the Management of Naturally Occurring Radioactive Material (NORM) in the Oil & Gas Industry. Report No. 412. London.

IAEA, 2005. Regulations for the Safe Transport of Radioactive Material – 2005 Edition. IAEA Safety Standards Series No. TS-R-1. International Atomic Energy Agency, Vienna.

URS Corp, 2010. Handling, Treatment and Disposing of NORMs Report 9/2010, for ADNOC. Abu Dhabi.

SAGE

ICRP
Annals of the ICRP

Use of effective dose in medicine

J. Harrison[a], P.O. Lopez[b]

[a]*Public Health England, Centre for Radiation, Chemical and Environmental Hazards, Chilton, Didcot, Oxon OX11 0RQ, UK; e-mail: john.harrison@phe.gov.uk*
[b]*International Atomic Energy Agency, Vienna*

Abstract–The protection quantity 'effective dose' was developed by the International Commission on Radiological Protection (ICRP) for use in the radiological protection of workers and the public. In this context, it is used as a risk-adjusted dosimetric quantity to optimise protection, comparing received or planned doses with constraints, reference levels, and limits expressed in the same quantity. Considering exposures incurred during medical procedures, effective dose can be of practical value for comparing: doses from different diagnostic examinations and interventional procedures; the use of similar technologies and procedures in different hospitals and countries; and the use of different technologies for the same medical examination, provided that the representative patients or patient populations for which the effective doses are derived are similar with regard to age and sex. However, as stated in ICRP *Publication 103*, '... risk assessment for medical diagnosis and treatment... is best evaluated using appropriate risk values for the individual tissues at risk and for the age and sex distribution of the individuals undergoing the medical procedures'. This topic was explored in a session of the First ICRP Symposium with arguments for and against the use of a new quantity referred to as 'effective risk', and examination of variations in estimated risk for different diagnostic procedures according to the age and sex of the exposed individuals. This paper restates the primary purposes of effective dose, and summarises estimates of variation in individual risk from medical procedures. The authors support the judicious use of effective dose as an indicator of possible risk, but caution against the use of effective risk as compared with the calculation of scientific best estimates of risk with consideration of associated uncertainties.

Keywords: Effective dose; Effective risk; Nominal risk coefficients

This paper does not necessarily reflect the views of the International Commission on Radiological Protection.

1. INTRODUCTION

Effective dose was originally introduced in the 1977 Recommendations of the International Commission on Radiological Protection (ICRP, 1977), at which time the over-riding concern was the control of occupational exposures. While the concept has remained essentially unchanged through the 1990 Recommendations (ICRP, 1991) to the 2007 Recommendations (ICRP, 2007), its use has been extended to members of the public of all ages, including in-utero exposures of the embryo and fetus. Particularly problematic is the use of effective dose in the assessment of medical exposures of patient populations of different ages, and where the requirement is for a measure of risk to individuals (Martin, 2007; Brenner, 2008).

Brenner (2008, 2012) proposed that effective dose represents flawed science, and should be replaced with a new quantity, ‘effective risk’, which could be used to provide estimates of risk to individuals from medical procedures, taking account of differences between males and females and at different ages at exposure. Wall et al. (2011) and Balonov and Shrimpton (2012) provided estimates of age- and sex-specific risk per unit effective dose for a range of medical diagnostic procedures to determine variations in comparison with the nominal risk coefficients derived by ICRP.

This paper outlines the purpose of effective dose and the underlying assumptions made in its calculation, and examines the proposal that it should be replaced by effective risk. Results obtained for the risk per unit effective dose for diagnostic x-ray procedures are summarised, with commentary on the use of effective dose as an indicator of risk.

2. EFFECTIVE DOSE

The ICRP protection quantities – equivalent and effective dose – enable the summation of doses from internal emitters and from external sources to provide a single number for comparison with dose limits, constraints, and reference levels that relate to stochastic effects of whole-body radiation exposure. Thus, the primary application of effective dose is in the planning and demonstration of compliance in various situations of exposure of workers and members of the public. The calculation of effective dose can be seen as a three-step process, starting with the calculation of absorbed dose to organs and tissues, in gray (Gy; J kg^{-1}). As radiation types differ in their ability to cause cancer per unit of absorbed dose, the second step is to multiply the calculated values of absorbed dose by radiation weighting factors that take account of the greater effectiveness of alpha particles and neutrons compared with beta particles and gamma rays. The results are termed ‘equivalent dose’, and expressed in sieverts (Sv). The final step is to sum the equivalent doses to individual organs and tissues, multiplying each by a tissue weighting factor that represents its contribution to total detriment from uniform whole-body irradiation. The intention is that the overall risk should be comparable, irrespective of the type and distribution of radiation exposure. Effective dose, in sieverts, is the well-known quantity that is often referred to simply as ‘dose’.

It is important to recognise that while effective dose is a risk-related construct, and is ideally suited for use in radiation protection, particularly in planning and optimising protection for workers and members of the public, it does not provide more than a rough estimate of dose and risk to individuals (Dietze and Menzel, 2004; ICRP, 2007; Dietze et al., 2009). Absorbed doses to organs and tissues are calculated in mathematical phantoms and used to provide sex-averaged values of effective dose for a 'reference person' (ICRP, 2007; Harrison and Day, 2008). Radiation and tissue weighting factors are chosen as simple sets of values that do not fully reflect our scientific understanding of radiation risks. For example, no account is taken of recognised differences between low-energy mammography x rays and Co-60 gamma rays (Hill, 2004). There is evidence that the relative effectiveness of different radiations is dependent on cancer type and may, for example, be greater for liver cancer than for leukaemia (Harrison and Muirhead, 2004). Tissue weighting factors are age- and sex-averaged values that conceal substantial differences between cancer risk estimates for males and females, and at different ages, dependent on cancer type (ICRP, 2007; Harrison and Day, 2008).

Effective dose and the tissue weighting factors used in its calculation relate to detriment from radiation-induced cancer and hereditary effects. Detriment is calculated as defined by ICRP and explained in detail in Annex A of *Publication 103* (ICRP, 2007). The main source of data on cancer risks is the follow-up studies of the Japanese atomic bomb survivors (A-bomb data), used to derive risk coefficients averaged over seven Western and Asian populations with different background cancer rates. The most recent ICRP (2007) calculations use cancer incidence data, adjusted for lethality, loss of quality of life, and years of life lost. In applying the risk factors obtained from epidemiological studies to exposures at lower doses and dose rates, ICRP applies a dose and dose-rate effectiveness factor of 2 for solid cancers, and uses a linear-quadratic model for leukaemia. Weighting for hereditary effects is based on estimates of disease occurring in the first two generations. ICRP has published nominal values of radiation detriment for an averaged world population, giving values for all ages (members of the public) and adults (workers).

The application of effective dose in the control of stochastic effects for protection purposes relies on a number of key assumptions, principally that:

- a linear non-threshold (LNT) relationship between dose and risk applies at low doses;
- acute and chronic exposures are equally effective at low doses; and
- external dose and internal dose from radionuclides deposited in body tissues can be summed, taking account of radiation quality through simple adjustments using radiation weighting factors.

3. EFFECTIVE RISK

In medical applications, effective dose can be used to compare doses from different diagnostic examinations and interventional procedures, the use of similar technologies, or the use of different technologies for the same procedure, when the patient populations are similar with regard to age and sex (ICRP, 2007; Fazel et al., 2009). However, as stated in the 2007 Recommendations, '... risk assessment for medical diagnosis and treatment... is best evaluated using appropriate risk values for the individual tissues at risk and for the age and sex distribution of the individuals undergoing the medical procedures'. Brenner (2008, 2012) proposed that effective dose should be replaced with a new quantity, 'effective risk', which could be used to provide estimates of risk to individuals from medical procedures, taking account of differences in risk between males and females and at different ages at exposure. Brenner (2012) referred to the risk estimates derived by NAS/NRC (2006) as published in the BEIR VII report, a selection of which is given in Table 1. These estimates were derived from the A-bomb data for the US population, and show that, in general, risk estimates are approximately double for irradiation in infancy compared with adults aged 20 years, and approximately five to six times greater for thyroid cancer. Risks at older ages are substantially lower in most cases.

Data of the type shown in Table 1 were also used by ICRP in their calculation of stochastic detriment and tissue weighting factors (see above), although with risks transferred to a composite of seven populations rather than a US population. Effective risk would not be a useful replacement for effective dose for its principal purpose of providing a single metric for control of exposures worldwide. An important argument against the effective risk quantity concerns the dose range to which this quantity would be applied. In radiation protection practice, the effective doses recorded are generally far below the 20-mSv limit for occupationally exposed persons, often as low as 10 μSv or even less. Dose measurements and assessments are well established in this dose range and, particularly for external exposures, quality

Table 1. Lifetime attributable risk of specific cancers after irradiation at different ages. Number of cases per 10^6 exposed to a single dose of 10 mGy. Selected data from BEIR VII (NRC/NAS, 2006).

	Age at exposure (years)					
	Males			Females		
Cancer site	0	20	60	0	20	60
Breast	–	–	–	1171	429	31
Colon	336	173	94	220	114	62
Liver	61	30	14	28	14	7
Lung	314	149	89	733	346	201
Thyroid	115	21	0.3	634	113	1
Leukaemia	237	96	82	185	71	57
All cancers	2563	977	489	4777	1646	586

assurance programmes ensure reliable measurements, even at such low doses. In addition, the operational dose quantities defined for external exposure and used for calibrating dosimeters are taken to provide sufficiently accurate assessments of effective dose for the purposes of radiation monitoring and dose recording. However, while doses can be measured with some precision at such low levels, the corresponding risks are uncertain or even unknown.

The protection system relies on the assumption of an LNT dose–response relationship. The LNT assumption is implicit in the addition of external and internal doses of different magnitudes, with different temporal and spatial patterns of delivery. However, a clear distinction should be drawn between the necessary and justifiable reliance on the LNT dose-response relationship for protection purposes and the assumption of its scientific validity extending to very low doses and all stochastic risks. ICRP (2007) concluded that the validity of the LNT dose–response assumption may prove to be beyond resolution for the foreseeable future. Hence, individual 'risk monitoring' at low doses can be judged to be unhelpful from a protection perspective, and scientifically misleading in implying greater knowledge of the risks of low-dose exposures than is currently warranted.

4. DOSES AND RISKS FROM DIAGNOSTIC PROCEDURES

Wall et al. (2011) and Balonov and Shrimpton (2012) estimated radiation risks from a range of medical x-ray examinations (radiography, fluoroscopy, and computed tomography) as a function of the age and sex of the patient, applying the risk models used by ICRP (2007) and UNSCEAR (2006). Risk estimates based on typical organ doses and age- and sex-specific risk factors for individual cancer types were compared with values derived from effective dose using ICRP nominal risk coefficients. Table 2 shows typical doses received in examinations in the UK, ranging from approximately 10 mSv effective dose for some computed tomography procedures to

Table 2. Typical doses from x-ray examinations of adults in the UK (from Balonov and Shrimpton (2012).

Type of examination		Effective dose (mSv) 1990	Effective dose (mSv) 2007	Highest organ dose: Organ	Highest organ dose: mGy
Radiography	Foot	0.0002	0.0002	Skin	0.007
	Head (skull)	0.05	0.07	Salivary glands	1.3
	Intravenous urography	2.3	2.1	Stomach	6.9
Fluoroscopy	Ba follow	1.5	1.3	Kidneys	6.1
	Coronary angiography	3.9	3.9	Lungs	15
Computed tomography	Head	1.6	1.4	Brain	45
	Chest + abdomen + pelvis	9.2	10	Thymus	15

microsievert doses for peripheral x-ray examinations. Lifetime risks of radiation-induced cancer, evaluated on the basis of typical organ doses, were shown to decrease with patient age for all examinations. However, the rate of change differed markedly between examinations and between males and females, reflecting differences in the changing radiosensitivity of organs and tissues. Overall, the analyses showed that using effective dose and nominal population risk coefficients, as published by ICRP, may underestimate risk for children (aged 0–9 years) by up to a factor of four, and overestimate risk for older patients (aged >60 years) by approximately one order of magnitude. Balonov and Shrimpton (2012) concluded that although effective dose was not intended to provide a measure of risk associated with medical x-ray examinations of individual patients, simple adjustments to nominal risk coefficients to take account of age and sex differences might make it a useful instrument when considering the justification of examinations.

5. DISCUSSION

Effective dose is a radiation protection quantity that is used to set limits, constraints, and reference levels that apply to reference workers or reference members of the public. It provides an elegant solution to the requirement for a single quantity that enables the summation of all radiation exposures, from external exposures and from radionuclides entering the body by inhalation or ingestion. Effective dose is used as a risk-related quantity for the optimisation of protection below constraints and reference levels. For protection purposes, the reasonable assumption is made of an LNT dose–response relationship down to very low doses of a few microsieverts or even less. However, while doses can be calculated with reasonable reliability at such low levels, the associated risk is uncertain.

Effective dose has proved to be a useful tool in controlling exposures received by patients undergoing medical diagnostic and interventional procedures. However, its use to provide estimates of risk to individual patients goes beyond its intended use (ICRP, 2007; Menzel and Harrison, 2012). Brenner (2008, 2012) suggested that effective dose should be replaced by effective risk as a more scientifically based quantity. Unfortunately, this approach ignores the uncertainties associated with the LNT dose–response relationship as applied to all cancer types, and gives a misleading impression of our ability to quantify risks at low doses.

A clear distinction should be drawn between the application of science in the definition of the risk-adjusted protection quantity, 'effective dose', which has proved to be of considerable practical utility, and the use of best scientific information to calculate risks to individuals and population groups. Risk estimation can be specific to individuals of known body mass and dimensions, applying age- and sex-specific risk factors derived for the most appropriate population group. An example referred to by Harrison (2015) in this issue is the National Aeronautics and Space Administration's estimates of risk to individual astronauts. Such scientific best estimates of risk should also consider associated uncertainties.

Although effective dose was not intended as a measure of risk to individuals, it might reasonably be used as an approximate indicator of risk. Considering a range of medical x-ray examinations and comparing best estimates of risk with values obtained using effective dose and nominal risk coefficients, Balonov and Shrimpton (2012) showed that effective dose might underestimate risk by up to a factor of four for children, and overestimate risk by approximately one order of magnitude for older adults. Making such simple adjustments to the risk associated with effective dose could help make informed judgements on the justification of examinations without implying greater knowledge of radiation risks at low doses than is warranted.

ICRP currently has a task group responsible for preparing a report on the use of effective dose (Harrison, 2015). This task group, led by Committee 2, has membership from Committees 1, 3, and 4, as well as external experts. Useful guidance on restrictions on the use of the quantity is provided in Annex B to the 2007 Recommendations (ICRP, 2007). This guidance needs to be expanded and proposals made for the control of exposures and risk management in situations where 'effective dose' should not be used. An important focus of the report will be medical exposures.

REFERENCES

Balonov, M.I., Shrimpton, P.C., 2012. Effective dose and risks from medical x-ray procedures. Ann. ICRP 41(3/4), 129–141.

Brenner, D.J., 2008. Effective dose: a flawed concept that could and should be replaced. Br. J. Radiol. 81, 521–523.

Brenner, D.J., 2012. We can do better than effective dose for estimating or comparing low-dose radiation risks. Ann. ICRP 41(3/4), 124–128.

Dietze, G., Harrison, J.D., Menzel, H-G., 2009. Comments on the paper of DJ Brenner 'Effective dose: a flawed concept that could and should be replaced'. Br. J. Radiol. 82, 348–351.

Dietze, G., Menzel, H-G., 2004. Dose quantities in radiation protection and their limitations. Radiat. Prot. Dosim. 112, 457–463.

Fazel, R., Krumhholz, H.M., Wang, Y., et al., 2009. Exposure to low-dose ionizing radiation from medical imaging procedures. N. Engl. J. Med. 361, 849–857.

Harrison, J.D., 2015. Overview of ICRP Committee 2: doses from radiation exposure. Ann. ICRP. DOI: 10.1177/0146645314559350.

Harrison, J.D., Muirhead, C.R., 2004. Quantitative comparisons of cancer induction in humans by internally deposited radionuclides and external radiation. Int. J. Radiat. Biol. 79, 1–13.

Harrison, J.D., Day, P., 2008. Radiation doses and risks from internal emitters. J. Radiol. Prot. 28, 137–159.

Hill, M.A., 2004. The variation in biological effectiveness of x-rays and gamma rays with energy. Radiat. Prot. Dosim. 112, 471–481.

ICRP, 1977. Recommendations of the International Commission on Radiological Protection. ICRP Publication 26. Pergamon Press, Oxford.

ICRP, 1991. 1990 Recommendations of the International Commission on Radiological Protection. ICRP Publication 60. Ann. ICRP 21(1–3).

ICRP, 2007. The 2007 Recommendations of the International Commission on Radiological Protection. ICRP Publication 103. Ann. ICRP 37(2–4).
Martin, C.J., 2007. Effective dose: how should it be applied to medical exposures? Br. J. Radiol. 80, 639–647.
Menzel, H-G., Harrison, J.D., 2012. Effective dose: a radiation protection quantity. Ann. ICRP 41(3/4), 117–123.
NAS/NRC, 2006. Health Risks from Exposure to Low Levels of Ionizing Radiation – BEIR VII. National Academies Press, Washington, DC.
UNSCEAR, 2006. Effects of Ionizing Radiation. United Nations Scientific Committee on the Effects of Atomic Radiation 2006 Report, Volume I, Annex A. United Nations, New York.
Wall, B.F., Haylock, R., Jansen, J.T.M., et al., 2011. Radiation Risks from Medical X-ray Examinations as a Function of the Age and Sex of the Patient. Report HPA-CRCE-028. Health Protection Agency, Chilton.

SAGE

ICRP
Annals of the ICRP

Radiological protection in computed tomography and cone beam computed tomography

M.M. Rehani

Massachusetts General Hospital and Harvard Medical School, Boston, USA; e-mail: madan.rehani@gmail.com

Abstract–The International Commission on Radiological Protection (ICRP) has sustained interest in radiological protection in computed tomography (CT), and ICRP *Publications 87* and *102* focused on the management of patient doses in CT and multi-detector CT (MDCT) respectively. ICRP forecasted and 'sounded the alarm' on increasing patient doses in CT, and recommended actions for manufacturers and users. One of the approaches was that safety is best achieved when it is built into the machine, rather than left as a matter of choice for users. In view of upcoming challenges posed by newer systems that use cone beam geometry for CT (CBCT), and their widened usage, often by untrained users, a new ICRP task group has been working on radiological protection issues in CBCT. Some of the issues identified by the task group are: lack of standardisation of dosimetry in CBCT; the false belief within the medical and dental community that CBCT is a 'light', low-dose CT whereas mobile CBCT units and newer applications, particularly C-arm CT in interventional procedures, involve higher doses; lack of training in radiological protection among clinical users; and lack of dose information and tracking in many applications. This paper provides a summary of approaches used in CT and MDCT, and preliminary information regarding work just published for radiological protection in CBCT.

Keywords: Computed tomography; Cone beam CT; Radiological protection; Patient dose; Radiation protection of patient

This paper does not necessarily reflect the views of the International Commission on Radiological Protection.

1. COMPUTED TOMOGRAPHY

Computed tomography (CT) is a unique imaging technique that has been growing in usage, particularly since the development of multi-detector CT (MDCT) which enables faster and higher resolution scanning. CT technology has been improving continuously, providing images of superb quality in two-, three-, or four-dimensional format. Back in 1998, there were no cases of skin injuries from CT examinations, and there was little interest in cancer risk estimate from CT scans. Instead, manufacturers were mainly competing with each other in terms of reducing scanning time rather than patient dose, as customers rarely asked about the radiation dose. At that time, the International Commission on Radiological Protection (ICRP) warned that the use of CT was increasing and there was potential for both single CT scans and repeat CT scans to result in high patient doses (ICRP, 2000; Rehani, 2012). A recent article summarised how the actions by ICRP and others led to an era of serious consideration of patient dose reduction in CT examinations, and competition among manufacturers to reduce the doses delivered by CT machines (Rehani, 2012). The article also summarised justification and optimisation principles of ICRP, as applied to CT scanning, as well as actions taken by the International Atomic Energy Agency (Rehani, 2012). The current paper summarises approaches used by ICRP in radiological protection in CT, and recommendations on radiological protection in cone beam CT (CBCT) which have been published recently (ICRP, 2015).

1.1. Approach to safety

Safety is best achieved when it is built into a system, rather than made optional for users. Realising this approach, *Publication 87* (ICRP, 2000) emphasised the actions that manufacturers may take on new features in CT equipment that can help users to manage patient dose. The best example of this approach is a 'collision avoidance system', which started within the automobile industry but has now been implemented in medical imaging equipment. If the gantry of the imaging machine just touches a person, the gantry immediately stops moving. When collisions have to be avoided through education, training, and instructions, the results cannot be the same. Both detection of the problem and avoidance of the effect should be automatic. Similar safety approaches have been implemented in the last decade through CT manufacturer use of automatic exposure control and tube current modulation to reduce patient radiation dose. In each rotation of the x-ray tube around the patient, the system detects attenuation and adjusts exposure automatically to achieve optimisation in dose and image quality. Technology holds the key to future advances in safety and optimisation of protection (Rehani and Vañó, 2011; Rehani, 2013).

1.2. Shift in emphasis from dose management to avoidance of overexposure

Between 2001 and 2007, the emphasis of radiological protection in CT was on management and optimisation of patient doses (ICRP, 2007a). Many publications used potential risk of cancer in CT exposure to draw attention to justification of exposure and optimisation of radiation protection. However, since the invention of

CT in 1972 until 2007, CT machines performed well across the world. Accidental overexposure resulting in radiation-induced skin injury, whether as a result of operator error or machine error, had not been reported. One report of temporary hair loss in CT brain perfusion [and digital subtraction angiography (DSA)] in 2005 was not attributed to error (Imanishi et al., 2005), but was attributed to overexposure that can potentially result in a typical situation.

In October 2009, the media highlighted a report of skin injury from overexposure to a child in 2008 due to operator error (Bogdanich, 2009). The report included information about administration of up to eight times the normal radiation dose to 206 possible stroke victims over an 18-month period during a procedure intended to obtain clearer images of the brain. The US Food and Drug Administration (FDA) in 2009 issued an initial notification regarding a safety investigation of facilities performing brain perfusion CT scans This alert indicated that US FDA had become aware of radiation overexposure during perfusion CT imaging performed to diagnose stroke at a single, particular facility. The incident reported above involved brain perfusion CT, which may be more prone to substantial radiation overexposure if performed incorrectly due to the cine nature of the acquisition.

US FDA has been working with the Medical Imaging and Technology Alliance on standardised dose reporting. The purpose is to ensure that necessary dose and patient information (e.g. size and age data) for x-ray imaging procedures are recorded in Digital Imaging and Communications in Medicine radiation dose structured reports, so they can be easily transferred to a dose registry.

To develop standards that define safeguards to help prevent CT scanners from delivering excessive radiation, the National Electrical Manufacturers Association and US FDA have collaborated to create new software called the 'Computed Tomography Dose Check'. The CT Dose Check notifications and alerts call the user's attention to what may be high-dose settings, and offer a 'time-out' opportunity for the user to confirm or change scan settings before proceeding. Further, there is a pop-up box that requires the user to type in their initials when over-riding the standard protocol. All new CT equipment must have this safety device.

In Europe, the Heads of the European Radiological Protection Competent Authorities have signed a Memorandum of Understanding with US FDA to share information on promoting radiation protection in CT imaging.

2. CONE BEAM CT

CBCT is a form of x-ray CT in which the x rays, in the form of a divergent cone, illuminate a wide-area detector for image capture. While conventional MDCT scanners acquire consecutive tomographic slices, in CBCT, two-dimensional projection images are acquired by an area detector and directly reconstructed into a three-dimensional dataset. Although first introduced in 1982, multiple teams in the 1990s worked to bring the system to a level of clinical utility that only gained popularity in the 2000s.

CBCT represents an emerging technology that enables high-resolution volumetric scanning of the anatomy under consideration. Just as in MDCT, use of CBCT is increasing steadily in clinical practice. Although it is a relatively new modality, CBCT is being used for a variety of clinical applications such as dental imaging, head and neck imaging (including sinus CT), high-resolution bone imaging, and intra-operative and interventional imaging. Mobile units represent a fast growing use of CBCT.

CBCT is also used in radiotherapy for pretreatment verification of patient position and target volume localisation.

In view of initial usage of CBCT in dental imaging, where doses were tiny, users are tempted to see CBCT as 'light' CT or consider it as 'low-dose CT'. However, at present, CBCT use, even in orthodentistry, can result in relatively high doses to the thyroid and lens of the eye. CBCT is widely used as C-arm mobile units for interventional applications, with much higher radiation doses that equal or exceed those from MDCT. As CBCT is a relatively new development in clinical practice, radiation doses and possible effects of CBCT are still being gathered and analysed. These units are poorly understood, with little or no training provided to the users, and poor dose tracking. Even at this early stage, however, studies indicate that there is room for optimisation to keep the radiation dose as low as reasonably achievable.

ICRP has produced a document on radiological protection issues in CBCT, and developed guidelines and recommendations (ICRP, 2015).

2.1. Differences between standard MDCT and CBCT

CBCT systems differ from 'standard' MDCT systems in several ways that affect image quality and radiological protection. Some key differences are listed below.

- Due to the cone beam nature of the irradiated field, and the associated non-uniformities in the primary and scatter radiation imparted to the scan volume, the standard dose metrics popularised by MDCT cannot be applied to CBCT.
- CBCT systems have superior spatial resolution for high-contrast objects (e.g. bone, lung), but inferior contrast resolution for low-contrast objects (e.g. soft tissue). A CBCT user can significantly influence the radiation dose imparted to the patient by judiciously deciding whether a 'high-dose' scan is needed, or if a 'low-dose' scan will suffice. A 'high-dose' scan is generally needed if soft tissue structures are the main diagnostic focus. A 'low-dose' scan may be sufficient for angiographic scans with arterial or venous contrast media, or for defining the position of interventional catheters.

2.2. Radiological protection in CBCT

2.2.1. Radiological protection of patients

Any attempt to match image quality of a thin-slice CBCT with a thick-slice MDCT will result in excessive dose.

In many CBCT scanners, the angular span over which the projection data are acquired can be customised. Most interventional and intraprocedural C-arm CBCT systems can scan an angular range spanning 180–240°, plus the cone angle of the x-ray beam. This allows keeping radiation-sensitive organs, such as the thyroid, eyes, female breast, and gonads, on the 'detector side' of the arc in order to achieve protection of these organs. The user should take advantage of this feature of CBCT, which is generally not available in MDCT.

The Commission in its *Publication 129* (ICRP, 2015) has recommended the provision of alerts when the dose is higher than specified, and has stopped exposure at levels that should not be exceeded.

Further recommendations include regular and continuous monitoring of radiation output throughout the examination, comparison with reference or desired levels, and provision of feedback to the system with automatic adjustment.

To date, as elaborated above, the radiological protection emphasis has been on dose management and, more recently, on avoidance of high-dose accidental exposures. However, there is another aspect that requires consideration in CBCT: risk management.

Many patients undergoing interventional procedures using CBCT are over 50 years of age, where the risk of stochastic effects is of lower consequence. Kothary et al. (2011) studied the impact of C-arm CT on radiation exposure of patients with hepatocellular carcinoma treated by chemoembolisation. They indicated that routine use of C-arm CT can increase stochastic risk, but decrease deterministic risk from DSA. The increase in stochastic risk is operator dependent; as such, with experience, it can be reduced. C-arm CT provides information that is not provided by DSA in many patients, while decreasing the use of iodinated contrast medium. Therefore, Kothary et al. (2011) recommended replacing DSA with C-arm CT based on risk management considerations.

Currently available CBCT scanners are not able to provide dose values in standardised dose indices for machine outputs or patients. There is a need to develop consensus and bring uniformity in CBCT dosimetry. Moreover, equipment used for both fluoroscopy and CBCT presents new challenges in dosimetry, and there is a need to develop methods that aggregate exposures to individual patients during entire procedures.

2.2.2. CBCT optimisation

Some points for consideration of future development are: collimation should not exceed detector size; laser guidance of needles in interventions; and minimisation of dose wastage by mechanical components.

ICRP provided recommendations dedicated to education and training in *Publication 113* (ICRP, 2009), and these are also applicable for CBCT. Essentially, the concept holds that 'the level of training in radiological protection should be commensurate with the level of expected radiation exposure'.

All personnel intending to use CBCT for diagnostic purposes should be trained in the same manner as for corresponding applications in diagnostic fluoroscopy and CT. All personnel intending to use CBCT for interventional purposes should be trained in the same manner as for interventional fluoroscopy and interventional CT.

2.2.3. Radiological protection of workers

Occupational radiation exposure is expected to be small in the case of clinic-based CBCT systems. When using a C-arm or other CBCT devices in an interventional suite or operating theatre, physicians, technologists, and other staff can protect themselves by using shielding devices. As required under national regulations in most countries, radiation workers must comply with regular individual dose monitoring requirements for managing radiation exposure, and keep a comprehensive dose record. Further, unless necessary, staff should move outside the fluoroscopy room when CBCT acquisition is taking place.

The principles of radiological protection of workers from ionising radiation are discussed in *Publication 75* (ICRP, 1997), in Paragraph 113 of *Publication 105* (ICRP, 2007b) and also in ICRP *Publication 120* (ICRP, 2013).

Interventionist physicians and other staff conducting CBCT should adhere to best practices in radiation protection, including use of protective shields and vests, maximising the distance from the x-ray source, and injecting contrast media using an automatic injector. While surgeons and interventionalists may be required to be close to the examined volume during two-dimensional fluoroscopy, patient condition permitting, all other staff should take extra shielding measures when acquiring CBCT scans. This may include leaving the room when possible, or using extra shielding when the clinical situation requires their presence in the room. The training of users, particularly those who use mobile C-arm and other mobile units with CBCT, has been covered in *Publication 129* (ICRP, 2015).

REFERENCES

Bogdanich, W., 2009. Radiation overdoses point up dangers of CT scans. The New York Times 16/10/2009. Available at: http://www.nytimes.com/2009/10/16/us/16radiation.html (last accessed 4 December 2013).

ICRP, 1997. General principles for the radiation protection of workers. ICRP Publication 75. Ann. ICRP 27(1).

ICRP, 2000. Managing patient dose in computed tomography. ICRP Publication 87. Ann. ICRP 30(4).

ICRP, 2007a. Managing patient dose in multi-detector computed tomography (MDCT). ICRP Publication 102. Ann. ICRP 37(1).

ICRP, 2007b. Radiological protection in medicine. ICRP Publication 105. Ann. ICRP 37(6).

ICRP, 2009. Education and training in radiological protection for diagnostic and interventional procedures. ICRP Publication 113. Ann. ICRP 39(5).

ICRP, 2013. Radiological protection in cardiology. ICRP Publication 120. Ann. ICRP 42(1).

ICRP, 2015. Radiological Protection in Cone Beam Computed Tomography (CBCT). ICRP Publication 129, Ann. ICRP 44(1).

Imanishi Y, Fukui A., Niimi H., et al., 2005. Radiation-induced temporary hair loss as a radiation damage only occurring in patients who had the combination of MDCT and DSA. Eur. Radiol 15, 41–46.

Kothary N., Abdelmaksoud M.H., Tognolini A., et al., 2011. Imaging guidance with C-arm CT: prospective evaluation of its impact on patient radiation exposure during transhepatic arterial chemoembolization. J. Vasc. Interv. Radiol 22, 1535–1543.

Rehani M.M. and Vañó E. 2011. Medical radiation protection in next decade. Radiat. Prot. Dosimetry 147, 52–33.

Rehani M.M. 2012. ICRP and IAEA actions on radiation protection in computed tomography. Ann ICRP 41(3/4), 154–160.

Rehani M.M. 2013. Challenges in radiation protection of patients for the 21st century. AJR. Am. J. Roentgenol 200, 762–764.

US FDA. Food and Drug Administration. CT Brain Perfusion Scans Safety Investigation: Initial Notification. Available at: http://www.fda.gov/Safety/MedWatch/SafetyInformation/SafetyAlertsforHumanMedicalProducts/ucm186105.htm (accessed 28 February 2015).

SAGE

ICRP
Annals of the ICRP

Radiation epidemiology and recent paediatric computed tomography studies

J.D. Boice Jr
Department of Medicine and Vanderbilt-Ingram Cancer Centre, Vanderbilt University School of Medicine, Vanderbilt University, Nashville, TN 37232, USA;
e-mail: john.boice@vanderbilt.edu

Abstract–Recent record-linkage studies of cancer risk following computed tomography (CT) procedures among children and adolescents under 21 years of age must be interpreted with caution. The reasons why the examinations were performed were not known, and the dosimetric approaches did not include individual dose reconstructions or account for the possibility for missed examinations. The recent report (2013) on children by the United Nations Scientific Committee on the Effects of Atomic Radiation concluded that the associations may have resulted from confounding by indication (also called 'reverse causation'), and not radiation exposure. The reported cancer associations may very well have been related to the patients' underlying health conditions that prompted the examinations. Reverse causation has been observed in other epidemiological investigations, such as a Swedish study of thyroid cancer risk following I-131 scintillation imaging scans, and in studies of brain cancer risk following Thorotrast for cerebral angiography. Epidemiological patterns reported in the CT studies were also inconsistent with the world's literature. For example, in a UK study, teenagers had a higher risk of brain tumour than young children; in an Australian study, cancers not previously linked to radiation were significantly elevated; and in a Taiwanese study, the risk of benign tumours decreased with age at the time of CT examination. In all studies, solid tumours appeared much earlier than previously reported. Remarkably, in the Australian study, brain cancer excesses were seen regardless of whether or not the CT was to the head, i.e. a significant excess was reported for CT examinations of the abdomen and extremities, which involved no radiation exposure to the brain. In the UK study, the significance of the 'leukaemia' finding was only because myelodysplastic syndrome was added to the category, and there was no significance for leukaemia alone. Without knowledge of why CT examinations were performed, any future studies will be equally difficult to interpret. It is noteworthy that two recent studies of children in France and Germany found no significant excess cancer risk from CT scans once adjustment was made for conditions that prompted the scan, family

This paper does not necessarily reflect the views of the International Commission on Radiological Protection.

history, or other predisposing factors known to be associated with increased cancer risk. Nonetheless, such studies have heightened awareness of these relatively high-dose diagnostic procedures, and the need to reduce unnecessary examinations and lower the dose per examination commensurate with the desired image quality.

Keywords: CT examinations; Radiation epidemiology; Reverse causation

1. INTRODUCTION

Computed tomography (CT) examinations for imaging have increased markedly over the last 20 years (NCRP, 2009). In the USA alone, over 85 million CT examinations were performed in 2012, and the overall use per year, while slowing, continues to rise (PRWeb, 2012). These examinations are associated with absorbed doses that are much higher than those associated with chest x-rays and most other diagnostic procedures. A CT examination to the abdomen could result in a stomach dose as high as 10 mGy, whereas a neonatal abdominal CT could result in a stomach dose of the order of 20 mGy (Brenner and Hall, 2007; Huda and Vance, 2007; Huda and Mettler, 2011). A single examination, however, only carries a small theoretical cancer risk, and this future risk must be balanced with the immediate clinical benefit. If an imaging scan is clinically warranted in young adults, the immediate benefits far outweigh a very small theoretical long-term risk (Zondervan et al., 2013). Diagnostic scans also reduce the number of invasive procedures, unnecessary hospital admissions, and the length of hospital stays. Nonetheless, because of the possibility that multiple CT examinations of children and young adults could result in organ doses in the range where epidemiological studies might be able to reveal risk, studies of CT examinations in childhood have the potential to provide new information on the risk of cancer following exposures in childhood (Boice, 2014).

Unfortunately, the methodological deficiencies of recent CT studies from the UK (Pearce et al., 2012), Australia (Mathews et al., 2013), and Taiwan (Huang et al., 2014) limit their relevance to both clinical practice and understanding low-dose radiation health effects. The critical problem is that the reasons for performing the CT examinations were not known, and these reasons (rather than the CT radiation dose) were the probable cause of the subsequent cancer diagnoses. This is called 'confounding by indication' or 'reverse causation'. No individual dosimetry was performed, no information on the actual CT scanners that exposed the studied populations was available, and the acknowledged problem of incomplete documentation of the number of CT examinations performed raises serious questions regarding the validity of the organ-specific dose estimates, and the purported dose–response relationships. These criticisms have been raised by scientific committees (NCRP, 2012; UNSCEAR, 2013) and others (Walsh et al., 2013, 2014). Recent CT studies from France and Germany confirm that confounding by indication can play a major role in distorting reported risk estimates if not properly taken into account (Journy et al., 2015; Krille et al., 2015).

2. THE UK STUDY

The UK CT study (Pearce et al., 2012) was a record-linkage study of leukaemia [and myelodysplastic syndrome (MDS)] and brain cancer incidence following CT scans to 178,000 individuals at 0–21 years of age. Collection of the scan data for individual patients was not possible, and thus average CT machine settings from two national surveys were used to estimate organ doses. In the absence of individual dosimetric information, the authors conducted dose–response analyses for leukaemia and MDS, and combined and estimated the excess relative risk per Gy to be 36, which is extremely high. Similarly, for brain cancer, a significant dose–response relationship was reported with an excess relative risk per Gy of 23. A recent United Nations Scientific Committee on the Effects of Atomic Radiation (UNSCEAR) report concluded, '...There are concerns about the risk estimates because of lack of information about indications for the CT scans and the consequent potential for 'reverse causation' (i.e. cancers may have been caused by the medical conditions prompting the CT scans rather than by the CT dose), and lack of individual dosimetry' (UNSCEAR, 2013). Similarly, the National Council on Radiation Protection and Measurements (NCRP) concluded that, 'Children who receive frequent examinations may have some underlying disability related to the outcome of interest. That is, a child who receives multiple CT examinations of the head may have a central nervous system disorder that is prompting such examinations' and it is these underlying disorders that are related to the cancer diagnosis and not the CT radiation dose (NCRP, 2012).

The risk of brain tumour was higher for young adults than for children at the time of CT examination among patients in the UK study (Pearce et al., 2012), and contrasted with the pattern seen in the Japanese atomic bomb survivor study (Table 1) where, 'The risk of glioma is highest at <5 years at irradiation and seems to largely

Table 1. Risk of brain tumour related to age at computed tomography examination among patients in the UK paediatric study (Pearce et al., 2012). Note that the Japanese atomic bomb survivor study had the opposite pattern: 'The risk of glioma is highest at <5 years at irradiation and seems to largely disappear at the age of 20 years or more at irradiation, suggesting that susceptibility decreases as brain development nears completion' (UNSCEAR, 2013).

Age at examination (years)	Brain tumour risk (ERR Gy^{-1})
0–4	5
5–9	28
10–14	37
15–21	41

ERR, excess relative risk.

disappear at the age of 20 years or more at irradiation, suggesting that susceptibility decreases as brain development nears completion' (UNSCEAR, 2013). Teenagers and young adults have been found to have small, if any, risk of radiation-induced tumours in other epidemiological studies.

The UK study excluded patients who developed a brain tumour within 5 years of their first CT examination, but this does not strictly address the issue of a short latency or possible dose-response relationship. The risk of brain tumour was concentrated among children who received multiple CT examinations, so brain tumours within 5 years of the last CT examination were included in the analysis, and may very well have distorted the results. Regardless, latency or time from exposure to occurrence was much shorter than observed in other studies (NA/NRC, 2006; UNSCEAR, 2008). Radiation-induced brain cancer is an example of the substantial variance with other studies of high-quality dosimetry and epidemiological rigor. The effect of age at CT exposure for radiation-induced brain tumour was the opposite of what has been observed in studies of Japanese atomic bomb survivors, Israeli children treated with radiation for ringworm of the scalp, and survivors of childhood cancer (Ron et al., 1988; Neglia et al., 2006; UNSCEAR, 2008). The risk following CT exposure in the UK increased with age at exposure, which is the opposite pattern seen in all previous studies where children under 5 years of age at exposure have the highest risk, and teenagers and young adults have minimal risk, if any.

The UK study appears to be the first radiation epidemiological investigation to combine leukaemia and MDS, and little if any justification was given for doing so. The UNSCEAR Committee noted, 'For 74 observed leukaemias, they found an excess relative risk Gy^{-1} of 36 . . . However, they included MDS with the leukaemias, and the MDS cases had an extremely high relative risk. Without the MDS cases, the estimated risk was still high, but no longer statistically significant' (UNSCEAR, 2013). NCRP stated, 'MDS accounted for 12% of the 74 leukemic conditions in the analysis'. MDS may be related to radiation, and this has recently been reported among long-term survivors from Nagasaki (Iwanaga et al., 2011). There is no evidence in any study that MDS occurs shortly after exposure. Further, in contrast to studies of atomic bomb survivors (Hsu et al., 2013), the leukaemia excess in the UK study was due to acute lymphocytic leukaemia, and no increase was seen for acute myelogenous leukaemia.

Similarly to the CT-induced risk of brain tumour, the non-significant excess relative risk per Gy of 19 for leukaemia (excluding MDS) was higher than seen in previous studies, including that of the atomic bomb survivors which reported an excess relative risk per Gy of 6.5 for children and young adults exposed under 20 years of age (Hsu et al., 2013). Interestingly, there are statements in UNSCEAR (2013) that may be slightly misleading regarding age at exposure effects among atomic bomb survivors. First, it was stated that very high risks were seen shortly after exposure among children under 10 years of age at exposure; however, this analysis was based on a statistical model that did not describe the incidence data, as well as the more recent analysis by Hsu et al. (2013) which showed a lower risk. Second, UNSCEAR (2013) stated that there was a substantial excess of childhood leukaemia cases among those exposed at less than 10 years of age and followed to 5–14 years of age: 10 observed against approximately 1.6 expected. Unfortunately, the expected value was based on Japanese general population rates for 1950 which have been judged to be inappropriate or, at best, misleading for comparison with the Life Span Study. Computations of observed over expected values from population rates have been found to give high-sided estimates of risk, and the elevations appear to be 'primarily due to non-radiation factors' (Grant et al., 2009).

3. THE AUSTRALIAN STUDY

The Australian CT study (Mathews et al., 2013) involved data linkages of 680,000 children aged 0–19 years who received CT scans and 10 million children with no record of such exposures. It was remarkable that excess risks and significant associations were reported for practically all cancers that were evaluated, except for breast cancer and acute lymphocytic leukaemia (two of the most highly radiosensitive sites) (Table 2).

Table 2. Cancers reported to be significantly increased in the data-linkage study of 680,000 Australian children (aged 0–19 years at exposure) who received computed tomography (CT) scans and 10,000,000 Australian children with no record of such exposures (Mathews et al., 2013).

Significant cancer excesses reported for:	No cancer excesses reported[‡] for:
Digestive organs	Breast[§]
Melanoma*	Leukaemia (lymphoid)[§]
Soft tissue	
Female genital	
Urinary tract	
Brain (after CT examinations of the head)[†]	
Brain (after CT examinations other than of the head)[†]	
Thyroid	
Ill-defined and unspecified sites	
Leukaemia (myeloid)	
Hodgkin's lymphoma*	
Other lymphoid cancers	
Myelodysplasias	

*These sites were not found to be increased following radiation exposure (UNSCEAR, 2008).

[†]Brain cancer excesses were seen regardless of whether or not the CT was to the head, i.e. significant excesses were reported for CT examinations of the abdomen and extremities that involved no radiation exposure to the brain.

[‡]All other sites were reported to have significant or non-significant excesses.

[§]These sites are highly radiosensitive and frequently found to be elevated in studies of childhood and young adult exposures.

Excesses were significantly elevated for melanoma and Hodgkin's lymphoma; cancers that are not inducible by ionising radiation (UNSCEAR, 2008). Remarkably, the associations with brain tumour were found regardless of whether the CT was to the head or to another part of the body. In other words, a significant association for brain tumour was found for CT scans of the abdominal region and extremities that resulted in no radiation exposure to the brain, clearly suggesting reverse causation (confounding by indication) where the conditions being screened or the characteristics of the patient were the cause of the CT examinations, rather than the CT examinations being the cause of the subsequent tumour.

UNSCEAR (2013) concluded that the Australian study has methodological flaws that resulted in implausible risks. Reverse causation was again raised as an explanation for the associations, i.e. that the cancers were caused by the medical conditions prompting the CT scans rather than by the CT dose. This potential bias could not be examined as no documentation was available on the indications for the CT scans. The UNSCEAR Committee went on to conclude that the exceptionally high-risk estimates for 'all cancers, excluding brain cancer after brain CT' risk (excess relative risk Sv^{-1}) were statistically incompatible with the data at comparable ages from the Japanese Life Span Study on atomic bomb survivors: 27 [95% confidence interval (CI) 17–37] vs 3 (95% CI 2–6).

The Australian findings were also at odds with previous studies in that two highly radiogenic sites were not elevated, i.e. breast cancer and acute lymphoid leukaemia were not increased. Breast cancer has been found to be elevated in several studies of childhood and young adult exposures to fractionated diagnostic radiation (Boice and Monson, 1977; Hoffman et al., 1989), and the absence of this tumour in contrast to the elevation of non-radiogenic sites raises caution in making any causal interpretations for the associations seen in this study (Walsh et al., 2013, 2014). Both acute lymphocytic leukaemia and acute myelogenous leukaemia were increased in the study of atomic bomb survivors (Hsu et al., 2013); in contrast, the UK CT study did not find acute myelogenous leukaemia to be increased, whereas the Australian CT study did not find acute lymphocytic leukaemia to be increased.

4. THE TAIWANESE STUDY

For completion, a recent record-linkage study was conducted in Taiwan to evaluate CT examinations of the head and subsequent development of benign brain tumour (Huang et al., 2014). Approximately 24,000 children with head CT were matched to four times this number of non-exposed children, i.e. children who did not receive a head CT. Any child with a reported diagnosis of cancer or conditions predisposing to brain tumour was excluded. A significant association with benign, but not malignant, brain tumour was reported. The reasons why the CT examinations were performed were not known. Apparently, children with benign tumours were not excluded, leaving open the real possibility that the results were confounded by indication, i.e. the symptoms prompting the CT scan were indications of an underlying brain tumour. This bias is supported by the peculiar epidemiological patterns that the risk decreased with time since CT examination, and that the oldest, not the youngest, subjects were at highest risk. Also, the study is limited by the small number of subsequent tumour cases (only five malignant and 14 benign tumours). The reported associations would essentially disappear if only a few of these brain conditions were present at the time of the CT examination. No dosimetry was performed. Similar to the UK and Australian studies, reverse causation implies that it is the early symptoms of undetected tumour, or of factors that predispose to tumour, that are the indications for the CT scans, rather than the CT scans per se that are causing the apparent excess risk of cancer.

5. CHALLENGES OF INDIVIDUAL DOSE ASSESSMENTS IN THE ABSENCE OF INDIVIDUAL DOSIMETRIC DATA

Collection of the scan data for individual patients and organ doses for individuals based on their exposure characteristics was not possible in any of the studies; therefore, average CT machine settings from national surveys were used. The resulting organ dose estimates made are highly uncertain and of questionable validity. A somewhat arbitrary year (2001) was chosen in the CT studies as the demarcation of the high exposures from CT examinations in previous years (conventional CT) and the lower exposures in use nowadays (helical CT) (Kim et al., 2012). This distinction is important because the scan time for paediatric patients with a conventional CT scanner could be from 10 to 35 min in the 1990s compared with 50–60 s nowadays with a helical CT scanner (Kaste et al., 1997).

For children examined during the early years of CT imaging, there is the likelihood that any movement during the examination, which could take up to 35 min for conventional CT scanners, would result in a blurred image and prompt a repeat examination (NCRP, 2012). Unfortunately, evidence of repeat examinations was not available from the electronic databases accessed in either the UK or Australian studies.

Accounting for chronological changes in CT dosimetry is challenging, complex, and complicated. Machines changed, time per examination changed, and exposure factors for children changed with time. The rapid and continued changes in CT technology are remarkable (Frush et al., 2013). These changes, however, were not simultaneous across nations, and not even in single hospitals. One important factor that influenced patient dose is when the exposure technique for children was lowered from that used for adults. The desired image quality was achieved at a much lower exposure level when the examinations were 'child-sized' (Goske et al., 2008, 2012). This change meaningfully affected the organ dose to a child, but unfortunately, no data are available for the actual exposure parameters for individuals in the studied populations.

Missed CT exposures also include any examinations occurring outside the age or calendar year ranges of the study design, i.e. CT examination occurring after the defined paediatric ages and calendar years of study eligibility would not be recorded, yet cancer diagnoses after the paediatric ages would be included. This combination of missing doses would tend to inflate the estimate of risk per unit dose.

6. EXAMPLES OF REVERSE CAUSATION (CONFOUNDING BY INDICATION)

An example of reverse causation (confounding by indication) is found in a Swedish epidemiological study of thyroid cancer following I-131 scans of the thyroid (Dickman et al., 2003). Over 35,000 adults were evaluated with I-131 scintillation scans for the suspicion of tumour and/or other thyroid conditions. The comprehensive study included abstracting all clinical data for all 35,000 patients, including thyroid size, I-131 activity administered, and, more importantly, the reason for the examination. The overall relative risk of thyroid cancer compared with the general population was significant (1.8) when all patients were included in the evaluation, including those examined because of suspicion of an underlying thyroid tumour (Table 3).

When the patients were divided into two examination categories, suspicion of tumour and other reasons, it was evident that the significant association was due entirely to the patients who were examined because of suspicion of tumour, with the relative risk overall being 3.5 (significant). For those patients who were examined for reasons other than suspicion of tumour, there was no evidence for an association between the I-131 scan and thyroid cancer (relative risk 0.9). The reverse causation bias lasted for more than 20 years after the initial I-131 examination, in that the suspicion of thyroid tumour was significantly associated with the occurrence of thyroid cancer even 20 years after the initial scintillation scan. Further, because the vast majority of these patients were over 20 years of age when examined, the observed high risk of thyroid cancer was not expected as the adult thyroid gland is relatively resistant to cancer induction by radiation (UNSCEAR, 2008).

Table 3. Reverse causation bias (confounding by indication) observed in a Swedish cohort study of thyroid cancer following I-131 scans (Dickman et al., 2003). Significant increases of thyroid cancer were observed only among patients examined with I-131 scans for suspicion of thyroid tumour, and this bias lasted for more than 20 years after the initial examination. The 'other reasons' for examination were mainly related to benign thyroid conditions such as hypothyroidism or hyperthyroidism.

Reason for scan (no. of thyroid cancers)	Relative risk of thyroid cancer by years after scan				
	2–4	5–9	10–20	>20	All
All reasons (105)	3.1*	2.5*	1.2	1.7*	1.8*
Suspicion of tumour (69)	6.3*	4.8*	2.3*	3.5*	3.5*
Other reasons (36)	1.3	1.5	0.6	0.9	0.9

*$P<0.05$.

Another example of reverse causation bias in radiation epidemiological studies is in: patients given radioactive Thorotrast (a colloid of thorium dioxide) as the contrast medium for cerebral angiography for suspicion of brain tumour (Dos Santos Silva et al., 2003; Travis et al., 2003). Patients in the USA (Travis et al., 2003) who had angiographic examinations because of suspicion of tumour were significantly more likely to develop brain cancer regardless of whether or not the examination included radioactive Thorotrast as the contrast medium [standardised mortality ratio (SMR) 33.6], or whether the examination involved a non-radioactive contrast medium (SMR 15.6). The relative risk of brain cancer comparing the Thorotrast group with the non-radioactive contrast medium group was 1.3 (95% CI 0.6–3.7). Similarly, in Portugal, the risk of dying from a nervous system disease was greatest in the subgroup of patients who received Thorotrast angiographic procedures because of suspicion of a nervous system condition (relative risk 7.6, 95% CI 1.22–317) (Dos Santos Silva et al., 2003). Again, it can be concluded that the preclinical symptoms of brain cancer prompted the clinician to request a diagnostic imaging scan, and it was the preclinical symptoms that led to the overt diagnosis of brain cancer, not the radiation dose received from Thorotrast.

A final example of confounding by indication cropping up in epidemiological studies involving diagnostic radiation is a recent report of meningioma associated with dental x rays (Claus et al., 2012). Putting aside for the moment the fact that the study was based on patient recall of dental x rays many decades before the diagnosis of meningioma, and that the high-dose dental examinations were associated with the lowest risk, confounding by indication was likely because of the association of jaw pain with both meningioma symptoms and presentation for dental x rays. Meningiomas can cause referred pain to the orofacial region. A patient with such pain may receive dental x rays during the course of his/her care. It may be that radiographs do not cause meningiomas, but rather the presence of tumours triggers the need for radiographs (White et al., 2013). In other words, the dental x rays did not cause the meningiomas; the meningiomas caused the dental x rays.

7. CONCLUSIONS

Epidemiological studies have value in drawing attention to diagnostic CT examinations and the need to reduce unnecessary examinations, as well as to reduce the dose per examination commensurate with desired image quality for clinical evaluation. Programmes and initiatives exist within the medical community to raise awareness about the need for dose reduction and for considering alternative modalities among both children and adults: Image Gently (Goske et al., 2008, 2012); Image Wisely (Brink et al., 2010); and Choosing Wisely (Rao and Levin, 2012, 2014). These efforts should be encouraged even in the absence of direct evidence that CT radiation exposures are causally linked to excess risk of cancer. The epidemiological evidence to date is limited as summarised below.

Current studies of paediatric CT examinations and subsequent cancer risk do not provide evidence that low doses are causally associated with cancers in children, nor are they of much clinical relevance. Statistical associations can result from study biases as apparent in the UK, Australian, and Taiwanese studies. Reverse causation (confounding by indication) is a likely reason for the observed associations, i.e. the condition being evaluated prompted the CT examinations, and was also the cause of the subsequent cancer diagnosis. It is thus noteworthy that two recent studies of nearly 70,000 children in France (Journy et al., 2015) and nearly 45,000 children in Germany (Krille et al., 2015) found no significant excess cancer risk from CT scans – once adjustment was made for conditions that prompted the scan, family history, or other predisposing factors known to be associated with increased cancer risk.

Inconsistency with the world's literature on radiation-induced cancers following childhood irradiation include: age at exposure effects being highest for exposures at older and not younger ages (most notably for brain tumour); latency or time from exposure to occurrence being much shorter than observed previously; tumour sites that have never been associated with radiation were increased whereas some of the most sensitive organs were not increased (Australian study); and the computed radiation risks per Gy were so extremely high as not to be believable (in the author's view).

Collection of the scan data and associated exposure information for individual patients was not done. Using national survey data to estimate scanner characteristics for individual procedures over a period of decades is of questionable validity. Doses from conventional and helical scanners are very different, and the type of scanner was not known for any individual patient. Exposure techniques changed over time, notably when it was recognised that adult techniques were providing unnecessary radiation dose to children of much smaller size. Excess relative risk per mGy is very difficult to interpret based on substantial uncertainties in organ doses associated with type of CT scanner, type of procedure, changes in scanners and technique over calendar year, missed examinations due to movement and blurred images prompting repeat examinations, and missed examinations that occurred outside the timeframe of the study age and calendar year eligibility criteria.

Unfortunately, these criticisms and methodological limitations have not discouraged recent media articles from claiming that epidemiology has now demonstrated directly that cancer risk is associated with CT examinations, concluding, '... we need to find ways to use them [CT examinations] without killing people in the process' (Redberg and Smith-Bindman, 2014). Such exaggerations based on highly questionable epidemiological observations serve little purpose, and may very well discourage patients from seeking the medical care needed for their children to diagnose life-threatening injuries, prevent unnecessary surgery and medications, and identify treatable causes of suffering (Ellenborgen et al., 2014). Theoretical future risks must always be balanced against the immediate benefits of these CT examinations. Epidemiology is yet to provide convincing evidence of future cancer risks following diagnostic CT examinations.

Unless the reasons for the examination can be determined and dosimetry improved in future studies (Einstein, 2012; Thierry-Chef et al., 2013; Walsh et al., 2014), they too will be equally ambiguous.

REFERENCES

Boice, J.D., Jr., 2014. Implications of radiation dose and exposed populations on radiation protection in the 21st century. Health Phys. 106, 313–328.

Boice, J.D., Jr., Monson, R.R., 1977. Breast cancer in women after repeated fluoroscopic examinations of the chest. J. Natl. Cancer Inst. 59, 823–832.

Brenner, D.J., Hall, E.J., 2007. Computed tomography – an increasing source of radiation exposure. N. Engl. J. Med. 357, 2277–2284.

Brink, J.A., Amis, E.S., Jr., 2010. Image Wisely: a campaign to increase awareness about adult radiation protection. Radiology 257, 601–602.

Claus, E.B., Calvocoressi, L., Bondy, M.L., et al., 2012. Dental x-rays and risk of meningioma. Cancer 118, 4530–4537.

Dickman, P.W., Holm, L.E., Lundell, G., et al., 2003. Thyroid cancer risk after thyroid examination with 131I: a population-based cohort study in Sweden. Int. J. Cancer 106, 580–587.

Dos Santos Silva, I., Malveiro, F., Jones, M.E., et al., 2003. Mortality after radiological investigation with radioactive Thorotrast: a follow-up study of up to fifty years in Portugal. Radiat. Res. 159, 521–534.

Einstein, A.J., 2012. Beyond the bombs: cancer risks of low-dose medical radiation. Lancet 380, 455–457.

Ellenbogen, P.H., Gerardi, M., Lee, S.I., et al., 2014. The uses and abuses of CT scans. New York Times 06/02/2014. Available at: http://www.nytimes.com/2014/02/07/opinion/the-uses-and-abuses-of-ct-scans.html?_r=0 (accessed 2 March 2015).

Frush, D., Denham, C.R., Goske, M.J., et al., 2013. Radiation protection and dose monitoring in medical imaging: a journey from awareness, through accountability, ability and action... but where will we arrive? J. Patient Saf. 9, 232–238.

Goske, M.J., Applegate, K.E., Boylan, J., et al., 2008. Alliance for Radiation Safety in Pediatric Imaging. Image Gently(SM): a national education and communication campaign in radiology using the science of social marketing. J. Am. Coll. Radiol. 5, 1200–1205.

Goske, M.J., Applegate, K.E., Bulas, D., et al., 2012. Image Gently 5 years later: what goals remain to be accomplished in radiation protection for children? AJR. Am. J. Roentgenol. 199, 477–479.

Grant, E.J., Shimizu, Y., Kasagi, F., et al., 2009. Radiation unlikely to be responsible for high cancer rates among distal Hiroshima A-bomb survivors. Environ. Health Prev. Med. 14, 247–249.

Hoffman, D.A., Lonstein, J.E., Morin, M.M., et al., 1989. Breast cancer in women with scoliosis exposed to multiple diagnostic x rays. J. Natl. Cancer Inst. 81, 1307–1312.

Hsu, W-L., Preston, D.L., Soda, M., et al., 2013. The incidence of leukemia, lymphoma and multiple myeloma among atomic bomb survivors: 1950–2001. Radiat. Res. 179, 361–382.

Huang, W.Y., Muo, C.H., Lin, CY., et al., 2014. Paediatric head CT scan and subsequent risk of malignancy and benign brain tumour: a nation-wide population-based cohort study. Br. J. Cancer 110, 2354–2360.

Huda, W., Vance, A., 2007. Patient radiation doses from adult and paediatric CT. AJR. Am. J. Roentgenol. 188, 540–546.

Huda, W., Mettler, F.A., 2011. Volume CT dose index and dose-length product displayed during CT: what good are they? Radiology 258, 236–242.

Iwanaga, M., Hsu, W.L., Soda, M., et al., 2011. Risk of myelodysplastic syndromes in people exposed to ionizing radiation: a retrospective cohort study of Nagasaki atomic bomb survivors. J. Clin. Oncol. 29, 428–434.

Journy, N., Rehel, J.L., Ducou Le Pointe, H., et al., 2015. Are the studies on cancer risk from CT scans biased by indication? Elements of answer from a large-scale cohort study in France. Br. J. Cancer 12, 185–193.

Kaste, S.C., Young, C.W., Holmes, T.P., et al., 1997. Effect of helical CT on the frequency of sedation in paediatric patients. AJR. Am. J. Roentgenol. 168, 1001–1003.

Kim, K.P., Berrington de González, A., Pearce, M.S., et al., 2012. Development of a database of organ doses for paediatric and young adult CT scans in the United Kingdom. Radiat. Prot. Dosim. 150, 415–426.

Krille, L., Dreger, S., Schindel, R., et al., 2015. Risk of cancer incidence before the age of 15 years after exposure to ionising radiation from computed tomography: results from a German cohort study. Radiat. Environ. Biophys. 54, 1–12.

Mathews, J.D., Forsythe, A.V., Brady, Z., et al., 2013. Cancer risk in 680 000 people exposed to computed tomography scans in childhood or adolescence: data linkage study of 11 million Australians. BMJ. 346, f2360.

NA/NRC, 2006. Health Risks from Exposure to Low Levels of Ionizing Radiation, BEIR VII Phase 2. National Academies Press, Washington, DC. Available at: http://www.nap.edu/catalog/11340/health-risks-from-exposure-to-low-levels-of-ionizing-radiation (accessed 2 March 2015).

NCRP, 2009. Ionizing Radiation Exposure of the Population of the United States. NCRP Report 160. National Council on Radiation Protection and Measurements, Bethesda, MD. Available at: http://www.ncrppublications.org/Reports/160 (accessed 2 March 2015).

NCRP, 2012. Uncertainties in the Estimation of Radiation Risks and Probability of Disease Causation. NCRP Report 171. National Council on Radiation Protection and Measurements, Bethesda, MD. Available at: http://www.ncrppublications.org/Reports/171 (accessed 2 March 2015).

Neglia, J.P., Robison, L.L., Stovall, M., et al., 2006. New primary neoplasms of the central nervous system in survivors of childhood cancer: a report from the Childhood Cancer Survivor Study. J. Natl. Cancer Inst. 98, 1528–1537.

Pearce, M.S., Salotti, J.A., Little, M.P., et al., 2012. Radiation exposure from CT scans in childhood and subsequent risk of leukaemia and brain tumours: a retrospective cohort study. Lancet 380, 499–505.

PRWeb, 2012. Latest IMV CT Survey Shows Hospitals Seek to Improve Productivity to Manage Increased Outpatient and Emergency CT Procedure Volume. Available at: http://www.prweb.com/releases/2012/6/prweb9578648.htm (accessed 2 March 2015).

Rao, V.M., Levin, D.C., 2012. The overuse of diagnostic imaging and the Choosing Wisely initiative. Ann. Intern. Med. 157, 574–576.

Rao, V.M., Levin, D.C., 2014. The Choosing Wisely initiative of the American Board of Internal Medicine Foundation: what will its impact be on radiology practice? AJR. Am. J. Roentgenol. 202, 358–361.

Redberg, R.F., Smith-Bindman, R., 2014. We are giving ourselves cancer. New York Times 31/01/2014. Available at: http://www.nytimes.com/2014/01/31/opinion/we-are-giving-ourselves-cancer.html?_r=0 (accessed 2 March 2015).

Ron, E., Modan, B., Boice, J.D., et al., 1988. Tumors of the brain and nervous system after radiotherapy in childhood. N. Engl. J. Med. 319, 1033–1039.

Thierry-Chef, I., Dabin, J., Friberg, E.G., et al., 2013. Assessing organ doses from paediatric CT scans – a novel approach for an epidemiology study (the EPI-CT study). Int. J. Environ. Res. Public Health 10, 717–728.

Travis, L.B., Hauptmann, M., Gaul, L.K., et al., 2003. Site-specific cancer incidence and mortality after cerebral angiography with radioactive Thorotrast. Radiat. Res. 160, 691–706.

UNSCEAR, 2008. United Nations Effects of Ionizing Radiation. UNSCEAR Report 2006. Volume I. Annex A. Epidemiological Studies of Radiation and Cancer. United Nations, New York. Available at: http://www.unscear.org/docs/reports/2006/07-82087_Report_Annex_A_2006_Web_corr.pdf (accessed 2 March 2015).

UNSCEAR, 2013. Sources, Effects and Risks of Ionizing Radiation. UNSCEAR Report 2013 to the General Assembly with Scientific Annexes. Volume II, Scientific Annex B: Effects of Radiation Exposure of Children. E.14.IX.2. United Nations, New York. Available at: http://www.unscear.org/docs/reports/2013/UNSCEAR2013Report_AnnexB_Children_13-87320_Ebook_web.pdf (accessed 2 March 2015).

Walsh, L., Shore, R., Auvinen, A., et al., 2013. Cancer risk in 680,000 people exposed to computed tomography scans in childhood or adolescence: data linkage study of 11 million Australians. BMJ. 346, f2360/rr/648506. Available at: http://www.bmj.com/content/346/bmj.f2360/rr/648506 (accessed 2 March 2015).

Walsh, L., Shore, R., Auvinen, A., et al., 2014. Risks from CT scans – what do recent studies tell us? J. Radiol. Prot. 34, E1–E5.

White, S.C., Hildebolt, C.F., Lurie, A.G., 2013. Dental x-rays and risk of meningioma. Cancer 119, 464.

Zondervan, R.L., Hahn, P.F., Sadow, C.A., et al., 2013. Body CT scanning in young adults: examination indications, patient outcomes, and risk of radiation-induced cancer. Radiology 267, 460–469.

SAGE

ICRP
Annals of the ICRP

Patient dose monitoring in Dubai in radiography and interventional procedures

J.S. AlSuwaidi[a], N.K. AlMazrouei[b], S. Pottybindu[c], M. Siraj[d], D. Mathew[d], A.A. Al Blooshi[d], V.P. Kuriakose[c]

[a]*Dubai Health Authority, Medical Education Department, Dubai Hospital, P.O. Box 7272, Dubai, United Arab Emirates; e-mail: jsalsuwaidi@dha.gov.ae*

[b]*Dubai Health Authority, Medical Physics Section, Radiation Protection Unit, United Arab Emirates*

[c]*Dubai Health Authority, Latifa Hospital, Radiology Department, United Arab Emirates*

[d]*Dubai Health Authority, Primary Health Care, Radiology Department, United Arab Emirates*

Abstract–This paper presents ongoing actions in Dubai on patient dose monitoring in digital radiographic examinations, mammography, interventional procedures, and dental radiological procedures. The aim of Dubai Health Authority (DHA) is to move towards the establishment of local diagnostic reference levels. DHA has participated in national and regional projects under the umbrella of the International Atomic Energy Agency. The need for local radiation protection educational programmes and wider patient dosimetry monitoring and recording emerged from this work.

Keywords: Radiation dose; Digital radiology; Dosimetry; Diagnostic reference levels; DRL

1. INTRODUCTION

Since 2005, the United Arab Emirates (UAE) has participated in the International Atomic Energy Agency's regional technical co-operation projects. In these projects, radiology and medical physics teams at the Dubai Health Authority (DHA)

This paper does not necessarily reflect the views of the International Commission on Radiological Protection.

managed to reduce patient radiation doses by up to 50% for common computed tomography (CT) examinations (head, chest, and combined abdomen and pelvis) (AlSuwaidi et al., 2013). Moreover, patient dose monitoring was initiated within conventional radiology practices, and was further enhanced with the introduction of digital radiology systems at DHA hospitals. This paper presents some DHA patient dosimetric data obtained from the practices of general radiology, mammography, interventional radiology, and dental radiology (both intra-oral and panoramic). The aim of this study was to evaluate radiation exposure levels to DHA patients, and to help to introduce steps towards establishing local diagnostic reference levels (DRLs) for common radiological examinations.

2. PATIENT DOSE MONITORING AND DIAGNOSTIC REFERENCE LEVELS

In relation to patient dosimetry, one of the major differences between digital and conventional film-based radiology is the fact that the radiographers receive immediate and direct feedback related to the patient's exposure in conventional radiology, by the white/dark appearance of the film. However, in digital systems, image brightness can be adjusted after processing, independent of the exposure level, and overexposure results in an image with less noise; as such, there may be no immediate indication of under- or overexposure. Underexposed images usually lead to the examination being repeated, which unintentionally increases exposure. On the other hand, the lack of negative impact on image appearance as a result of overexposure leads to a phenomenon known as 'dose creep'; the gradual increase of dose. It is therefore imperative that dose monitoring is undertaken.

At DHA hospitals, an integrated Picture Archiving and Communication System (PACS) and a Radiology Information System (RIS) were installed for the radiology and nuclear medicine departments in 2008. The method of using PACS and the information available within the DICOM header to monitor CT dosimetry data was established at DHA hospitals to keep a record of patient radiation doses (AlSuwaidi et al., 2011a). The establishment of a local or national patient dose registry involves tracking all radiation doses, across all imaging modalities, and thereby aims to enhance the quality of radiology management.

Dose registry provides a dynamic tool for the management of patient exposure. Moreover, in digital radiology systems, cumulative doses can be tracked. However, the advantages of digital radiology systems are coupled with some difficulties related to the manual part that requires operator input, such as patient weight. In some systems, the radiographer has to enter the patient dose from the image modality to the RIS. Hence, automatic transmission of radiation doses from imaging modalities to PACS (and regular calibration of dose quantities) is essential for the establishment of a dose database.

In the UAE and the Gulf region, the actions taken to establish DRLs are very limited. In addition to improvements in patient safety, this dose assessment limitation has provided new opportunities for applied research.

3. METHODS AND RESULTS

This paper presents DHA's experiences in dose monitoring and local DRLs, along with comparisons with published international values. This DHA study, and those presented jointly between DHA and other healthcare organisations in the UAE (AlSuwaidi et al., 2011b), are the first published papers on patient dosimetry from the UAE.

3.1. General radiology

Paediatric and adult dose surveys on digital general x-ray machines are in progress at three primary healthcare hospitals and a children and women's hospital (LH) within DHA. Chest radiography and full digital radiography using a built-in dose area product (also known as 'kerma area product') meter are used for general radiology at the children and women's hospital and the primary healthcare hospitals, respectively. Dose variations between measured and displayed values, through measurements on phantoms, were examined for these general x-ray systems. The variation ranged from 2% to 29%. The authors aimed to collect dosimetric data for 20 patients in each age group who underwent common radiological examinations [chest, abdomen, pelvis, combined abdomen and pelvis, lumbar spine (antero-posterior and lateral)]. As indicated in Fig. 1, fewer than 20 patients were included in the abdomen and pelvis groups. Adult patient dose data from the MIZ clinic (primary healthcare hospital) are presented in Fig. 1 in comparison with UK values (Hart et al., 2012). Paediatric dosimetric data at the primary healthcare hospitals were not

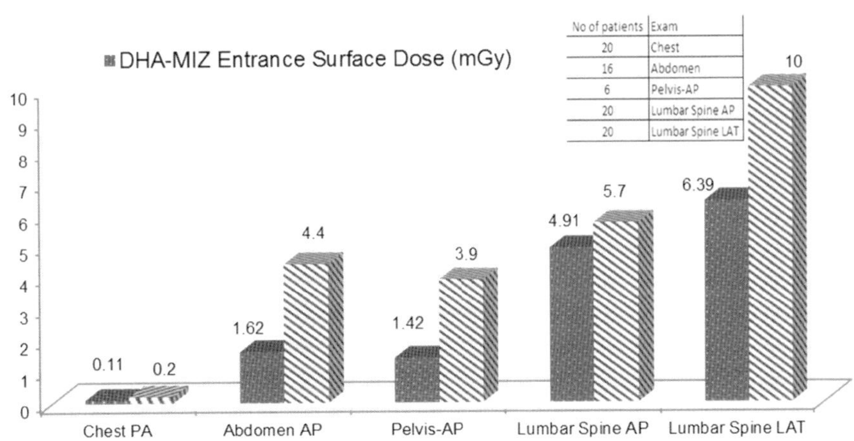

Fig. 1. Entrance surface dose (mGy) at the MIZ clinic (primary healthcare hospital) compared with adult data from Hart et al. (2002).

sufficient for analysis at this stage of the study. At the children and women's hospital, dosimetric data were obtained from 175 patients who underwent chest and abdomen x-ray examinations. In paediatric patients, entrance surface doses given in Table 1 are within the same range as those cited in *Publication 121* (ICRP, 2013) and Hart et al. (2002).

3.2. Mammography

Variations between measured and displayed dose values, through measurements on phantoms, were performed for two mammography units within DHA (H-1 and H-2, Siemens MAMMOMAT Inspiration and Hologic Lorad Selenia). Variation was 9–20% for H-1 and 19–45% for H-2. Radiation dose, in terms of average glandular dose and entrance dose, was collected from the DICOM header for 82 patients referred for mammography. Patient and image parameters (patient's name and healthcare number, date of birth, imaging view, number of views, kV, mAs, anode and filter combination, compression, and breast thickness) were recorded for each individual exposure. The average glandular dose and entrance dose values were grouped based on breast thickness: 20–<30 mm, 30–<40 mm, 40–<45 mm, 45–<50 mm, 50–<60 mm, 60–<70 mm, and ≥70 mm (Table 2). DHA's mammography dose results are within the range of European data (Perry et al., 2006). The sample of data that was analysed is small, and it is not yet possible to obtain local DRLs. DHA is in the process of collecting more data, and aims to combine this data with that from other hospitals in the UAE to establish national DRLs.

3.3. Interventional procedures

The Philips biplane flat detector system used for interventional procedures (IPs) at Dubai Hospital is fitted with a built-in kerma area product meter. Data were collected for 72 paediatric patients who underwent cardiac procedures. The majority of the cases were therapeutic (occlusion of patent ductus arteriosus, valvuloplasty, and left pulmonary artery ballooning and stenting). DHA's paediatric results are shown in Fig. 2. Average fluoroscopy times for the frontal C-arm (Plane A) were 7.13, 4.02, and 6.42 s for age groups 0–<1 year, 1–<5 years, and 5–<10 years, while for the lateral C-arm (Plane B), corresponding results were 7.14, 2.81, and 2.22 s. The adult data were based on 30 patients (10 patients for each cardiac procedure) for coronary angiography, coronary angioplasty, and peripheral angioplasty. The dose values from coronary angiography are presented in Fig. 3, along with other published IP dose values (Kuipers et al., 2012) for comparison purposes. DHA's coronary angiography sample was mixed (femoral and radial access), while all data from Kuipers et al. (2012) were for radial access. UNSCEAR presented IP dosimetric values in the range of 12.7–147.43 Gy cm^2 (UNSCEAR, 2008). The wide range of IP dose data is mainly due to the complexity of IPs. The average fluoroscopic time for coronary angiography was approximately 2 min. The doses from the other two adult cardiac procedures showed a wide range of values. Hence, further IP dosimetric data collection is in progress as the present data are not sufficient to propose representative local DRLs.

Table 1. Incident air kerma (μGy) for chest (CXR) and abdomen (AXR) x-ray examinations in children at the children and women's hospital.

Age	0–<1 month	1 month–<1 year	1–<5 years	5–<10 years	0–<1 month	1 month–<1 year	1–<5 years	5–<10 years	10–<15 years
Examination	CXR-AXR	AXR	AXR	AXR	CXR	CXR	CXR	CXR	CXR
Average	56.40	65.30	94.88	284.95	54.69	68.87	45.84	64.30	136.44
Minimum	42.72	49.24	50.63	79.87	39.65	57.31	19.22	35.59	44.69
Maximum	76.28	96.09	238.54	1107.51	73.14	92.84	86.50	294.45	373.74
Third quartile	61.14	70.21	114.30	324.88	59.08	72.03	61.78	54.89	118.33
No. of patients	17	21	22	20	20	24	25	21	5
Standard age*	Infant (0)	–	1 year	5 years	Infant (0)	–	1 year	5 years	10 years
ICRP (2013) min–max (μGy)	80–3210	–	–	60–2920	20–1000	–	–	20–1350	20–1160
Hart et al. (2002) min–max (μGy)	–	–	270–400	380–780	30–90	–	60–90	50–230	60–80

*Standard age as given in ICRP (2013) and Hart et al. (2002).

Table 2. Average glandular dose (AGD) (in mGy) at Hospitals 1 (H-1) and 2 (H-2) within Dubai Health Authority.

Breast thickness (mm)	H-1 AGD (mGy)	H-2 AGD (mGy)	H-1 Max (mGy)	H-2 Max (mGy)	European reference level for AGD (mGy) (breast thickness) (Perry et al., 2006)
20–<30	0.67	1.32	0.98	1.42	<1 (20 mm)
30–<40	0.76	1.67	0.86	1.77	<1.5 (30 mm)
40–<45	0.84	1.82	0.9	1.96	<2 (40 mm)
45–<50	0.99	1.67	1.06	1.73	<2.5 (45 mm)
50–<60	1.12	1.80	1.2	1.88	<3 (50 mm)
60–<70	1.29	1.83	1.38	1.92	<4.5 (60 mm)
≥ 70	1.59	1.77	1.78	1.85	<6.5 (70 mm)

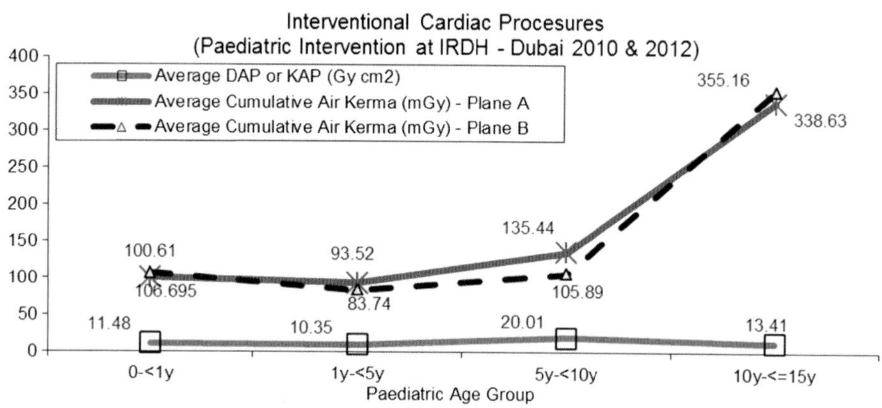

Fig. 2. Dubai Health Authority interventional procedure doses in children. DAP, dose area product; KAP, kerma area product.

3.4. Dental radiology

Sixty intra-oral machines (22 conventional film-based intra-oral x-ray machines and 38 digital intra-oral x-ray machines) and nine panoramic (OPG) dental radiology units were included in DHA's survey. This study presents the dosimetric results of digital intra-oral and OPG systems. Radiation exposures from intra-oral machines were measured using an electronic dental dosimeter (Multi-O-Meter, UNFORS, Billdal, Sweden). The dose survey for the OPG machines was performed using a CT cylindrical ionisation chamber. Exposure parameters implemented in this work reflected those used in clinical situations. The exposures for the intra-oral machines were measured in air at the end of the spacer cone; this estimated the patient entrance

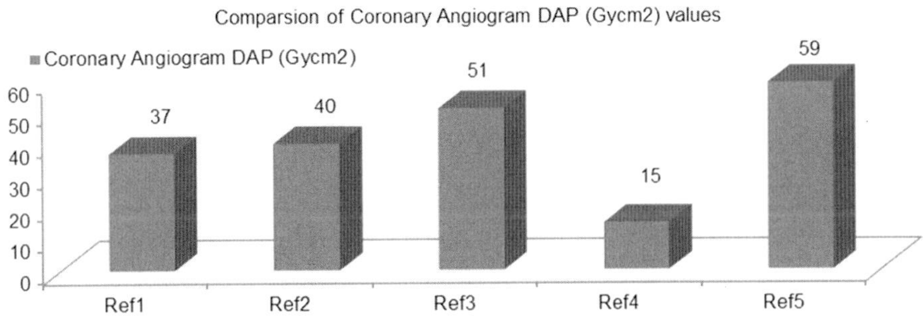

Fig. 3. Dubai Health Authority (DHA) interventional procedure doses in adults. Ref1, DHA, 2013, Dubai; Ref2, Kuipers et al. (2012), Netherlands; Ref3, Sandborg et al. (2004); Ref4, Lange et al. (2006); Ref5, Brasselet et al. (2008), all cited in Kuipers et al. (2012). DAP, dose area product.

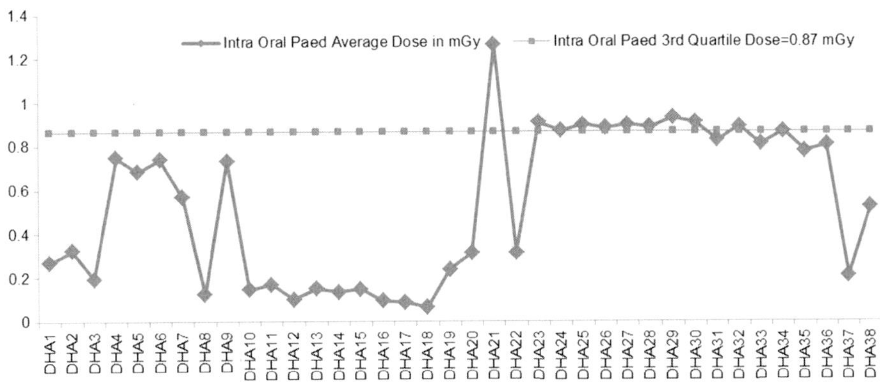

Fig. 4. Dubai Health Authority's intra-oral dose survey (posterior view) in paediatric patients.

dose. Measurements were taken for common intra-oral views: apical (anterior view), pre-molar, molar (posterior view), and bitewing views. This study presents the intra-oral molar (posterior view) dosimetric data in order to compare the results with a UK review undertaken in 2010 and other international references (EC, 2004; Hart et al., 2012). These results are shown in Figs 4–6. Exposures for the OPG systems were measured as the dose at the surface of the image receptor; these results are shown in Fig. 7.

4. DISCUSSION

This article has presented DHA's patient dose monitoring data; this is the first such study in the UAE. The radiation dose levels compare favourably with those published in the literature. For adult patient doses in general digital practice, doses

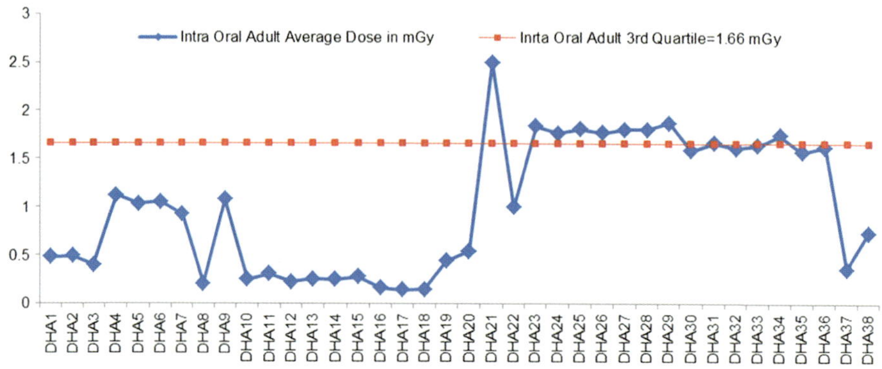

Fig. 5. Dubai Health Authority's intra-oral dose survey (posterior view) in adult patients.

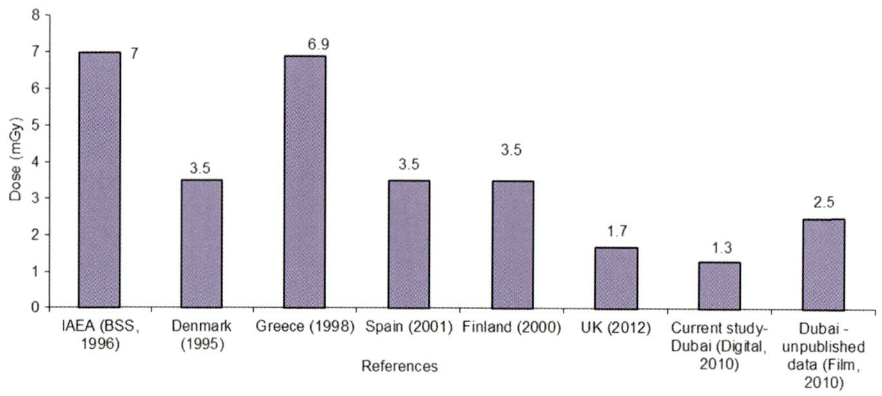

Fig. 6. Comparison of intra-oral dose values from Dubai Health Authority with international references* for adults. *References: IAEA (1996), Denmark (1995), Greece (1998), Spain (2001), and Finland (2000) are cited in EC (2004); UK (2010) is cited in Hart et al. (2012).

are lower than those reported by Hart et al. (2012). Furthermore, paediatric entrance surface dose results in this study are within the same range as those mentioned in *Publication 121* (ICRP, 2013) and Hart et al. (2002). The literature indicates large variations in paediatric DRLs, and there is no standardisation in quoting these values. The authors have not been able to obtain adequate data for the 10–15-year age group, mainly because the paediatric age group used by DHA is up to 13 years (the age group in this study was up to 15 years to correspond with the common paediatric age group used worldwide). Patient data collection for adult and paediatric groups is currently in progress. DHA mammography results are lower than those reported in a European publication, and the range of dose levels among different breast thicknesses also show lower variation (Perry et al., 2006). However, the results for H-2 did not show the expected increase in dose level with increasing breast thickness. The phantom results of H-2 showed a large degree of variation. As the

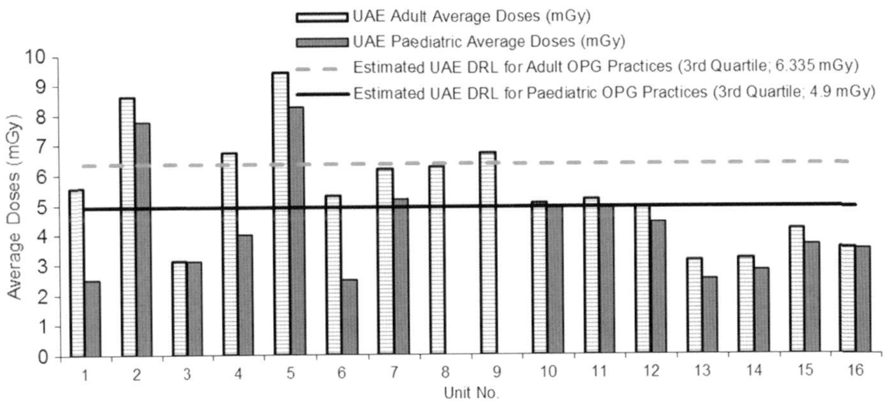

Fig. 7. Dubai Health Authority's panoramic (OPG) dental radiology unit dose survey, 2010–2011, in children and adults. Thanks to Jacek Janaczek and Fatima Al Kaabi, Tawam Hospital, SEHA, UAE for OPG data from seven machines (10–16).

same individual performed the work at H-1 and H-2, and the results of H-2 were repeated for verification purposes, the variation is likely to be due to the conversion factor used by the software in the mammography system. The conversion factor is not mentioned in the DICOM header. DHA's results for IP dosimetric monitoring are limited, and the authors intend to obtain further dose data and cover further IPs. The dental dosimetric survey in this study included dental systems that existed at DHA between 2010 and 2011. However, as the dental data were collected as digital dental radiology was introduced, and new dental systems have since been purchased, it is worth undertaking a dose survey review. The authors aim to continue to collect data, and emphasise the need to collect further data on patient dose in order to obtain significant population samples for adult and paediatric patient groups.

Recording and audit of patient radiation exposure are considered to represent integral elements of a quality assurance programme (RCSI, 2011); however, to date, they have not been fully implemented by DHA. Standardisation of automated dose evaluation by manufacturers is required, and this will help to improve patient radiation safety. Automated DICOM dose extraction techniques are essential to handle large samples of patient dosimetric data. A lack of qualified experts and medical physicists in the study region hinders progress in research work related to patient radiation dosimetry and safety.

In conclusion, the authors believe that the following issues are of great importance in patient radiation safety in the UAE:

- support of educational programmes;
- education and instruction of radiology teams to avoid overexposure and 'dose creep' in digital radiology practices;
- recommendation of simple methods for patient dose evaluation, and emphasis on dose recording within patients' medical reports; and

- a uniform, accurate, and calibrated dose registry at local and national levels.

This study suggests that there is a need to develop further radiation protection and education programmes, along with patient dosimetric monitoring and recording, within the UAE. These radiation protection requirements are considered to be an essential prerequisite to a radiation safety culture within the healthcare community. An educational programme, DHA's Radiation Protection Educational Programme, has now been established; this is delivered on a regular basis (every 3–4 months), and is attended by radiologists, radiographers, medical physicists, biomedical engineers, and quality administrative officers from various areas within the UAE. It is hoped that this will facilitate extensive national patient dose monitoring in the future.

REFERENCES

AlSuwaidi, J., Bayoumi, M., Al Shibli, N., et al., 2011a. Utilisation of PACS to monitor patient CT doses. Radiat. Prot. Dosim. 150, 1–4.

AlSuwaidi, J., AlAwadhi, H., Rahanjam, A., et al., 2011b. National Survey on Medical Exposure to Patients Undergoing Diagnostic Radiology Examinations: an Initial Overview on CT Doses at UAE Hospitals (MECBME). 1st Middle East Conference on Biomedical Engineering, Sharjah, UAE. Available at: http://ieeexplore.ieee.org/Xplore/home.jsp (accessed 4 March 2015).

AlSuwaidi, J., AlBalooshi, L., AlAwadhi, H., et al., 2013. Continuous monitoring of CT dose indexes at Dubai Hospital. AJR, Am. J. Roentgenol. 201, 858–864.

EC, 2004. European Guidelines on Radiation Protection in Dental Radiology. The Safe Use of Radiographs in Dental Practice. Issue N° 136. Radiation Protection 136. European Commission, Luxembourg City.

Hart, D., Hiller, M., Wall, B., 2002. Doses to Patients from Medical X-ray Examinations in the UK 2000 Review. NRPB-W14. Chilton: National Radiological Protection Board (NRPB).

Hart, D., Hillier, M., Shrimpton, P., 2012. Doses to Patients from Radiographic and Fluoroscopic X-ray Imaging Procedures in the UK. Health Protection Agency (HPA) 2010 Review. Chilton: Health Protection Agency (HPA).

Kuipers, G., Delewi, R., Velders, X., et al., 2012. Radiation exposure during percutaneous coronary interventions and coronary angiograms performed by the radial compared with the femoral route. J. Am. Coll. Cardiol. Intv. 5, 752–757.

IAEA, 1996. Basic Safety Standards (BSS). Safety Series No. 115. International Atomic Energy Agency, Vienna.

IAEA, 2007. Dosimetry in Diagnostic Radiology: an International Code of Practice. IAEA Technical Reports Series No. 457. International Atomic Energy Agency, Vienna.

ICRP, 2013. Radiological protection in paediatric diagnostic and interventional radiology. ICRP Publication 121. Ann. ICRP 42(2).

Perry, N., Broeders, M., de Wolf, C., et al., 2006. European Guidelines for Quality Assurance in Breast Cancer Screening and Diagnosis. 4th edn. European Commission, Luxembourg.

RCSI, 2011. Guidelines for the Implementation of a National Quality Assurance Programme in Radiology – Version 2.0. Royal College of Surgeons in Ireland, Dublin.

UNSCEAR, 2008. Sources and Effects of Ionizing Radiation, 2010. United Nations Scientific Committee on the Effects of Atomic Radiation Report to the General Assembly, Volume I. United Nations, New York.

SAGE

Annals of the ICRP

Trends in radiation protection of positron emission tomography/computed tomography imaging

A. Alenezi, K. Soliman

Prince Sultan Military Medical City, Department of Medical Physics, P.O. Box 7897, Riyadh 11159, Saudi Arabia; PH: +966 (11) 4777714; FAX: +966 (11) 2063001; e-mail: aalenezi@psmmc.med.sa

Abstract–Over the past decade, the number of positron emission tomography/computed tomography (PET/CT) imaging procedures has increased substantially. This imaging technique provides accurate functional and anatomical information, particularly for oncological applications. Separately, both PET and CT are considered as high-dose imaging modalities. With the increased use of PET/CT, one could expect an increase in radiation doses to staff and patients. As such, major efforts have been made to reduce radiation dose in PET/CT facilities. Variations in working techniques have made it difficult to compare published results. This study aimed to review the literature on proposed methods to reduce patient and staff dose in clinical PET/CT imaging. A brief overview of some published information on staff and patient doses will be analysed and presented. Recent trends regarding radiation protection in PET/CT imaging will be discussed, and practical recommendations for reducing radiation doses to staff and patients will be discussed and summarised. Generally, the CT dose component is often higher in magnitude than the dose from PET alone; as such, focusing on CT dose reduction will decrease the overall patient dose in PET/CT imaging studies. The following factors should be considered in order to reduce the patient's dose from CT alone: proper justification for ordering contrast-enhanced CT; use of automatic exposure control features; use of adaptive statistical iterative reconstruction algorithms; and optimisation of scan parameters, especially scan length. The PET dose component can be reduced by administration of lower activity to the patient, optimisation of the workflow, and appropriate use of protective devices and engineered systems. At the international level, there is wide variation in work practices among institutions. The current observed trends are such that the annual dose limits for radiation workers in PET/CT imaging are unlikely to be exceeded.

Keywords: PET/CT; Radiation protection; Staff dose; Patient dose; Radiation dose reduction techniques

1. INTRODUCTION

Over the last decade, there have been significant developments in medical imaging techniques. One of the main evolutions is the combination of positron emission tomography (PET) and computed tomography (CT) on a single gantry as a hybrid PET/CT scanner. This provides both anatomical and functional images of the patient as a fused image. Since its introduction in 1998 (Beyer et al., 2000), PET/CT has become an important imaging modality worldwide. PET/CT is one of the fastest-growing medical imaging modalities, with more than 5000 PET/CT systems installed worldwide (Beyer et al., 2011). PET/CT is widely recognised as the main imaging modality used in oncology for early diagnosis, staging, restaging, and monitoring the therapy of several types of tumours. Oncological applications account for more than 90% of PET/CT procedures, whereas other applications, such as cardiology, neurology, and psychiatry, account for 4% (IMV Medical Information Division, 2011). Due to the increased use of PET/CT with increased patient throughput, one could expect an increase in radiation doses to workers. Radiation protection issues in PET/CT have been addressed adequately, and significant progress has been made in reducing occupational radiation doses in PET/CT facilities. Applied measurement techniques and different working practices increase the variations in published data. Other factors, such as the level of experience of staff, type of PET/CT scanner, work practice, administered activity, procedure used during dose dispensing, time spent by worker close to patient, distance from the patient, workload, and CT parameters, are the causes of these variations.

There is a need to review the current literature and produce updated recommendations. The main aim of this study was to present and analyse published information on recent trends in radiation protection at PET/CT facilities, and discuss state-of-the-art radiation reduction techniques for staff and patients.

Separately, both PET and CT are considered as high-dose imaging modalities. However, a properly justified diagnostic CT examination must be taken as part of a PET/CT examination in order to avoid the radiation dose received subsequently from an additional CT examination.

Experience has shown real difficulty regarding CT procedure standardisation due to the wide variation in radiology practice worldwide. Most CT studies have adequate image quality, allowing the radiologist to make a proper diagnosis. Therefore, it is always possible to optimise the patient radiation dose by accepting a lower image quality. However, this task is extremely challenging to implement.

Primary tumour staging often requires a good-quality diagnostic CT scan. For cancer patients, the benefit from a CT examination is much higher than the risk associated with the radiation dose received during imaging.

2. CT DOSE COMPONENT IN PET/CT IMAGING

Data on the doses delivered during PET/CT imaging to patients and staff are available in the scientific literature. Implementation of the optimisation principles in PET/CT requires assessment of patient doses, exposed workers, and the operational aspects of radiation protection in the clinical environment.

It has been reported that low-dose CT used for attenuation correction (AC) and anatomical orientation may only account for 50% of the whole PET/CT dose, with an average PET dose of 7 mSv and an average low-dose CT dose of 2–3 mSv (total 9–10 mSv) (Mattson and Soderberg, 2012).

It has been reported that only 73% of PET/CT facilities use dedicated low-dose CT acquisition for AC of PET/CT studies (Beyer et al., 2011). The remaining 27% of PET/CT facilities should be encouraged to implement low-dose CT in routine practice in order to reduce the patient radiation dose.

Three different CT acquisition protocols were compared using a humanoid (Alderson-Rando) phantom equipped with thermoluminescent dosimeters (TLD-100) during a whole-body PET/CT scan; the CT dose component accounted for 54–81% of the total combined dose. The calculated effective dose from PET/CT scanning with a diagnostic CT protocol and an administered ^{18}F-FDG activity of 370 MBq was approximately 32 mSv for a whole-body diagnostic scan (Huang et al., 2009).

Brix et al. (2005) measured the effective dose from PET/CT examinations with four diagnostic CT protocols, and reported a range of 23.7–26.4 mSv. Wu et al. (2004) reported that the effective dose from diagnostic CT and PET was 18.9 mSv and 10.7 mSv, respectively.

When full CT is required along with whole-body PET, the advantage of conducting the CT scan before the PET scan is that the CT data can be used for AC, and misalignment artefacts will be minimised. On the other hand, if the CT scan is conducted after the PET scan, a smaller field of view can be selected and an optimised CT examination can be performed. Also, if the patient has an allergic reaction to the contrast media, this will not affect the PET study, but there may be more image misalignment artefacts.

3. JUSTIFICATION OF CT EXAMINATION IN PET/CT IMAGING

One important concern in diagnostic radiology is that all examinations should be justified to avoid unnecessary radiation exposure of the individual patient. To prevent unjustified examinations, close cooperation between radiologists, other clinicians, and medical physicists is required. In a recent survey about the justification of performing CT examinations in Sweden, it was reported that 20% of all CT examinations were unjustified, and that a major population dose reduction could be achieved by avoiding such unjustified examinations (Swedish Nuclear Safety Authority, 2009).

Efficient cooperation between nuclear medicine and radiology departments is necessary to use PET/CT investigations most appropriately. Efforts should be made to better inform referring physicians about various alternatives to radiological examinations and the criteria for their use. Radiologists and nuclear medicine physicians should establish evidence-based imaging parameters as a function of clinical indications for PET/CT imaging. The development of clear referral criteria for imaging will decrease the number of inappropriate examinations, and hence reduce collective dose to patients.

To reduce radiation dose, low-dose CT in combination with ultrasound and/or magnetic resonance imaging for the assessment of anatomy is often preferred to a full-dose examination (Biermann et al., 2013). Guidelines are difficult and expensive to produce and maintain. Such decision support must be comprehensive, user-friendly, relevant, and interactive. The current challenge in the area is the fact that the proposed guidelines must be based on sound methodology. In conclusion, there is a need for imaging referral guidelines in a useable format.

4. CT DOSE OPTIMISATION IN PET/CT IMAGING

CT technology is used in PET/CT to perform AC and anatomical reference localisation for the PET scan data set. Low-resolution CT imaging using multi-slice CT scanners, usually 64-slice scanners, is usually requested.

In CT imaging, proper selection of exposure parameters (e.g. kVp, mAs, scan length, pitch, and gantry rotation speed) is required. Dose reduction features such as automatic exposure control, conformal three-dimensional dose modulation, and new image reconstruction algorithms (e.g. iterative reconstruction techniques) must be applied when available. The use of these technologies will empower the end user to achieve patient and staff dose reduction and optimisation.

The relationship between justification and optimisation principles is complex; there is a need to establish proper clinical justification criteria for high-resolution or contrast-enhanced CT (CECT) examinations in PET/CT. There is also a need to avoid repeated prescriptions because these will increase patient dose.

In CT procedures, radiation dose can be estimated using the dosimetric quantities available as part of the image DICOM file, such as the CT dose index and the dose length product. The dose may also be measured using established reference dosimetry methods with body phantoms and pencil dosimeters coupled with electrometers. At the national level, a patient radiation dose survey will require application of a standardised acquisition protocol for all surveyed scanners, and adequate time for performance of the survey.

In order to evaluate the successful implementation of dose optimisation techniques, the standard imaging protocols in use must be examined. Medical institutions may customise the suggested optimised protocols to fit their scanners. However, training and awareness of technical and medical staff performing and reporting the PET/CT scan are vital. Vendors' cooperation during the implementation stage is also important.

4.1. Contrast-enhanced CT

CT scanners used in PET/CT imaging system are identical to any standard diagnostic CT scanner (modern multi-slice CT). These CT scanners are used to produce scans specifically for AC of the PET data. CT-based AC (CTAC) scans are usually lower-dose scanning parameters. The CTAC scans must not be used for other diagnostic purposes. On the other hand, a high-quality diagnostic scan can also be used for AC. Diagnostic scans should only be used if more clinical indications are required other than AC of PET images. In such case, diagnostic CT examination will replace the CTAC scan to avoid additional patient dose. CECT studies have shown some advantages, such as improved delineation of anatomical structures, increased sensitivity for detection of pathological lesions, and improved accuracy in lesion characterisation (Antoch et al., 2004).

In a study to evaluate low-dose non-enhanced CT (LDCT) and full-dose CECT, PET/CECT was found to be a more accurate imaging modality with higher confidence for assessing ovarian cancer recurrence than PET/LDCT (Kitajima et al., 2012).

Another study has clearly shown the usefulness of PET/CT for better evaluation and management of patients with fibroblast proliferative disease. Accurate determination of the level of inflammation was significantly improved when PET/CT and CECT were performed in the same study, and used for better delineation of areas of residual inflammation (Liu et al., 2012). When performing PET/CT, optimal detection and characterisation of liver lesions required the use of a fused CECT scan in patients with colorectal cancer (Ozkan et al., 2012).

The use of iodinated contrast media for small-animal PET imaging has been found to significantly improve tumour delineation and diagnostic performance, without significant alteration of small animal PET quantitative accuracy and National Electrical Manufacturers Association (NEMA) image quality parameters (Lasnon et al., 2013).

Other studies have shown that there is little advantage in using CECT (e.g. in staging malignant melanoma) in comparison with LDCT (Elstrom et al., 2008; Pfluger et al., 2011). Yoshida et al. (2009) reported that routine CECT is not justified for lesion detection in staging primary head and neck cancers with PET/CT.

While contrast CT examinations have good diagnostic value in PET/CT imaging studies, there is some disagreement on their role in AC. Clinical PET/CT studies have shown that contrast agents may result in overestimation of standardised uptake values on images corrected for attenuation with attenuation maps derived from CT (Berthelsen et al., 2005; Yau et al., 2005; Mawlawi et al., 2006; Bunyaviroch et al., 2008; Aschoff et al., 2012).

The mean and maximum standardised uptake values were significantly increased in contrast-enhanced PET/CT compared with non-enhanced PET/CT at each anatomical site (Prechtel et al., 2012). On the other hand, it has been reported that intravenous contrast does not have an adverse effect on the sensitivity and specificity of PET/CT (Hall et al., 2011). Non-AC PET images can also be used to solve potential image artefacts arising from CECT AC PET images in PET/CT.

4.2. Image reconstruction algorithms

Adaptive statistical iterative reconstruction (ASIR) has been reported to save dose in comparison with the known filtered back projection algorithm. The images produced have the appearance of noise-free images and may not be easily accepted by radiologists. A blended solution exists, composed of a linear combination of filtered back projection and ASIR, with a natural appearance and less noise. These preliminary results support body CT dose index reductions of 32–65% when ASIR is used. Studies with larger statistical samples are needed to confirm these findings (Hara et al., 2009). Using the iterative reconstruction technique, it has been shown that the temporal bone dose can be reduced by 50% in comparison with the conventional filtered back projection algorithm (Niu et al., 2012). Most CT vendors have now incorporated the iterative reconstruction technique into their scanners. Technical evaluations of the proposed iterative reconstruction algorithms showed reduced noise levels and increased signal-to-noise ratio in images while maintaining the same spatial resolution. The immediate benefit from applying this new reconstruction technique is the ability to use it extensively on obese patients, in paediatric examinations, on patients requiring multiple imaging studies for follow-up, and on patients unable to use magnetic resonance imaging as an alternative imaging modality (Silva et al., 2010).

Continuous effort from major CT vendors in the area of reconstruction algorithm development is demonstrated by introduction of the model-based iterative reconstruction technique, which has shown good prospective achievement of image noise and patient dose reduction while maintaining good image quality (Pickhardt et al., 2012; Vardhanabhuti et al., 2013). This method can be applied to reconstruct low-dose acquired CT images (Katsura et al., 2012). The only drawback of the method is its longer computing time.

4.3. Optimisation of CT scan parameters

In PET/CT imaging, the CT scan is used mainly for AC of the PET images; such correction is achieved by using low-dose CT. Dose reduction in CT can be achieved by using adjusted examination acquisition parameters such as lower kVp and mAs, larger pitch for spiral acquisition, and, when possible, shorter scan length. A large number of studies covering the previously mentioned parameters have been published (Graham et al., 2011). The use of automatic tube current modulation has been recommended (Alessio et al., 2009).

Optimisation of scanning parameters is now possible due to numerous published studies using simulation tools, and such tools have shown that parameter optimisation will lead to reduced patient dose without affecting image quality. The simulation works simply by adding an artificial level of noise to the CT raw data in order to simulate a lower dose examination. As mAs is directly related to dose, a lower mAs examination will definitively produce a lower radiation dose scan (Kumar et al., 2012b). A specific level of image quality is desired by clinicians in order to acquire valid CT studies that will answer the clinical question under investigation. Other reported parameters affecting patient dose are kVp, slice thickness, and scan length.

Automatic exposure control techniques are now integrated in all new CT scanners, and a large number of studies have shown their usefulness in patient dose reduction.

One important parameter affecting patient dose from CT is the scan length, as the effective dose is related to the dose length product using published conversion factors relating the two parameters. Proper scan length can be estimated easily from the internationally published dose reference levels, usually expressed using the CT dose index and dose length product.

Hence, administered activity for a PET scan cannot be reduced substantially because it is related to the patient's weight. The easiest way to decrease the patient radiation dose burden from PET/CT imaging is to optimise and adjust the CT scan acquisition parameters for each patient. There is a need for well-established clinical guidelines for CT examination requirements as part of the PET/CT examination.

5. PET DOSE COMPONENT IN PET/CT IMAGING

PET/CT procedures include many tasks that contribute significantly to the occupational radiation exposure of staff. These tasks include: (1) dispensing and preparation of individual patient dose; (2) dose administration; (3) measurement of residual activity; (4) patient contact during uptake period; (5) escorting the patient to the PET/CT scanner; (6) positioning the patient on the PET/CT couch and helping the patient to rise from the couch; (7) scanning of the patient (worker will not be exposed to x rays during the CT examination and the patient is exposed to external radiation); and (8) escorting the patient out of the department at the end of the PET/CT examination. The total radiation dose depends on the skill of the worker to implement the radiation protection principles of time, distance, and shielding. Therefore, reducing radiation dose to workers in a PET/CT facility is a complicated issue, and more effort is required to implement the ALARA (as low as reasonably achievable) principles and International Commission on Radiological Protection (ICRP) recommendations.

ICRP has recommended that the occupational dose limit for radiation exposure is 20 mSv year^{-1} averaged over 5 years, and a maximum of 50 mSv in any single year (ICRP, 1991). According to the literature, the annual occupational dose to a PET/CT worker is always well below the 20 mSv dose limit. Seierstad et al. (2007) estimated the occupational dose to workers in PET/CT, and attributed approximately 60% of the occupational dose to the technologist handling the radioactive materials and 40% to close contact with the patient. This indicates that handling radioactive materials before and during injection, and dealing with the patient after the injection, are important aspects and must be addressed when attempting to reduce occupational radiation dose in PET/CT facilities.

6. PET DOSE REDUCTION TECHNIQUES

6.1. Administered activity

It has been reported that only 44% of PET/CT facilities employ weight-based radioactivity dose injection (Beyer et al., 2011). This indicates that the remaining PET/CT facilities should implement weight-based radioactivity dose injection in their practice in order to optimise radiation dose to patients. Given the availability of the three-dimensional acquisition mode, administered activity could be reduced without affecting image quality. For a standard adult patient (75 kg), the European Association of Nuclear Medicine recommends using 380 MBq for the two-dimensional mode and 190 MBq for the three-dimensional mode (Boellaard et al., 2010).

All new commercial PET/CT scanners offer time-of-flight (TOF) technology, which can help to overcome poor signal from large patients. TOF accurately measures the actual time difference between detection of the two annihilation photons, and this leads to improved image contrast and higher sensitivity. El Fakhri et al. (2011) reported that TOF reduces the injected dose, and therefore reduces the exposure of patients and workers to radiation. Etard et al. (2012) reported that new TOF technology can decrease average specific activity by approximately 20% (from 4.3 MBq kg^{-1} to 3.5 MBq kg^{-1}) while maintaining image quality.

Hydration and voiding are also important for patient preparation during PET/CT. The patient is encouraged to drink water after the ^{18}F-FDG injection and before scanning in order to limit the radiation dose to the bladder. The bladder is the organ that receives the largest radiation dose during PET/CT procedures. The estimated absorbed radiation dose to the bladder is 0.16 mGy MBq^{-1}, 0.21 mGy MBq^{-1}, and 0.32 mGy MBq^{-1} for an adult, a 15-year-old, and a 5-year-old, respectively (ICRP, 1998).

6.2. Extremity dose

Staff preparing and administering unsealed ^{18}F-FDG doses may receive a significant dose to their hands, especially their fingers. The location of the highest exposure to the worker's hand should be measured and monitored. ICRP has recommended that skin dose should be measured at a depth of 0.07 mm (the skin's basal cell layer is considered to be 70 μm), and should not exceed the annual dose limit of 500 mSv (ICRP, 2007). ICRP also recommends that ring dosimeters should be worn at the base of the middle finger, with the TLD facing the palm, and recommends a factor of 3 to estimate the highest extremity dose at the fingertip if the TLD faces the palm and a factor of 6 when the TLD faces the dorsal side of the hand (ICRP, 2008). The extremity dose to the fingers of workers dealing with ^{18}F-FDG is more than three orders of magnitude higher than the dose for workers handling ^{99m}Tc (3.0 μGy MBq^{-1} for ^{18}F and 0.0012 μGy MBq^{-1} for ^{99m}Tc) (Leide-Svegborn, 2012).

The skin dose due to contact with 5 ml of ^{18}F is eight times higher than the skin dose from ^{99m}Tc (2.88 mSv MBq h^{-1} for ^{18}F and 0.354 mSv MBq h^{-1} for ^{99m}Tc) (Delacroix et al., 2002). The skin dose limit is applied to dose averaged over an area of 1 cm^2 (ICRP, 1991). Therefore, it should be measured accurately at the location that represents the highest skin dose. This is considered as the extremity dose and should be measured at the tip of the fingers or the thumb. Published studies for extremity dose indicate that there is a significant variation in the measured extremity dose in PET/CT imaging using ^{18}F-FDG. These variations are mainly due to position of the dosimeter (palmer or dorsal), location of the dosimeter (base of ring finger or wrist), choice of hand (dominant or non-dominant), exposure time (condition of the patient and experience of the worker), amount of activity, individual skills and habits of the workers, number of patients handled (or number of measurements per worker), total number of workers, distance between finger and radioactive source, and use of syringe shields. Exposure time may be affected by the inconvenience of TLD location on the finger and the experience (or skill) of the worker. The skill of the worker plays an important role in the dose received (ICRP, 2007). Carnicer et al. (2011) found that good working habits were more important than experience. More experienced PET/CT workers should be responsible for dealing with paediatric, disabled, and elderly patients.

Most hospitals do not perform extremity dose monitoring (Kemerink et al., 2012). Based on the ORAMED (Optimisation of RAdiation protection of MEDical staff) study, nearly 20% of nuclear medicine staff are likely to receive a skin dose that exceeds the annual skin dose limit of 500 mSv (ORAMED, 2011). This is because it is impractical to place the dosimeter at the fingertip, and inconvenient for the worker. The fingertip is the part of the hand that is likely to receive the highest radiation exposure due to its proximity to the radiation source. Therefore, in routine practice, extremity dose is measured by placing the dosimeter at the base of the ring finger (ring dosimeter) or at the wrist (wrist dosimeter), which underestimates the actual fingertip dose due to distance from the fingertip. There is a large discrepancy between finger doses received by the dominant and non-dominant hands. Withdrawal of activity can lead to a greater finger dose discrepancy between the dominant and non-dominant hands than injection and waste handling (Leide-Svegborn, 2012). This is likely to be due to use of the non-dominant hand to support the syringe during the removal of air bubbles, and to support the needle during withdrawal of activity from the vial into the syringe.

As a result, in routine daily practice, the majority of workers use ring dosimeters because they are convenient to wear. However, a corrective factor should be applied to estimate the highest finger (fingertip) dose. Correction factors of 6 (Carnicer et al., 2011) and 9 (Covens et al., 2007) have been reported in the literature. This indicates that there is a need to change the practice of wearing finger dosimeters from the wrist to the base of the index finger of the non-dominant hand (facing the palm) in order to give a more accurate estimate of the maximum finger dose (Carnicer et al., 2011).

Covens et al. (2007) studied the highest skin dose for seven workers (right-handed) performing ^{18}F manipulation (kit preparation, dispensing, and injection), and reported that the highest dose location was situated on the tip of the ring finger of the left hand. The highest skin dose for one worker was 850 μSv per handled GBq of activity when manipulating ^{18}F-FDG without radiation protection measures. This worker would easily exceed the annual dose limit. Carnicer et al. (2011) reported skin doses of 930 μSv GBq^{-1} and 1200 μSv GBq^{-1} for the administration and preparation of ^{18}F-FDG, respectively, and showed that the dose to the hands might surpass the dose limit of 500 mSv. Leide-Svegborn (2012) reported that the finger dose might reach 420 mSv, which is close to the 500 mSv annual finger dose limit, and suggested that much has to be done in order to reduce the extremity dose from ^{18}F-FDG procedures.

Although use of a syringe shield is a practical solution to reduce radiation dose to fingers during manual injection, other efficient systems such as automatic (or semi-automatic) dispensing/injection systems have been recommended to limit the finger dose. Covens et al. (2010) studied one of these automated dispensing/injection systems, and found that extremity dose can be reduced by more than 95% with the use of an automated dispensing/injection system. Lecchi et al. (2012) studied the use of a combined dispensring/injector system, and reported that the extremity dose to the physician (when administration of ^{18}F-FDG was performed by a physician) was reduced by 93%. Prevot et al. (2012) reported that collective finger doses were reduced by 39% with the use of an automated dispensing/injection system.

6.3. Automated dispensing/injection systems

Most of the occupational radiation dose is received during dispensing and injection of ^{18}F-FDG. The extremity dose is mainly received during handling of the syringe. ^{18}F-FDG is often dispensed and injected manually. Recently, some manufacturers have developed automated combined dedicated dispensing/injection systems. Covens et al. (2010) studied one of these systems and reported that the dose was reduced by 20% during the ^{18}F injection phase and by up to 50% for the whole-body dose.

Schleipman and Gerbaudo (2012) studied occupational dose to workers from both automated infusion and manual injection protocols. Their study discovered a10-fold reduction in staff extremity and body dose with use of an automated infusion device compared with manual injection of ^{18}F-FDG radiotracers. Prevot et al. (2012) reported that collective whole-body exposure was decreased by 25% in 2 years with the use of an automated dispensing/injection system. Lecchi et al. (2012) reported that the whole-body radiation dose was reduced significantly by 38% with the use of a combined dispensing/injection system. These systems have some disadvantages, such as high capital cost, cost of consumables, cost of multi-dose vials from off-site cyclotrons, unfamiliarity with the system, and space requirement. Kumar et al. (2012a) studied dose rate to staff during different phases (injection, positioning, and scanning of the patient), and found that injected dose made a greater contribution to the total radiation dose. All these studies indicate the need to consider the use of automated injection systems, particularly in PET/CT facilities with high patient throughput.

6.4. Shielding

Use of primary syringe shields (dedicated for 511 keV) can reduce the average dose per injection by approximately 44% (from 6.8 nSv MBq^{-1} with unshielded syringe to 3.8 nSv MBq^{-1} with shielded syringe) (Roberts et al., 2005). The wrist dose was reduced by approximately 50% by the use of syringe shields (15 nSv MBq^{-1} with unshielded syringe to 7.5 nSv MBq^{-1} with shielded syringe) (Amaral et al., 2007). Reductions of 35% in the left hand and 33% in the right hand were obtained by shielding the syringe (Demir et al., 2010). It has been reported that use of a syringe shield can reduce finger doses, yet the finger doses for some workers are still high. This is due to improper use of syringe shields, where workers position their fingers around the needle during injection with a syringe shield (Carnicer et al., 2011).

6.5. Other techniques to minimise radiation exposure in PET/CT

The main principles of radiation protection (time, distance, and shielding) should be followed when working with ^{18}F-FDG radiopharmaceuticals and injected patients. Explaining the procedure to the patient before the injection is considered to be a practical method to reduce radiation exposure to workers. After ^{18}F-FDG injections, the patient should be viewed and communicated with remotely through a closed-circuit television camera and two-way intercom systems when possible. This will reduce contact time and increase distance from the patient. The staff or family member who escorts the patient when leaving the PET/CT facility at the end of the procedure should keep a distance of >0.5 m in the first 2 h after the injection, and the patient should stay away from children and pregnant women for at least 6 h (Demir et al., 2011). There are some PET/CT facilities where one technologist performs all tasks, and this significantly increases the whole-body dose to the worker. Sharing patient injections with other staff, such as nurses and doctors, can reduce the technologist's dose. Staff education on radiation protection issues on a regular basis with updates on radiation dose reduction techniques can promote safer practices. New workers should learn how to perform dispensing and injection procedures with dummy sources in order to become more familiar and improve their handling technique. Rotation of the work tasks between staff can help to reduce occupational dose. For individual members of staff, reducing the workload can help to reduce staff dose.

7. PET/CT FACILITY WITH HIGH PATIENT THROUGHPUT

Clearly, increasing the number of PET/CT procedures will lead to proportional increase in occupational dose to workers. The annual dose to workers should be estimated and monitored before increasing the number of PET/CT procedures. Orlacchio et al. (2012) reported that the maximum number of PET/CT examinations should be 14 examinations day^{-1}. Based on the data presented by Guillet et al. (2005), the effective whole-body dose to a technologist handling all tasks was 2.95 μSv $patient^{-1}$ (per 309 MBq administered activity), and using the assumption of 14 patients day^{-1}, 5 days $week^{-1}$, and 43 weeks $year^{-1}$, the annual occupational dose for an individual worker would be approximately 8.88 mSv. This value is still below the annual occupational dose limit (20 mSv). Demir et al. (2010) reported that the whole-body dose will exceed the ICRP limit after 4300 patients (working 12 h day^{-1}, five technologists, and administered activity of 518 MBq). Seierstad et al. (2007) reported that the annual body dose limit is reached after handling approximately 3000 patients or after injection of 1 TBq. The previous assumption did not consider variations in administered activity.

With regard to skin dose, assuming a finger dose of approximately 2–3 μSv MBq^{-1} for 350 MBq for 500 patients day^{-1} as reported by Leide-Svegborn (2012), the annual finger dose limit will be reached after handling approximately 2400 patients year^{-1} (Mattsson et al., 2011). Demir et al. (2010) reported that the skin radiation dose limit will be exceeded after 3000 episodes. This indicates that the maximum number of ^{18}F-FDG injections should not exceed 2400 patients year^{-1} during 8 working h/day (for manual injection only) in order to remain below the annual finger dose limit.

With the use of semi-automatic injectors, finger doses can be reduced from 2–3 μSv MBq^{-1} to 0.2–0.6 μSv MBq^{-1} (Guillet et al., 2005). This will enable PET/CT facilities to increase the number of PET/CT procedures performed per year without exceeding skin dose limits.

It is recommended that each PET/CT facility should define their workload, taking into consideration the administered activity and the use of automated dispensing/injection systems. Additional staff should be considered to manage the extra radiation dose from high workloads.

8. RESULTS

Table 1 summarises the recommendations suggested to optimise radiation protection practice in PET/CT imaging.

9. CONCLUSIONS

Occupational radiation dose can be decreased by reducing the time that workers spend near the radioactive source (this includes the patient after administration), by increasing the distance from the radioactive source, and by using dedicated shielding tools. Extremity dose for workers in PET/CT is an issue that needs to be considered carefully in routine daily practice, and the highest radiation dose should be estimated. Use of automated dispensing/injection systems significantly reduces the extremity dose of workers. All manufacturers should invest in research that focuses on improving the sensitivity and efficiency of PET/CT scanners.

There is also a need to focus on educating and training PET/CT staff to pay more attention to practical aspects of radiation protection. There is an urgent need for international efforts to standardise and optimise the use of PET/CT imaging techniques in order to keep the radiation dose as low as reasonably achievable while optimising image quality. The international effort should focus on accurate monitoring of extremity doses, and increased awareness of high extremity doses.

Table 1. Practical radiation dose reduction techniques in positron emission tomography/computed tomography (PET/CT) imaging.

Preparation phase	• More communication between the referring physician and the nuclear medicine physician to reduce unjustified procedures • Patient instructions should be given to the patient prior to injection of the radiopharmaceutical • Establish clear intravenous access before the injection to avoid the need for repeat injection • Workers must wear an extremity dosimeter in PET/CT imaging practice
Dose injection	• Administered activity of the radiopharmaceutical should be minimal • Use unit doses instead of multi-dose • Use distance tools such as forceps • Use automated dispensing (or/and injection) system • Use radiation-shielded tools (syringe shields, syringe carriers, carts, waste containers, etc.)
Uptake phase	• Minimise time near the patient after the injection • CCTV and intercom facility should be used in the uptake room to minimise contact time with the patient • During uptake time, the patient should be encouraged to drink fluids and void frequently in order to reduce the radiation dose to the urinary bladder
Imaging phase	• Maximisation of distance when escorting and positioning the patient • Increase the acquisition time per bed position • During positioning of the patient on the PET/CT scanner, workers are encouraged to stand to the right of the patient and to minimise close contact time • Positioning time of the patient on the PET/CT table should be short
CT dose optimisation	• Use low-dose non-enhanced CT for AC • Use dose optimisation techniques by adjusting scanning parameters • Use automatic exposure control and dose modulation techniques • Use iterative reconstruction techniques • Encourage medical organisations and involved groups to develop CT examination prescription guidelines for PET/CT
Patient discharge	• Use other personnel for escorting the patient • For nursing mothers, expressed milk should be discarded for 2 h after the scan • Injected patients should keep a distance of >0.5 m from others in the first 2 h after the injection, and stay away from children and pregnant women for at least 6 h

Staff-related issues	• Fluorodeoxyglucose preparation and administration by a skilled worker • Training and rotation scheme within workers (divide responsibilities between staff) • Practical continuing education of PET/CT staff to reduce the extremity dose • Increase the number of PET/CT technologists in a department with high throughput • Optimisation of the individual workload is often used to restrict staff doses • Harmonising a competency-based programme for the certification of personnel in radiation protection (dual certificate on radiation safety in PET/CT imaging) • Workers should gain experience and improve their skills in handling radioactive material (e.g. more practice with non-radioactive materials)
Technology-related issues	• Vendors, workers, and inventors should collaborate to develop equipment, software, and protocols that deliver less radiation • Use of three-dimensional PET acquisition mode and time-of-flight technology • Perform regular quality control and preventive maintenance of PET/CT scanner

CCTV, closed-circuit television; AC, attenuation correction.

REFERENCES

Alessio, A.M., Kinahan, P.E., Manchanda, V., et al., 2009. Weight-based low dose paediatric whole-body PET/CT protocols. J. Nucl. Med. 50, 1570–1577.

Amaral, A., Itie, C., Bok, B., 2007. Dose absorbed by technologists in positron emission tomography procedures with FDG. Braz. Arch. Biol. Technol. 50, 129–134.

Antoch, G., Freudenberg, LS., Beyer, T., et al., 2004. To enhance or not to enhance? ^{18}F-FDG and CT contrast agents in dual-modality 18F-FDG PET/CT. J. Nucl. Med. 45 (Suppl. 1), 56S–65S.

Aschoff, P., Plathow, C., Beyer, T., et al., 2012. Multiphase contrast-enhanced CT with highly concentrated contrast agent can be used for PET attenuation correction in integrated PET/CT imaging. Eur. J. Nucl. Med. Mol. Imag. 39, 316–325.

Berthelsen, A.K., Holm, S., Loft, A., et al., 2005. PET/CT with intravenous contrast can be used for PET attenuation correction in cancer patients. Eur. J. Nucl. Med. Mol. Imag. 32, 1167–1175.

Beyer, T., Townsend, D.W., Brun, T., et al., 2000. A combined PET/CT scanner for clinical oncology. J. Nucl. Med. 41, 1369–1379.

Beyer, T., Czernin, J., Freudenberg, L.S., 2011. Variations in clinical PET/CT operations: results of an international survey of active PET/CT users. J. Nucl. Med. 52, 303–310.

Biermann, M., Schwarzlmüller, T., Fasmer, KE., et al., 2013. Is there a role for PET-CT and SPECT-CT in paediatric oncology? Acta Radiol. 54, 1037–1045.

Boellaard, R., O'Doherty, M.J., Weber, W.A., et al., 2010. FDG PET and PET/CT: EANM procedure guidelines for tumour PET imaging: version 1.0. Eur. J. Nucl. Med. Mol. Imag. 37, 181–200.

Brix, G., Lechel, U., Glatting, G., et al., 2005. Radiation exposure of patients undergoing whole-body dual-modality 18F-FDG PET/CT examinations. J. Nucl. Med. 46, 608–613.

Bunyaviroch, T., Turkington, T.G., Wong, T.Z., et al., 2008. Quantitative effects of contrast enhanced CT attenuation correction on PET SUV measurements. Mol. Imag. Biol. 10, 107–113.

Carnicer, A., Sans-Merce, M., Baecheler, S., et al., 2011. Hand exposure in diagnostic nuclear medicine with 18F- and 99mTc-labelled radiopharmaceuticals – results of the ORAMED project. Radiat. Measur. 46, 1277–1282.

Covens, P., Berus, D., Vanhavere, F., et al., 2010. The introduction of automated dispensing and injection during PET procedures: a step in optimisation of extremity doses and whole-body doses of nuclear medicine staff. Radiat. Prot. Dosimetry 140, 250–258.

Covens, P., Berus, D., Buls, N., et al., 2007. Personal dose monitoring in hospitals: global assessment, critical applications and future needs. Radiat. Prot. Dosimetry 124, 250–259.

Delacroix, D., Guerre, J.P., Leblanc, P., et al., 2002. Radionuclide and radiation protection data handbook. 2nd edn. Radiat. Prot. Dosimetry 98, 1–168.

Demir, M., Demir, B., Sayman, H., et al., 2011. Radiation protection for accompanying person and radiation workers in PET/CT. Radiat. Prot. Dosimetry 147, 528–532.

Demir, M., Demir, B., Yaşar, D., et al., 2010. Radiation doses to technologists working with F-FDG in a PET center with high patient capacity. NUKLEONIKA 55, 107–112.

El Fakhri, G., Surti, S., Trott, C., et al., 2010. Improvement in lesion detection with whole-body oncologic time-of-flight PET. J. Nucl. Med. 52, 347–353.

Elstrom, R.L., Leonard, J.P., Coleman, M., et al., 2008. Combined PET and low-dose, non-contrast CT scanning obviates the need for additional diagnostic contrast-enhanced CT scans in patients undergoing staging or restaging for lymphoma. Ann. Oncol. 19, 1770–1773.

Etard, C., Celier, D., Roch, P., et al., 2012. National survey of patient doses from whole-body FDG PET-CT examinations in France in 2011. Radiat. Prot. Dosimetry 152, 334–338.

Graham, M., Badawi, R., Wahl, R., 2011. Variations in PET/CT methodology for oncologic imaging at U.S. academic medical centers: an imaging response assessment team survey. J. Nucl. Med. 52, 311–317.

Guillet, B., Quentin, P., Waultier, S., et al., 2005. Technologist radiation exposure in routine clinical practice with 18F-FDG PET. J. Nucl. Med. Technol. 33, 175–179.

Hall, L.T., Struck, A.F., Guglielmo, C.G., 2011. Does the use of IV contrast enhanced CT for attenuation correction affect clinical interpretation of head and neck PET/CT? Open Nucl. Med. J. 3, 12–18.

Hara, A.K., Paden, R.G., Silva, A.C., et al., 2009. Iterative reconstruction technique for reducing body radiation dose at CT: feasibility study. AJR Am. J. Roentgenol. 193, 764–771.

Huang, B., Law, M.W., Khong, P.L., 2009. Whole-body PET/CT scanning: estimation of radiation dose and cancer risk. Radiology 251, 166–174.

IMV Medical Information Division, Inc., 2008. PET Market Summary Report. Available at: http://www.marketreserch.com/IMV-Medical-Information-Division-Inc-v229/ (accessed 20 September 2013).

ICRP, 1991. The Recommendations of the International Commission on Radiological Protection. ICRP Publication 60. Ann. ICRP 21(1–3).

ICRP, 1998. Radiation dose to patients from radiopharmaceuticals (Addendum to ICRP Publication 53). ICRP Publication 80. Ann. ICRP 28(3).

ICRP, 2007. The 2007 Recommendations of the International Commission on Radiological Protection. ICRP Publication 103. Ann. ICRP 37(2–4).
ICRP, 2008. Radiation Dose to Patients from Radiopharmaceuticals - Addendum 3 to ICRP Publication 53. ICRP Publication 106. Ann. ICRP 38 (1–2), Annex E.
Katsura, M., Matsuda, L., Akahane, M., et al., 2012. Model-based iterative reconstruction technique for radiation dose reduction in chest CT: comparison with the adaptive statistical iterative reconstruction technique. Eur. Radiol. 22, 1613–1623.
Kemerink, G., Vanhavere, F., Barth, H., et al., 2012. Extremity doses of nuclear medicine personnel: a concern. Eur. J. Nucl. Med. Mol. Imag. 39, 529–532.
Kitajima, K., Ueno, Y., Suzuki, K., et al., 2012. Low-dose non-enhanced CT versus full-dose contrast-enhanced CT in integrated PET/CT scans for diagnosing ovarian cancer recurrence. Eur. J. Radiol. 81, 3557–3362.
Kumar, S., Pandey, A.K., Sharma, P., et al., 2012a. Instantaneous exposure to nuclear medicine staff involved in PET-CT imaging in developing countries: experience from a tertiary care centre in India. Jpn. J. Radiol. 30, 291–295.
Kumar, S., Pandey, A.K., Sharma, P., et al., 2012b. Optimization of the CT acquisition protocol to reduce patient dose without compromising the diagnostic quality for PET-CT: a phantom study. Nucl. Med. Commun. 33, 164–170.
Lasnon, C., Quak, E., Briand, M., et al., 2013. Contrast-enhanced small-animal PET/CT in cancer research: strong improvement of diagnostic accuracy without significant alteration of quantitative accuracy and NEMA NU 4–2008 image quality parameters. EJNMMI Res. 3, 5.
Lecchi, M., Lucignani, G., Maioli, C., et al., 2012. Validation of a new protocol for ^{18}F-FDG infusion using an automatic combined dispenser and injector system. Eur. J. Nucl. Med. Mol. Imag. 39, 1720–1729.
Leide-Svegborn, S., 2012. External radiation exposure of personnel in nuclear medicine from 18F, 99mTcC and 131I with special reference to fingers, eyes and thyroid. Radiat. Prot. Dosimetry 149, 196–206.
Liu, X.F., He, B.M., Ou-Yang, X.H., et al., 2012. Different imaging findings of inflammatory myofibroblastic tumour of the liver. World J. Gastroenterol. 18, 5821–5825.
Mattson, S., Soderberg, M., 2012. Radiation dose management in CT, SPECT/CT and PET/CT techniques. Radiat. Prot. Dosimetry 147, 13–21.
Mawlawi, O., Erasmus, J.J., Munden, R.F., et al., 2006. Quantifying the effect of IV contrast media on integrated PET/CT: clinical evaluation. AJR Am. J. Roentgenol. 186, 308–319.
Niu, Y.T., Mehta, D., Zhang, Z.R., et al., 2012. Radiation dose reduction in temporal bone CT with iterative reconstruction technique. AJNR Am. J. Neuroradiol. 33, 1020–1026.
ORAMED, 2011. International Workshop on Optimization of Radiation Protection of Medical Staff. Radiat. Measur. 46, 1195–1334.
Orlacchio, A., Ciarrapico, A.M., Schillaci, O., et al., 2012. Relevant factors and optimization in the management of a PET/CT facility. Open J. Radiol. 2, 105–109.
Ozkan, E., Soydal, C., Araz, M., et al., 2012. Serum carcinoembryonic antigen measurement, abdominal contrast-enhanced computed tomography, and fluorine-18 fluorodeoxyglucose positron emission tomography/computed tomography in the detection of colorectal cancer recurrence: a correlative study. Nucl. Med. Commun. 33, 990–994.
Pfluger, T., Melzer, H.I., Schneider, V., et al., 2011. PET/CT in malignant melanoma: contrast-enhanced CT versus plain low-dose CT. Eur. J. Nucl. Med. Mol. Imag. 38, 822–831.

Pickhardt, P.J., Lubner, M.G., Kim, D.H., et al., 2012. Abdominal CT with model-based iterative reconstruction (MBIR): initial results of a prospective trial comparing ultralow-dose with standard-dose imaging. AJR Am. J. Roentgenol. 199, 1266–1274.

Prechtel, H.W., Verburg, F.A., Palmowski, M., et al., 2012. Different intravenous contrast media concentrations do not affect clinical assessment of 18F-fluorodeoxyglucose positron emission tomography/computed tomography scans in an intraindividual comparison. Invest. Radiol. 47, 497–502.

Prevot, S., Vrigneaud, J.J., Riedinger, J., et al., 2012. Minimising occupational exposures: a 2-year radiation protection strategy to achieve the goal. Annual Congress of the EANM 12, Abstract P0232.

Roberts, F., Gunawardana, D., Pathmaraj, K., et al., 2005. Radiation dose to PET technologists and strategies to lower occupational exposure. J. Nucl. Med. Technol. 33, 44–47.

Schleipman, A.R., Gerbaudo, V.H., 2012. Occupational radiation dosimetry assessment using an automated infusion device for positron-emitting radiotracers. J. Nucl. Med. Technol. 40, 244–248.

Seierstad, T., Stranden, E., Bjering, K., et al., 2007. Doses to nuclear technicians in a dedicated PET/CT centre utilising 18F fluorodeoxyglucose (FDG). Radiat. Prot. Dosimetry 123, 246–249.

Silva, A.C., Lawder, H.J., Hara, A., et al., 2010. Innovations in CT dose reduction strategy: application of the adaptive statistical iterative reconstruction algorithm. AJR Am. J. Roentgenol. 194, 191–199.

Swedish Nuclear Safety Authority, 2009. National Survey on Justification of CT Examinations in Sweden. Report Number: 2009:03. Report number: 2009:03 ISSN: 2000-0456; Available at www.stralsakerhetsmyndigheten.se. (accessed 23 October 2013)

Vardhanabhuti, V., Ilyas, S., Gutteridge, C., et al., 2013. Comparison of image quality between filtered back-projection and adaptive statistical and novel model-based iterative reconstruction techniques in abdominal CT for renal calculi. Insights Imag. 4, 661–669.

Wu, T.H., Huang, Y.H., Lee, J.J., et al., 2004. Radiation exposure during transmission measurements: comparison between CT- and germanium-based techniques with a current PET scanner. Eur. J. Nucl. Med. Mol. Imag. 31, 38–43.

Yau, Y.Y., Chan, W.S., Tam, Y.M., et al., 2005. Application of intravenous contrast in PET/CT: does it really introduce significant attenuation correction error? J. Nucl. Med. 46, 283–291.

Yoshida, K., Suzuki, A., Nagashima, T., et al., 2009. Staging primary head and neck cancers with (18)F-FDG PET/CT: is intravenous contrast administration really necessary? Eur. J. Nucl. Med. Mol. Imag. 36, 1417–1424.

SAGE

ICRP
Annals of the ICRP

Referral criteria and clinical decision support: radiological protection aspects for justification

M. del Rosario Pérez

World Health Organization, Department of Public Health, Environmental and Social Determinants of Health, Radiation Programme 20, ave Appia, 1211 Geneva 27, Switzerland; PH: +41 (22) 719 5027; FAX: +41 (22) 791 4123; e-mail: perezm@who.int

Abstract—Advanced imaging technology has opened new horizons for medical diagnostics and improved patient care. However, many procedures are unjustified and do not provide a net benefit. An area of particular concern is the unnecessary use of radiation when clinical evaluation or other imaging modalities could provide an accurate diagnosis. Referral criteria for medical imaging are consensus statements based on the best-available evidence to assist the decision-making process when choosing the best imaging procedure for a given patient. Although they are advisory rather than compulsory, physicians should have good reasons for deviation from these criteria. Voluntary use of referral criteria has shown limited success compared with integration into clinical decision support systems. These systems support good medical practice, can improve health service delivery, and foster safer, more efficient, fair, cost-effective care, thus contributing to the strengthening of health systems. Justification of procedures and optimisation of protection, the two pillars of radiological protection in health care, are implicit in the notion of good medical practice. However, some health professionals are not familiar with these principles, and have low awareness of radiological protection aspects of justification. A stronger collaboration between radiation protection and healthcare communities could contribute to improve the radiation protection culture in medical practice.

Keywords: Radiological protection; Medical exposures; Justification

This paper does not necessarily reflect the views of the International Commission on Radiological Protection.

1. INTRODUCTION

The system of radiological protection aims to control radiation risks to provide an adequate level of protection without unduly limiting the potential benefits for individuals and society. Achieving this balance in medical exposures is particularly challenging. The use of ionising radiation for diagnosis and treatment of human diseases has increased worldwide (UNSCEAR, 2010). Benefits have gained recognition, and advanced imaging technology has opened new horizons to medical diagnostics and improved patient care. While modern technology improves the safety of new applications, inappropriate or incorrect use can lead to unnecessary or unintentional radiation exposures with associated radiation risks for patients and staff. As such, there is a need for public health policies that recognise the multiple health benefits that can be obtained, while addressing and minimising health risks.

It has been reported that a substantial proportion of medical imaging procedures (>20% in some areas) may be inappropriate (Malone et al., 2012). An area of particular concern is the unnecessary use of ionising radiation when clinical evaluation or other imaging modalities could provide an accurate diagnosis (justification of procedures). In addition to the scientific evidence, justification of an examination must rely on professional evaluation of comprehensive patient information, such as relevant clinical history, prior imaging, laboratory tests and treatment. Once the procedure has been justified, methods for dose reduction should be applied to manage the radiation dose commensurate with the medical purpose (optimisation of protection). Radiation protection in medicine is built on these two principles: justification and optimisation. However, some health professionals are not familiar with these principles and have a low level of awareness of radiation doses and potential risks (Shiralkar et al., 2003; Lee et al., 2004; Carpeggiani et al., 2012). There is a need to establish a radiation protection culture in medical practice in order to ensure that patients benefit from the medical use of radiation, and allowing more cost-effective allocation of health resources. This paper addresses the radiological protection aspects of justification of medical imaging, and discusses the concept of referral criteria and decision support systems, and their potential role to improve justification of medical imaging procedures.

2. JUSTIFICATION OF MEDICAL EXPOSURES IN THE CURRENT SYSTEM OF RADIOLOGICAL PROTECTION

In *Publication 103*, the International Commission on Radiological Protection (ICRP) proposed a system of radiological protection based on the characteristics of the radiation exposure situations (planned, existing, or emergency exposure situations) and the categories of exposure (occupational, medical, or public exposure) (ICRP, 2007a). The principles of justification and optimisation apply universally to all three exposure situations and categories of exposure. In this context, the principle of justification implies that any decision that alters the radiation exposure situation should do more good than harm; in other words, the benefits to individuals and society of introducing or continuing the activity should outweigh the harm (including radiation detriment) resulting from the activity. ICRP recognises that radiological protection in medicine requires an approach different from that applied in other exposure situations. One major difference is that radiation exposure in medicine is intentional and for the direct benefit of the patient.[1] Neither dose limits nor dose constraints are therefore recommended for patients, because they might reduce the effectiveness of the procedure, thus doing more harm than good. The emphasis is on justification of the procedure and optimisation of the protection, noting that medical exposure of patients calls for a different and more detailed approach to the process of justification than any other category of exposure. In this case, justification often lies with health professionals rather than the government or the competent regulatory authority. The responsibility for justification of a particular procedure falls on the relevant medical practitioners, who need to have special training in radiological protection (ICRP, 2007a). The ICRP recommendations for radiological protection and safety in medicine are given in *Publication 105* (ICRP, 2007b), which explains that the principle of justification applies at three levels in medicine as described below.

- Level 1: the proper use of radiation in medicine is accepted as doing more good than harm to society.
- Level 2: a specified procedure is justified for a group of patients showing relevant symptoms, or for a group of individuals at risk for a clinical condition that can be detected and treated.
- Level 3: the application of a specified procedure to an individual patient is justified if that particular application is judged to do more good than harm to the individual patient.

3. JUSTIFICATION OF MEDICAL EXPOSURES IN THE INTERNATIONAL RADIATION SAFETY STANDARDS

Co-sponsored by eight international organisations,[2] the International Basic Safety Standards (BSS) for Protection against Ionising Radiation and for the Safety of Radiation Sources (IAEA, 2014) provide the leading international benchmark for safety standard setting, policy making and decision making. The International BSS establish specific responsibilities for health professionals related to radiation protection and safety in medical exposures. The introductory chapter notes that application of the principle of justification to medical exposures requires a special approach. In addition to general justification of the use of ionising radiation in medicine (Level 1),

[1]Although risks and benefits associated with a medical procedure usually accrue to the same person, there may be other considerations such as occupational exposure of health workers, and screening programmes that may benefit the population rather than every screened individual.
[2]European Commission, Food and Agriculture Organization, International Atomic Energy Agency, International Labour Organization, Organization for Economic Co-operation and Development/Nuclear Energy Agency, Pan American Health Organization, United Nations Environment Programme, and World Health Organization.

there is a need for generic justification of a given radiological procedure that applies to the justification of new technologies and techniques as they evolve (Level 2). Finally, justification of a radiological procedure for a given individual (Level 3) includes consideration of the purpose of the exposure, the clinical circumstances, and the characteristics of the individual involved. Referral criteria developed by professional bodies and health authorities have to be taken into account. These introductory concepts are further developed in Chapter 3 of the International BSS, which is devoted to planned exposure situations. The government or the regulatory body, as appropriate, should ensure that provision is made for the justification of any type of practice and for review of the justification, as necessary, and should ensure that only justified practices are authorised. It is emphasised that medical exposures should be justified by weighing the expected diagnostic or therapeutic benefits against the potential radiation detriment, with the benefits and the risks of available alternative techniques that do not involve exposure to radiation taken into account. While the responsibility for generic justification is assigned to the health authority in conjunction with appropriate professional bodies, the individual justification should be performed through consultation between the radiological medical practitioner and the referring medical practitioner, as appropriate, with consideration of appropriateness, urgency, type of exposure, patient conditions, and relevant previous information, with relevant referral guidelines taken into account. Justification of radiological procedures in asymptomatic populations as part of a health screening programme should be undertaken by the health authority in conjunction with appropriate professional bodies. However, specific individual justification by the radiological medical practitioner and the referring medical practitioner is required if the procedure is to be performed on an asymptomatic individual for the early detection of disease, but not as part of an approved health screening programme. The individual should be informed of the expected benefits, risks, and limitations of the procedure. Special considerations regarding pregnancy and breast feeding are provided in the International BSS. Procedures should be in place for ascertaining the pregnancy status of a female patient of reproductive capacity before the performance of any radiological procedure that could result in a significant dose to the embryo or fetus, so that this information can be considered in the justification and optimisation processes. Justification of nuclear medicine procedures that could result in a significant dose to a breastfed infant is addressed explicitly. Specific requirements for justification of exposure of volunteers as part of a programme of biomedical research are included in the International BSS. Critical review of implementation of the principle of justification should be part of radiological audits performed in medical radiation facilities.

Safety requirements concerning the justification of medical exposures represent a major component of the new European Commission Council Directive 2013/59/ Euratom of December 2013 (EC, 2013), laying down the BSS for protection against the dangers arising from exposure to ionising radiation. Availability of clinical imaging guidelines is mandatory in the Euratom BSS, which are legally binding. In contrast, the International BSS are not legally binding; however, the implementation of their safety requirements, including those related to justification, can be subject to inspection by national regulatory bodies.

4. DOES THE INCREASE IN THE FREQUENCY OF RADIOLOGICAL IMAGING PROCEDURES IMPLY 'OVERUSE'?

The increasing trend in the use of radiation in medical imaging cannot be taken *per se* as evidence of 'overuse'. Considerable disparities exist between and within countries with respect to the use of radiation technologies (e.g. high-, medium-, and low-income countries; rural vs urban areas). While most developing countries still lack adequate capacity and resources to provide access to radiology services on a widespread basis, developed countries are increasingly facing the risk of overuse of computed tomography and other sophisticated technologies.

Health service delivery is one of the building blocks of health systems. Good health services deliver safe and effective health interventions to those that need them, when and where needed, with minimum waste of resources (WHO, 2007). In this context, the appropriate use of radiation in health care contributes to the strengthening of health systems.

The ultimate purpose of healthcare systems is to deliver the best care to every person, any time, and everywhere. Poor-quality care may come in the form of overuse (i.e. giving people care that they do not need), underuse (i.e. failing to give people the care that they need), and misuse (i.e. making errors that can damage people). All three problems should be addressed to improve the quality and safety of health care.

In medicine, the term 'overuse' refers to the application of a procedure where/when it is unlikely to improve patient outcome (i.e. unnecessary or unjustified). In this context, the 'overuse' of radiological medical procedures represents an issue that goes beyond the boundaries of radiation protection, and falls within the notion of good medical practice.

When choosing a procedure using ionising radiation, the benefit/risk balance must be considered carefully. Even if benefits outweigh risks, there is unnecessary use of radiation when clinical evaluation or other imaging modalities could provide an accurate diagnosis. Cost, local expertise, available resources, accessibility, and patient values have to be considered, as well as efficacy.

The benefit outweighs the risk when the procedure is prescribed appropriately and performed properly. This is not the case if there is no clinical indication or the radiation dose is higher than necessary for the clinical purpose (e.g. adult protocols used to image children). A medical imaging examination is useful if its outcome, either positive or negative, influences the management of the patient or strengthens confidence in the diagnosis. When indicated and available, imaging modalities that do not use ionising radiation (e.g. ultrasound or magnetic resonance imaging) are preferred, especially for children. The possibility of deferring imaging to a later time if/when the patient's condition may change must also be considered. The final decision may also be influenced by cost, expertise, availability of resources, and/or patient values.

Table 1. Possible referral incentives for radiological imaging procedures.

Low awareness of radiation doses and associated health risks
Lack of availability of imaging referral guidelines
Reliance on personal experience rather than evidence-based medicine
No access to alternative modalities (e.g. ultrasound, magnetic resonance imaging) due to lack of expertise and/or technology
Pressure on physicians for performance (e.g. patient throughput)
Pressure from referring clinicians or other specialists
Pressure from promotion and marketing of health technology
Fragmentation of health care (e.g. too frequent or unnecessary repeated procedures)
Lack of dialogue/consultation between referrers and radiologists
Concern about malpractice litigation (defensive medicine)
Self-referral, financial incentives
Consumer's demand (patient's and/or family's expectations)
Physicians' low confidence in clinical diagnosis and/or over-reliance on imaging tests
Insufficient, incorrect, or unclear clinical information provided for justification

Various driving forces may act as referral incentives for radiological examinations (see Table 1).

Depending on the national healthcare system, commercial interests may influence medical imaging referral as radiological examinations may be a source of income to radiology departments. Such a situation may create referrals that exceed the needs of good medical practice, confer unjustifiable risk on patients, and are inconsistent with principles of radiological protection and medical ethics (ICRP, 2007b).

In some countries, health professionals have concerns about possible malpractice litigation. Physicians may respond to this threat of litigation by ordering more referrals and more tests, some of which may be wasteful and even harmful. This so-called 'defensive medicine' often results in overuse of radiation for medical imaging, and this is particularly relevant in trauma centres and emergency departments (Hadley et al., 2006; Sierzenski et al., 2014). It has been demonstrated that such defensive behaviour can be modified towards evidence-based behaviour by training physicians in the use of referral guidelines (Rohacek et al., 2012).

5. REFERRAL CRITERIA FOR MEDICAL IMAGING

Faced with a clinical presentation, the referring physician makes a decision for an individual patient based on overall best practice (Level 3). However, complexities and rapid advances in medical imaging render it difficult for referrers to follow changes in evidence-based standards of care. Therefore, generic guidance for justification of imaging (Level 2) is usually provided by professional societies in conjunction with health authorities as referral criteria for medical imaging. These criteria are consensus statements based on the best-available evidence to assist the decision-making process when choosing the best imaging procedure for a given patient. Although these criteria are advisory rather than mandatory, physicians should have good reasons for deviation from these criteria.

Imaging referral guidelines provide physicians with information regarding which procedure is most likely to yield the most informative results, and whether another modality is equally or more effective, and therefore more appropriate. As decision-aiding tools for good medical practice, these guidelines can improve health service delivery. These guidelines support the practice of evidence-based medicine and form a foundation to guide appropriateness in prescribing diagnostic imaging services.

Examples of referral guidelines include the Appropriateness Criteria of the American College of Radiology (ACR, 2014), 'iRefer: Making the Best Use of Clinical Radiology' (RCR, 2013), 'Diagnostic Imaging Pathways' (Government of Western Australia, Department of Health, 2014), and 'Guide de bon usage d'imagerie médicale' (SFR/SFMN, 2013).

The use of imaging referral guidelines could reduce the number of unjustified examinations (RCR, 1993; Oakeshott et al., 1994; Hadley et al., 2006). It was recently demonstrated by clinical audit that, even when awareness of the availability of referral guidelines is relatively low, scrutinizing based on imaging referral guidelines and amending or returning those found inappropriate enables a high level of appropriate imaging (95%), thus making the best use of clinical radiology (Remedios et al., 2014).

Evidence-based imaging referral guidelines have gained widespread global acceptance, and many countries are now considering the adoption or adaptation of one of the existing evidence-based referral guidelines. Given the frequency of similar pathologies, it is not surprising to find comparable guidelines in different regions of the world. Despite various formats (e.g. tabulated vs flow charts) and media (e.g. hard copy, electronic copy, interactive web-based, smart phone-based, etc.), the common philosophy behind the use of these guidelines is to promote a questioning attitude among physicians before making decisions about medical imaging. An example of this approach is the Socratic questions[3] proposed in iRefer, summarised in Table 2.

6. DECISION SUPPORT SYSTEMS

Voluntary use of referral criteria has shown limited success compared with integration into clinical decision support (CDS) systems. These systems support good medical practice, can improve health service delivery, and foster safer, more efficient, fair, cost-effective care, thus contributing to the strengthening of health systems. The importance of having such systems in place to assist referring physicians in making the best decisions has been acknowledged over the past few years (Health Council of Canada, 2010).

[3]Classical method to stimulate erudite thought: http://www.ncbi.nlm.nih.gov/pubmed/21075021.

Table 2. Examples of questions proposed for referring clinicians (adapted from RCR, 2013).

Which are the preventable situations that may result in wasteful medical exposures?	Which questions should the referring physician answer to prevent such situations?
Unnecessarily repeating investigations that have been done previously	Has it been done already?
Undertaking investigations when results are unlikely to affect patient management or over-investigating	Do I need it?
Investigating too early	Do I need it now?
Performing the wrong investigation	Is this the best investigation?
Failing to provide appropriate clinical information and questions that the imaging investigation should answer	Have I explained the problem?

Formerly called 'computerised physician order[4] entry' systems, these CDS systems facilitate the integration of imaging referral guidelines into the daily workflow, with provision of interactive reminders. Such CDS systems should be available at the point of care, implemented through clinical workflow, and preferably executed in real time. At the time of entry into the CDS system, the referring doctors receive feedback on the degree of appropriateness of their choices relative to the referral guidelines (e.g. warnings and coloured codes) (Kawamoto et al., 2005).

The integration of CDS into radiology requesting systems embedded in patient care pathways has proven to slow down the increasing use of computed tomography, to be acceptable to clinicians and to improve appropriateness of imaging referral, particularly in the emergency department (Sistrom et al., 2009; Ip et al., 2012).

Potential caveats related to the use of CDS systems include technical challenges of connectivity and interfacing with existing clinical information systems, inability to cover all possible clinical presentations, limited applicability of their guidance to individual patients, and possibility of bypassing 'soft stops' in the system. Nevertheless, it can be concluded that CDS systems are useful tools to facilitate implementation of evidence-based imaging referral guidelines at the time of referral.

7. DISCUSSION

The principle of appropriateness in imaging can be encapsulated into three words: 'best test first'. A number of international, regional, and national initiatives are being

[4]The term 'order' is not appropriate because it gives the false impression that referring physicians are entering 'orders' while, in fact, they are entering 'requests' for advice from the imaging specialists. As such, the abbreviation 'CPOE' is not recommended.

conducted to increase appropriateness, reduce unnecessary radiation exposures, and thus prevent unnecessary radiation risks.

ICRP currently has a Working Party on Justification of Medical Imaging composed of members of ICRP Committee 3 on Protection in Medicine as well as external experts. In the framework of its Global Initiative on Radiation Safety in Health Care Settings, the World Health Organization (WHO) promotes availability and use of imaging referral guidelines at a global level, and is currently conducting a project on justification of medical imaging of asymptomatic people for individual health assessment (Lau et al., 2011a,b). In the framework of its International Action Plan for Radiological Protection of Patients, the International Atomic Energy Agency (IAEA) launched a '3-A's campaign' promoting the concept of awareness, appropriateness, and audit to improve justification (Malone et al., 2012). The European Commission, its Working Party on Medical Exposures of Article 31 of the Euratom Treaty,[5] and the association of Heads of European Radiological Protection Competent Authorities[6] complement each other's efforts to foster appropriateness and enhance justification in Europe. Large-scale campaigns are being pursued intensively by professional societies (e.g. Image Gently, Image Wisely, Eurosafe[7]). National initiatives on this topic are numerous.

An international conference on radiation protection in medicine was organised by IAEA in cooperation with WHO in December 2012, hosted in Bonn by the Government of Germany. More than 500 participants from approximately 80 countries and 16 organisations reviewed advances, challenges, and opportunities. The main outcome of this conference was a list of 10 priority actions to improve radiation safety in health care in the next decade.[8] The first priority action of the so-called 'Bonn Call for Action' refers to enhancing implementation of justification by:

- introducing and applying awareness, appropriateness, and audit to facilitate and enhance justification in practice;
- developing harmonised evidence-based criteria to strengthen the appropriateness of clinical imaging, including diagnostic nuclear medicine and non-ionising radiation procedures, and involving all stakeholders in this development;
- implementing clinical imaging referral guidelines globally, keeping local and regional variations in mind, and ensuring regular updating, sustainability, and availability of these guidelines;
- strengthening the application of clinical audit in relation to justification, and ensuring that justification becomes an effective, transparent, and accountable part of normal radiological practice;

[5]See: http://ec.europa.eu/energy/nuclear/radiation_protection/medical/applications_en.htm
[6]See: http://www.herca.org/
[7]See: http://www.imagegently.org/; http://wwwimagewisely.org/; http://wwweurosafeimaging.org/
[8]See: http://www.who.int/ionizing_radiation/about/med_exposure/en/index3.html and https://rpop.iaea.org/RPOP/RPoP/Content/News/bonn-call-for-action-joint-position-statement.htm

- introducing information technology solutions, such as decision support tools, in clinical imaging, and ensuring that these are available and freely accessible at the point-of-care; and
- further developing criteria for justification of health screening programmes for asymptomatic populations (e.g. mammography screening), and for medical imaging of asymptomatic individuals who are not participating in approved health screening programmes.

Justification of procedures and optimisation of protection, the two pillars of radiological protection in health care, are implicit in the notion of good medical practice. However, some health professionals are not familiar with these principles, and have low awareness of radiological protection aspects of justification.

Global partnerships should be expanded, and stronger collaboration between radiation protection and healthcare communities should be fostered to improve radiation protection culture in the medical sector. The most effective means to decrease radiation dose associated with medical imaging without compromising the benefits is to eliminate unnecessary or inappropriate referrals. Application of the principle of justification is therefore a key measure to control unnecessary radiation exposures in health care.

REFERENCES

ACR, 2014. Appropriateness Criteria. American College of Radiology, Reston. Available at: http://www.acr.org/Quality-Safety/Appropriateness-Criteria (last accessed 19 July 2014).

Carpeggiani, C., Kraft, G., Caramella, D., et al., 2012. Radioprotection (un)awareness in cardiologists, and how to improve it. Int. J. Cardiovasc. Imag. 28, 1369–1374.

EC, 2013. Council Directive 2013/59/Euratom of 5 December 2013, Laying Down Basic Safety Standards for Protection Against the Dangers Arising from Exposure to Ionising Radiation, and Repealing Directives 89/618/Euratom, 90/641/Euratom, 96/29/Euratom, 97/43/Euratom and 2003/122/Euratom. European Commission, Brussels. Available at: http://eur-lex.europa.eu/LexUriServ/LexUriServ.do?uri=OJ:L:2014:013:0001:0073:EN:PDF (last accessed 15 July 2014).

Government of Western Australia, Department of Health, 2014. Diagnostic Imaging Pathways (DIP), a Clinical Decision Support Tool and Educational Resource for Diagnostic Imaging Endorsed by the Royal Australian and New Zealand College of Radiology. Government of Western Australia, Perth. Available at: http://www.imagingpathways.health.wa.gov.au/ (last accessed 14 July 2014).

Hadley, J.L., Agola, J., Wong, P., 2006. Potential impact of the American College of Radiology Appropriateness Criteria on CT for trauma. Am. J. Radiol. 186, 937–942.

Health Council of Canada, 2010. Decisions, Decisions: Family Doctors as Gatekeepers to Prescription Drugs and Diagnostic Imaging in Canada. Health Council of Canada, Toronto. Available at: http://www.healthcouncilcanada.ca/rpt_det.php?id=154#sthash.yGqF0KJU.dpuf (last accessed 12 July 2014).

IAEA, 2014. Radiation Protection and Safety of Radiation Sources: International Basic Safety Standards. IAEA Safety Standards Series GSR Part 3. International Atomic

Energy Agency, Vienna. Available at: http://www-pub.iaea.org/MTCD/Publications/PDF/Pub1578_web-57265295.pdf (last accessed 24 July 2014).

ICRP, 2007a. The 2007 Recommendations of the International Commission on Radiological Protection. ICRP Publication 103. Ann. ICRP 37(2–4).

ICRP, 2007b. Radiological protection in medicine. ICRP Publication 105. Ann. ICRP 37(6).

Ip, I.K., Schneider, L., Hanson, R., et al., 2012. Adoption and meaningful use of computerized physician order entry (CPOE) with an integrated clinical decision support system for radiology: ten-year analysis in an urban teaching hospital. J. Am. Coll. Radiol. 9, 129–136. Available at: http://www.ncbi.nlm.nih.gov/pubmed/22305699 (last accessed 14 July 2014).

Kawamoto, K., Houlihan, C., Balas, E.A., et al., 2005. Improving clinical practice using clinical decision support systems: a systematic review of trials to identify features critical to success. BMJ 330, 765. Available at: http://www.bmj.com/content/330/7494/765 (last accessed 18 July 2014).

Lau, L.S., Pérez, M.R., Applegate, K.E., et al., 2011a. Global quality imaging: emerging issues. J. Am. Coll. Radiol. 8, 508–512.

Lau, L.S., Pérez, M.R., Applegate, K.E., et al., 2011b. Global quality imaging: improvement actions. J. Am. Coll. Radiol. 8, 330–334.

Lee, C.I., Haims, A.H., Monico, E.P., et al., 2004. Diagnostic CT scans: assessment of patient, physician, and radiologist awareness of radiation dose and possible risks. Radiology 231, 393–398.

Malone, J., Guleria, R., Craven, C., et al., 2012. Justification of diagnostic medical exposures: some practical issues. Report of an International Atomic Energy Agency Consultation. Br. J. Radiol. 85, 523–538.

Oakeshott, P., Kerry, S.M., Williams, J.E., 1994. Randomized controlled trial of the effect of the Royal College of Radiologists' guidelines on general practitioners' referrals for radiographic examination. Br. J. Gen. Pract. 44, 427e8.

RCR, 1993. Influence of Royal College of Radiologists' guidelines on referral from general practice. BMJ 306, 110e1.

RCR, 2013. iRefer: Making the Best Use of Clinical Radiology. Royal College of Radiologists, London. Available at: http://www.rcr.ac.uk/content.aspx?PageID=995 (last accessed 11 September 2014).

Remedios, D., Drinkwater, K., Warwick, R., 2014. National audit of appropriate imaging. Clin. Radiol. 69, 1039–1044.

Rohacek, M., Buatsi, J., Szucs-Farkas, Z., et al., 2012. Ordering CT pulmonary angiography to exclude pulmonary embolism: defense versus evidence in the emergency room. Intens. Care Med. 38, 1345–1351.

SFR/SFMN, 2013. Guide du Bon Usage des examens d'imagerie médicale de la Societé Francaise de Radiologie (SFMN) et la Societé Française de Médecine Nucléaire (SFMN). 2nd edn. Société Française de Radiologie et Société Française de Médecine Nucléaire, Paris. Available at: http://gbu.radiologie.fr/ (last accessed 13 July 2014)

Shiralkar, S., Rennie, A., Snow, M., et al., 2003. Doctors' knowledge of radiation exposure: questionnaire study. BMJ 327, 371–372.

Sierzenski, P.R., Linton O.W., Amis, E.S. Jr., et al., 2014. Applications of justification and optimization in medical imaging: examples of clinical guidance for computed tomography use in emergency medicine. Ann. Emerg. Med. 63, 25–32.

Sistrom, C.L., Dang, P.A., Weilburg, J.B., et al., 2009. Effect of computerized order entry with integrated decision support on the growth of outpatient procedure volumes: seven-year

time series analysis. Radiology 251, 147–155. Available at: http://radiology.rsna.org/content/251/1/147.long (last accessed 17 July 2014).
UNSCEAR, 2010. United Nations Scientific Committee on the Effects of Atomic Radiation Report to the General Assembly. Annexe A: Medical Radiation Exposures. United Nations Scientific Committee on the Effects of Atomic Radiation, New York. Available at: http://www.unscear.org/docs/reports/2008/09-86753_Report_2008_Annex_A.pdf (last accessed 12 July 2014).
WHO, 2007. Strengthening Health Systems to Improve Health Outcomes – WHO's Framework for Action. World Health Organization, Geneva. Available at: http://www.who.int/healthsystems/strategy/everybodys_business.pdf (last accessed 16 July 2014).

SAGE

ICRP
Annals of the ICRP

ICRP's approach to protection of the living environment under different exposure situations

R.J. Pentreath[a], C-M. Larsson[b], D. Copplestone[c]

[a]*Plymouth Marine Laboratory, Prospect Place, The Hoe, Plymouth PL1 3DH, UK; e-mail: janpentreath@yahoo.co.uk*

[b]*Australian Radiation Protection and Nuclear Safety Agency, Miranda, NSW, Australia*

[c]*School of Natural Sciences, University of Stirling, Stirling, UK*

Abstract–The International Commission on Radiological Protection's (ICRP) system to protect the living components of the environment is designed to provide a broad and practical framework across all exposure situations. The objectives of ICRP are therefore also set in fairly broad terms, recognising that national and local environmental protection requirements may need to be set within them. The framework recognises the need to be able to demonstrate an adequate level of protection in relation to planned exposure situations, whilst also providing an ability to manage existing situations and accidents, as well as emergency situations, in a rational way. The objects of protection are always real biota in real exposure situations, and the scientific basis for their protection needs to be based on data originating from studies on the relationships between exposure and dose, dose and effects, and effects and consequences in real animals and plants. The framework that has been developed has therefore had to take such realities into account to make the optimum use of the data currently available, whilst being sufficiently flexible to accommodate new scientific information as it arises without having to alter the framework as a whole.

Keywords: Biota; Environmental protection; Reference Animals and Plants

1. INTRODUCTION

The need to protect the living components of the environment (the biota) by taking either proactive or retrospective action is now widely accepted as part of the regulation and management of any large-scale industrial activity (ICRP, 2003). In the framework of radiological protection developed by the International

This paper does not necessarily reflect the views of the International Commission on Radiological Protection.

Commission on Radiological Protection (ICRP), this necessity applies to three exposure situations which, in the context of environmental protection, essentially consist of the following: planned exposure situations, which involve the discharge and disposal of radioactive waste, the decommissioning of installations, and activities related to the eventual remediation and decontamination of sites; existing exposure situations, where the source already exists and a decision on control has to be taken, including prolonged exposure situations after emergencies; and emergency exposure situations that may occur as a result of accidents, malicious acts, or any other unexpected situation in which control of the source has been lost, and urgent actions are required in order to avoid or reduce undesirable consequences.

ICRP's aims for the protection of animals and plants in their natural environmental settings apply to these situations. The aims are those of '. . . preventing or reducing the frequency of deleterious radiation effects to a level where they would have a negligible impact on the maintenance of biological diversity, the conservation of species, or the health and status of natural habitats, communities and ecosystems' (ICRP, 2007).

The questions that naturally arise from these two starting points (different types of exposure situations, and general aims relating to the protection of biota that could be affected by them) therefore include the following:

- Which biological effects of radiation are of relevance?
- To which components of the environment (i.e. which animals and plants) do they apply?
- How would one know what doses these animals and plants have received?
- How can one know what the effects of radiation are on these animals and plants?
- How would one know that the original objectives had been met?

This paper briefly addresses these questions and summarises ICRP's approach to environmental protection from a high-level perspective. Full documentation has been published in *Publications 108*, *114,* and *124* (ICRP, 2008, 2009, 2013); further – and more specific – detail is given in the Proceedings of the First ICRP Symposium on the International System of Radiological Protection (ICRP, 2012).

2. BIOLOGICAL EFFECTS OF RELEVANCE

Given the general aims of ICRP, it is evident that the principal focus is that of the protection of 'populations' or large groups of individuals. This is quite different from the aims in relation to the protection of human beings, which relate to the risks of radiation affecting individuals. However, radiation effects on 'populations' are mediated through effects on individuals. The biological endpoints of interest to individuals that could have a consequence at a population level are therefore those of:

- early mortality (leading to changes in age distribution, death rate, and population density);

- some forms of morbidity (that could reduce 'fitness' of the individuals, making it more difficult for them to survive in a natural environment);
- impairment of reproductive capacity by either reduced fertility or fecundity (affecting birth rate, age distribution, number, and density); and
- induction of chromosomal damage.

ICRP has therefore summarised the information available on the effects of radiation to certain types of animals and plants under these categories, rather than as stochastic or non-stochastic (deterministic) effects as it has for human radiological protection.

3. APPLICATION TO ACTUAL EXPOSURE SITUATIONS

In order to apply ICRP's recommendations in practice, and to meet its overall objective, a framework would ideally include all of the following elements.

- Clearly stated local environmental protection objectives that relate to a specific environmental exposure situation.
- Knowledge of the effects of radiation, at different dose rates, to different tissues, organs, and life stages of the relevant biota relating to such objectives.
- Estimates of the dose likely to be received by the relevant biota under those environmental exposure situations, in terms of the tissues, organs, and life stages most likely to be at risk with regard to the relevant biological endpoints.
- Number of individuals, or fraction of the relevant population, that would be likely to receive such dose rates, and when.
- Actions, or choice of actions, that could be taken to optimise the level of protection of the relevant biota relating to radiation exposure, bearing in mind other possible threats to the same population.

Due to the immense variety of biota and their presumed response to radiation, any credible system also needs to have some key points of reference that provide some form of auditable trail which links the basic elements of the framework together, or at least could do so if further data were forthcoming, and it is feasible to obtain such data. The advantage of such a systematic approach is that, as the needs for change to any component of the system arise (as in the acquisition of new scientific data, changes in societal attitudes, or simply from experience gained in its practical application), it is possible to consider what the consequences of such a change may be elsewhere within the system, and upon the system as a whole.

Regarding the biota that represent the actual targets of protection in any specific exposure situation, these may already be identified because of existing environmental legislation. Such legislation may relate to the type of habitat or to certain components of, or indeed all of, the local fauna or flora. This protective need may have arisen because of nature conservation or commercial (e.g. fisheries) concerns. Thus, in planned exposure situations, some selection and dialogue with other relevant regulatory or 'stakeholder' groups may be necessary. In existing exposure situations, however, the objects of concern are those components of the biota that are within the contaminated area, or otherwise associated with it. However, even for the purposes of nature conservation, it is usually impossible to know about all of the species within any ecosystem, and the 'health' of such areas is thus often assessed by studying a subset of its components. Therefore, for the purposes of assessing the actual or potential effects of radiation on an environmental area, it is recommended that Representative Organisms (ROs) should be identified to serve as representatives of a particular species, or a group of organisms, in relation to a site-specific assessment, taking account of their assumed location with respect to the source.

4. ESTIMATING DOSE RATES AND ASSESSING THE POTENTIAL FOR RADIATION EFFECTS

In existing exposure situations, dose rates will usually be made based on direct observations (e.g. external dose rates), or calculated from measured concentrations of radionuclides within, and external to, the relevant biota. In some cases, however, dose rates may need to be estimated from the use of models, either to extrapolate backwards or forwards in time, or to areas that are not readily accessible. For planned (and emergency) exposure situations, however, dose rates are most likely to be obtained from modelling. Obviously the best-available models and databases should be used. Once estimates of dose rates have been obtained, the potential effects of such dose rates need to be assessed.

There is clearly a limit to the amount of information available on the effects of radiation on different types of biota, and a limit to the extent that such information could ever be obtained as a result of laboratory experimentation. Even for those biota that have been the subject of study, the dose rates used are high, and there is no equivalent of the linear, no-threshold model to allow extrapolation from effects at high doses and dose rates to lower doses and dose rates. In general, there is little difference in response across a range of dose rates for mammals, and this may also apply to birds (because they are also homothermic vertebrates with high metabolic rates), but there are insufficient data at present to draw firm conclusions. For other vertebrates, generalisations are again difficult to make because their lower metabolic rates (that are also temperature dependent) are seldom taken into account.

ICRP has therefore used a set of Reference Animals and Plants (RAPs) to serve as points of reference to relate exposure to dose, and dose to effect, for different types of biota. The RAPs were selected on the basis of a number of criteria, including the fact that a reasonable amount of radiobiological information was already available on them, including data on probable radiation effects and/or that they were amenable to future research in order to obtain the necessary missing or imprecise data, particularly with regard to radiation effects. A set of RAPs was therefore identified by ICRP (ICRP, 2008), but there is nothing sacrosanct about the set; other biotic types could have been chosen. They were all considered to be organisms that are 'typical' of different environments, in the sense that one might expect to find them there. The set is also essentially one of 'wild' animals and plants rather than domesticated organisms. Quite clearly, in order to serve as points of reference, the RAPs needed to be defined fairly precisely and are thus described to the generality of the taxonomic level of Family, with defined anatomical, physiological, and life-history properties.

In order to use existing databases on the effects of radiation, the only pragmatic approach therefore seemed to be to organise the existing databases for each type of RAP into bands of dose within which certain effects have been noted, or might be expected, and then to select a single band for each RAP to serve as a derived consideration reference level (DCRL; i.e. a 'reference level' for effects, transparently derived, that can be considered as the starting point for decision making, depending on the purpose for making decisions under a defined exposure situation). Tables were therefore constructed to cover dose rate ranges, in bands, from <0.1 mGy day^{-1} to >100 mGy day^{-1} (ICRP, 2008). Dose rates >1 Gy day^{-1} are essentially of no environmental relevance. The resultant set of DCRLs can therefore be used as points of reference to inform on the appropriate level of effort that should be expended on environmental protection, dependent on the overall management objectives, the exposure situation, the actual fauna and flora present, and the numbers of individuals thus exposed. The DCRLs have been defined in terms of bands of dose rates spanning one order of magnitude, and are relevant to a large and a small terrestrial mammal, a bird, an amphibian, a marine fish and a freshwater fish, a crustacean, an annelid, a tree, a small plant, and a seaweed.

5. MEETING ENVIRONMENTAL OBJECTIVES

As the databases, particularly on radiation effects, are so limited, it would obviously be an advantage – given a choice – to select ROs that are as similar to RAPs as possible. Where this is not possible, differences between ROs and the RAPs should nevertheless be quantifiable, in relation to their basic biology, dosimetry, or radiation effects, and such differences need to be noted and taken into account. The extent to which such factors need to be applied, and their relevant impact on the final management decision, will depend on the nature of the implementation and application of the planning process relevant to protection of the environment. As other regulatory bodies are likely to be involved, such as those responsible for wildlife management, it is essential to have a clearly set out logical link between any radioactive releases and any potential risk of biological effects (for which the RAP framework should be a starting point), and a clearly laid out strategy by which the relevant stakeholders can be engaged in the decision-making process.

From a practical point of view, and to help create a set of reference points, *Publication 108* (ICRP, 2008) also included reference data sets (dose conversion factors) by which concentrations of radionuclides inside or outside the RAPs could be converted into dose rates at an approximate whole-body level, and a further publication (ICRP, 2009) provided reference data sets (concentration ratios) by which concentrations in the ambient media, under equilibrium conditions, could be related to whole-body concentrations for the relevant RAPs.

It is therefore considered that ICRP's original objectives in relation to protection of the environment would be met by reference to the bands of DCRLs. Thus, in planned exposure situations, the lower boundary of the relevant DCRL band should be used as the appropriate reference point for protection of different types of biota within a given area during the planning of controls to be applied to a source. As the DCRL bands apply to animals and plants within a given location, the extent of such an area needs to be determined in advance relative to the overall conservation objectives. The area is unlikely to be sufficiently large to involve an entire 'population' of the relevant biota. In the case of multiple sources of exposure (e.g. from historical discharges or multiple sites), these other sources should be taken into account in comparison with the DCRLs when assessing protection options. If the assessed dose rates are below the appropriate reference point, the level of control is determined by selection of the most reasonable protective action.

Planned exposure situations relating to the management of long-lived wastes are especially difficult with regard to protection of the environment, because the biosphere is likely to change, and may even change substantially, over the long time frames that are considered in such waste disposal. Such changes may entail alterations that are natural or are enhanced, or perturbed, through human action. The default case for protection, and protective actions, should therefore be the set of RAPs, bearing in mind that this set was chosen deliberately because its components are considered to be 'typical' biotic types of the major environmental domains of land, sea, and fresh water.

Planned exposure situations may also involve consideration of the possibilities and magnitude of the consequences of potential events or accidents (potential exposures). Thus, during planning phases, there may be a need to consider different siting options for a specific source (such as placing an outlet into a river, an estuary, or the sea), with regard to the potential environmental impact of accidental releases of radionuclides into these different media. It may also include a need to consider the potential impact of accidental releases (such as an accidental release into the aquatic medium or the atmosphere) from a defined source on the different surrounding media (terrestrial, freshwater, or marine environments) and the mitigating measures that may be available. The DCRL bands again may serve as a point of reference in such assessments, and be used as a mechanism to compare impacts in the overall siting and emergency planning activities.

6. CONCLUSIONS

As the nuclear industries move into an era of 'new build', decommissioning, waste disposal, and coping with the aftermath of (hopefully rare, but always possible) accident and emergency situations, it is essential for them to be regulated within a framework that takes account of the potential impact upon the environment, irrespective of any potential impact upon humans. Confidence in the nuclear industries by the public, and their political representatives, will ensure that this is the case. In the past, such confidence has been lacking because neither a conceptual nor practical framework existed within which the subject of protection of the environment, with respect to ionising radiation, could be handled in a logical and transparent way. ICRP has, over the last decade, developed such a framework and it remains for it to be applied in a variety of ways in order to mature into a robust basis for future decision making.

REFERENCES

ICRP, 2003. A framework for assessing the impact of ionising radiation on non-human species. ICRP Publication 91. Ann. ICRP 33(3).

ICRP, 2007. The 2007 Recommendations of the International Commission on Radiological Protection. ICRP Publication 103. Ann. ICRP 37(2–4).

ICRP, 2008. Environmental protection: the concept and use of reference animals and plants. ICRP Publication 108. Ann. ICRP 38(4–6).

ICRP, 2009. Environmental protection: transfer parameters for reference animals and plants. ICRP Publication 114. Ann. ICRP 39(6).

ICRP, 2012. Proceedings from the First ICRP Symposium on the International System for Radiological Protection. Ann. ICRP 41(3/4).

ICRP, 2013. Protection of the environment under different exposure situations. ICRP Publication 124. Ann. ICRP 43(1).

SAGE

ICRP
Annals of the ICRP

Establishing relationships between environmental exposures to radionuclides and the consequences for wildlife: inferences and weight of evidence

J. Garnier-Laplace[a], F. Alonzo[b], C. Adam-Guillermin[b]

[a]*Institute for Radioprotection and Nuclear Safety, IRSN/PRP-ENV/SERIS, Bld 159, Cadarache, 13115 Saint-Paul-lez-Durance, Cedex, France; e-mail: jacqueline.garnier-laplace@irsn.fr*

[b]*IRSN/PRP-ENV/SERIS/LECO, Laboratory of Ecotoxicology of Radionuclides, France*

Abstract–Ecological risk assessments for radioactive substances are based on a number of inference rules to compensate for knowledge gaps, and generally require the implementation of a weight-of-evidence approach. Until recently, dose (rate)–response relationships used to derive radioprotection criteria for wildlife have mainly relied on laboratory studies from a limited number of species as representatives of biodiversity. There is no doubt that additional knowledge, combined with advanced conceptual and mathematical approaches, is needed to develop general rules and increase confidence when extrapolating from test species to complex biological/ecological systems. Moreover, field data sets based on robust sampling strategies are still needed to validate benchmark values derived from controlled laboratory tests, and to indicate potential indirect ecological effects, if any. This paper illustrates, through several examples, the need for implementing a combined laboratory–field-model approach to obtain science-based benchmark doses (or dose rates) (e.g. screening benchmarks for ecological risk assessments or derived consideration reference levels), based on robust meta-analysis of dose–effect relationships covering ecologically relevant exposure time scales, species, and endpoints.

Keywords: Ecological risk assessment; Extrapolations; Radionuclides; Effect benchmarks

This paper does not necessarily reflect the views of the International Commission on Radiological Protection.

1. INTRODUCTION

1.1. International background in brief

Over the last 15 years, there has been considerable international effort to investigate the issue of protection of the environment from radioactive substances. In Europe, much of the focus has been on collating relevant information and developing different approaches to enable regulatory assessments and protective actions to be undertaken [FASSET (Williams, 2004), ERICA (Larsson, 2008), PROTECT (Howard et al., 2010)]. These developments were well aligned with the European framework for managing environmental risk from chemicals. The major operational outcomes were: an ecological risk assessment (ERA)-type tiered approach and associated tool for characterising and managing the environmental risk for radioactive substances [the ERICA Integrated Approach (Larsson, 2008)]; the derivation of screening ecological benchmarks needed for implementation of the tiered approach (Garnier-Laplace et al., 2008); and the FREDERICA database collating primary data on effects of ionising radiation on non-human species (Copplestone et al., 2008). In the meantime, the United Nations Scientific Committee on the Effects of Atomic Radiation (UNSCEAR) reviewed the literature to update its estimates of effects of ionising radiation on non-human biota (UNSCEAR, 2008). Through *Publication 108*, the International Commission on Radiological Protection (ICRP) developed an approach for radiological protection of the environment, consistent with the approach for humans (ICRP, 2008). This publication defines the underlying basic concepts and data requirements for Reference Animals and Plants (RAPs), including derived consideration reference level (DCRL) values, defined as the 'band of dose rates within which there is likely to be the same chance of deleterious effects occurring to individuals of such type of organism'. The International Atomic Energy Agency has revised the International Basic Safety Standards, which now include references to radiological protection of the environment (IAEA, 2011), and the revised version of the Euratom Basic Safety Standards was adopted consistently at the end of 2013 (Council of the European Union, 2013).

1.2. Basic definitions and extrapolation issues

Broadly, ERAs estimate the probability and magnitude of detrimental effects that are likely to occur in exposed biota at different levels of biological organisation (individuals, populations, communities, ecosystems). The method used to characterise the risk is based on comparing the doses (or the dose rates) to which living species of interest are exposed during a defined period with critical effect values (including 'no-effect' values). These effect benchmarks are defined for specific RAPs in ICRP's methodology, for specific wildlife groups when referring to the UNSCEAR's estimates, and for the structure and function of an ecosystem when using the ERICA-PROTECT's screening benchmark associated with the ERICA Integrated Approach (Table 1). Whichever method is

Table 1. Published effects benchmarks (dose rate in μGy h^{-1}) for chronic exposure situations.

	UNSCEAR (2008)	ERICA – PROTECT*	ICRP *Publication 108* (2008)
Terrestrial ecosystems		10	
Plants	400		
Reference Pine Tree†			4–40
Reference Wild Grass			40–400
Animals	100		
Reference Bee			400–4000
Reference Earthworm			400–4000
Reference Duck			4–40
Reference Deer			4–40
Reference Rat			4–40
Aquatic ecosystems		10	
Freshwater organisms	400		
Reference Frog			4–40
Reference Trout			40–400
Marine organisms	400		
Reference Crab			400–4000
Reference Flatfish			40–400
Reference Brown Seaweed			40–400

*Garnier-Laplace et al. (2010). Statistical method applied to the best set of chronic radiotoxicity critical values (dose rate giving 10% effect in various species and population-relevant endpoints). Meta-analysis indicated that 80 μGy h^{-1} would protect 95% of the species in a generic ecosystem (i.e. there will be less than 10% effect in the population-relevant endpoints for 95% of the species). In a precautionary approach to account for data limitations, a safety factor of 5 was applied to obtain the screening benchmark for ecological risk assessment purposes in the ERICA Integrated Approach.

†Reference 'organism type' refers to the International Commission on Radiological Protection's Reference Animals and Plants (RAP). For any RAP, derived consideration reference levels are given in the last column as a band of dose rates within which there is likely to be the same chance of deleterious effects occurring to individuals of such type of organism.

used for derivation, a number of extrapolation issues that arise in undertaking an ERA need to be dealt with, such as from acute to chronic exposures, from external to internal irradiation, between biological endpoints, between species, and from individual to population levels (Garnier-Laplace et al., 2008). This paper reviews three of these issues – extrapolation from individual to population, between species, and from laboratory to field – and discusses them in terms of basic knowledge, inference methods used, and main lessons learnt.

2. MATERIALS AND METHODS

2.1. Overview of the method for estimating population effects for a given species by combining Leslie matrices and dose-rate–effects relationships

Generally, protection of the environment targets populations and ecosystems. However, most laboratory effect tests are performed at individual and subindividual levels. One of the most important challenges is to extrapolate measured effects on individual endpoints to the population. Focusing solely on individual endpoints can lead to inaccurate estimates of risk to populations due to the complexity and non-linearity of the relationship between effects on individual survival, reproduction, or growth, and population dynamics. The proposed method is based on the use of a Leslie matrix, where the population of a species is represented as an age-structured vector $N(t)$ containing the numbers $n_i(t)$ of individuals in each age class i at time t, with i being the individual age ranging from 1 to i_{max}. The population at $t + \Delta$ is obtained from the equation $N(t + \Delta) = A.N(t)$, where A, the transition matrix constructed on the basis of the species' life-cycle features, collected from the literature, is defined by the following variables:

- the survival rate P_i, i.e. the probability that individuals of age class i survive to the next age class over one time step; and
- the fecundity rate, F_i, for each age class i during one time step.

Based on dose-rate–effect relationships established for individual endpoints, responses to ionising radiation are expressed as reduction coefficients applied to each variable (Fig. 1). Such dose-rate–response relationships were built according to a well-defined procedure explained in detail in Garnier-Laplace et al. (2010). The method, initially proposed by Lance et al. (2012), has been generalised to representative wildlife species (Alonzo et al., 2013). Here, only two terrestrial species, a mammal (*Mus musculus*) and a soil invertebrate (*Eisenia fetida*), are used as examples to compare the chronic external gamma radiosensitivity at the individual level using the EDR_{10} (dose rate giving 10% change in observed effect) on survival or fecundity with two population-relevant endpoints modelled as the lowest dose rate inducing significant change in the population growth rate λ-*loEDR* and the individual reproductive rate R_0-*loEDR* (number of offspring per individual over a lifetime) (Lance et al., 2012).

2.2. Overview of the method for estimating the range of variation of radiosensitivity among species

The method aims to represent the range of variation in radiosensitivity among species using a statistical extrapolation model named 'species sensitivity distribution' (SSD). Using the FREDERICA database, focusing on terrestrial species and endpoints directly relevant to population demography, the analysis was restricted to chronic external exposure situations acquired under controlled conditions, and to chronic effect data derived from field studies within the Chernobyl exclusion zone.

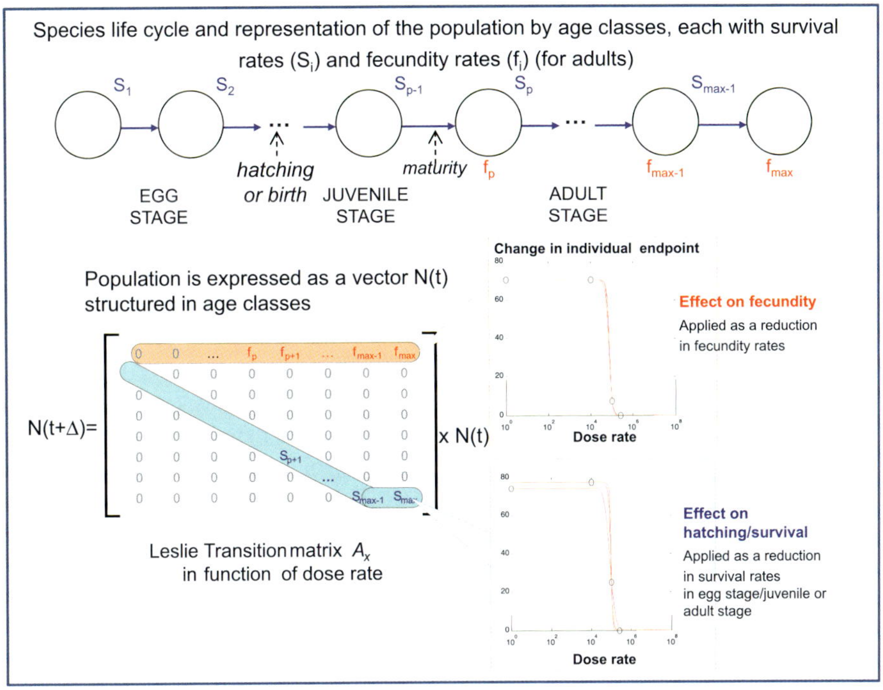

Fig. 1. Extrapolation approach proposed for predicting population-level response by combining dose-rate–effect relationships and an age-structured population matrix.

For each species and endpoint, the dose-rate–effect relationship was built, and the EDR_{10} was estimated. EDR_{10} values were allocated into two consistent data sets of critical ecotoxicity data: one documenting effects from chronic exposure to external gamma irradiation under controlled conditions, and the other documenting effects from chronically exposed wildlife within the Chernobyl exclusion zone in the last 15 years. Curve fitting on quality-assessed data subsets was performed with the 'drc' add-on package (Ritz and Streibig, 2005) of R software (R Development Core Team, 2010).

A statistical distribution was fitted to the selected sets of EDR_{10} (one minimum value per species) and allowed to compare interspecies variation in radiosensitivity between controlled and field exposure conditions. A direct weighted bootstrap method was used to build the SSD confidence intervals. The method is described in full in Garnier-Laplace et al. (2010).

3. RESULTS AND DISCUSSION

3.1. Extrapolation from individual to population level for two species

The population response to chronic ionising radiation was explored in *E. fetida* and *M. musculus* on the basis of a modelling exercise examining how population growth rate declined as a result of combined changes in several life-history traits. For the two selected species – and for a series of others in Alonzo et al. (2013) – evidence was found to show that population-level effects depended on how key variables were affected by exposure to ionising radiation at the individual level, and on how population dynamics responded to such combination of individual effects (Fig. 2). The simulations supported the view that the most influential individual endpoint for population dynamics was not necessarily the most sensitive to radiation (Alonzo et al., 2013). Combining available effects observed in each species on distinct life stages and individual endpoints yielded significant changes in both individual R_0 and/or population λ at minimum dose rates of 2610 μGy h^{-1} and 140 μGy h^{-1}, respectively, in the soil invertebrate (*E. fetida*) and the terrestrial mammal (*M. musculus*). Significant reduction in individual R_0 or population λ resulting from a combination of slight effects (<10% reduction) on several life-history traits was predicted at dose rates below any EDR_{10} derived for *E. fetida* at the individual level (lowest EDR_{10} of 3400 μGy h^{-1}), whereas the individual level for *M. musculus* (lowest EDR_{10} of 26 μGy h^{-1}) was clearly more radiosensitive than the population level. These observations supported the idea that a species might not be protected against a significant change below the EDR_{10} derived for the most sensitive individual endpoint, if several life-history traits are affected simultaneously. As a consequence for ICRP DCRL bands for the corresponding RAP (ICRP, 2008), the recommended DCRL band for the Reference Rat (i.e. 0.1–1 mGy day^{-1} or 4–40 μGy h^{-1}) fits with data reported here for the individual level, but may be overly cautious at the population level (in that case, the DCRLs would be 1–10 mGy day^{-1} or 40–400 μGy h^{-1}). For Reference Earthworm, the DCRL of 10–100 mGy day^{-1} (or 400–4000 μGy h^{-1}) is appropriate for both individual and population levels.

3.2. Extrapolation from one species to another, and from laboratory exposure conditions to the field

Based on laboratory data on radiation effects, species radiosensitivity varied over six orders of magnitude, with vertebrates being among the most radiosensitive organisms (Garnier-Laplace et al., 2010). This suggested that research on primary mechanisms of interactions between ionising radiation and living organisms (from biomolecules up to individuals) is still needed to extrapolate effects in terms of quality and intensity on a rational basis among the wide biodiversity. Additionally, two sets of chronic radiotoxicity data in terrestrial non-human species (one acquired under controlled external gamma irradiation in the laboratory or in the field, and another acquired from field studies in the Chernobyl exclusion zone characterised by a much more complex exposure situation) were compared. This comparison, discussed in detail in Garnier-Laplace et al. (2013), indicated that the best estimate of the median value of the distribution established for field conditions at Chernobyl (approximately 100 mGy h^{-1}) was eight times lower than that from controlled experiments (approximately 850 mGy h^{-1}), suggesting that organisms in their natural environmental were

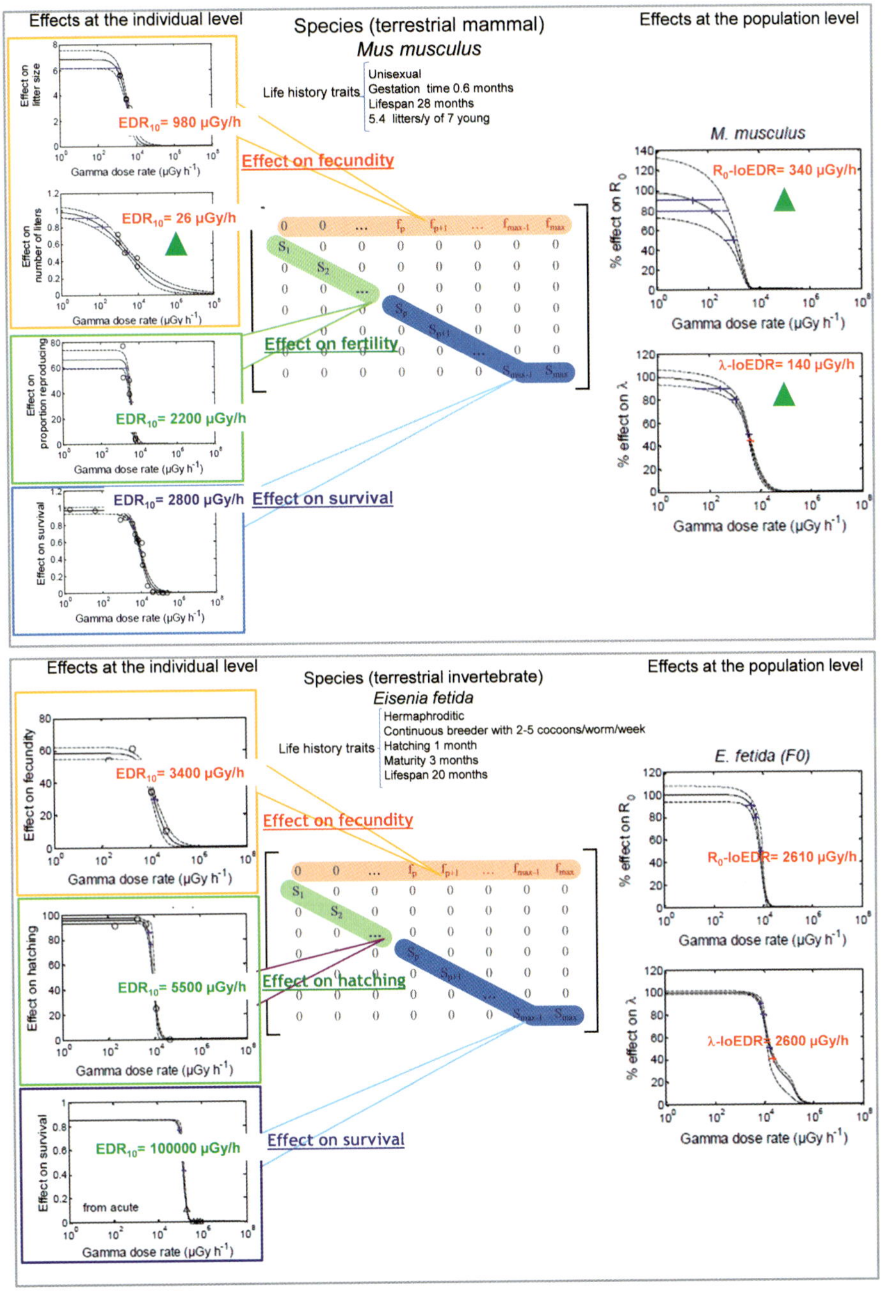

Fig. 2. Extrapolation approach proposed for predicting population-level response by combining dose-rate–effect relationships derived from chronic toxicity data and Leslie matrices (top: results for *Mus musculus*; bottom: results for *Eisenia fetida*).

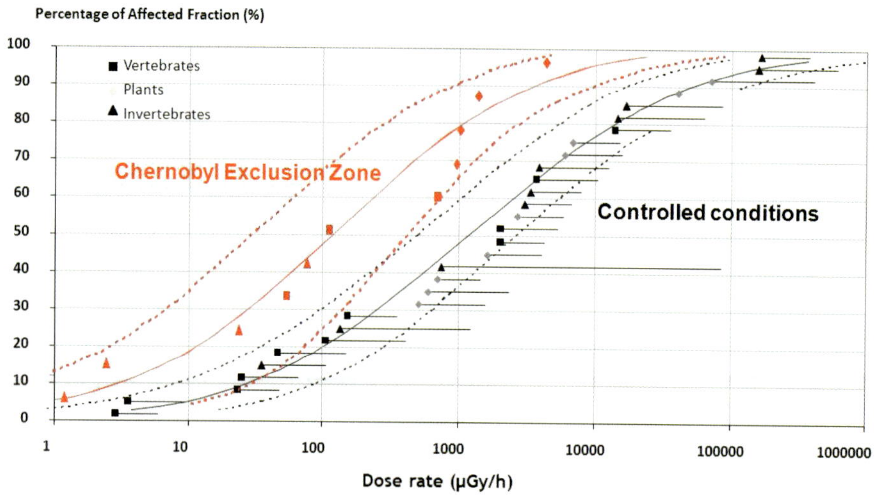

Fig. 3. Extrapolation approach proposed for modelling interspecies radiosensitivity variation. Right: species sensitivity distribution (SSD) fitted to minimum values of EDR_{10} (dose rate giving 10% change in observed effect) for terrestrial species exposed under controlled conditions to external gamma irradiation; left: SSD fitted on EDR_{10} data set acquired in real field conditions from the Chernobyl exclusion zone.

more sensitive (Fig. 3). However, this first comparison highlighted a lack of mechanistic understanding and confusion resulting from field sampling strategies that did not account sufficiently for confounding factors. The field exposure–effects relationship may be modified due to the combination of radiotoxicity effects on growth rate/reproduction and geographic gene diversity, competition, predation, and abiotic factors including pollutants other than radionuclides. Until more field data are acquired, this study calls for a precautionary approach when deriving any benchmark values.

4. CONCLUSIONS

Population models proposed in this study integrate available radiation effect data measured at the individual level, and infer effects at the population level. They can be used to derive dose rates above which population growth is significantly affected when chronically exposed to ionising radiation. This modelling approach is useful to identify critical endpoints for population sustainability in any species, and to prioritise future research. Additionally, the discrepancy between controlled tests and Chernobyl effects data on wildlife is intriguing, mainly due to a shift to a greater radiosensitivity of wildlife in the field. This strongly highlights the need for fundamental research to understand elementary mechanisms.

REFERENCES

Alonzo, F., Vives I Battle, J., Hertel-Aas, T., et al., 2013. Life History Traits, Radiosensitivity and Population Modeling: Methods to Extrapolate from Individual Endpoints to Population Dynamics. STAR Contract Number: Fission-2010-3.5.1-269672, Deliverable 5.2.

Copplestone D., Hingston J., Real A., 2008. The development and purpose of the FREDERICA radiation effects database. J. Environ. Radioact 99, 1456–1463.

Council of the European Union, 2013. Directive 2013/59/EURATOM of 5 December 2013 laying down basic safety standards for protection against the dangers arising from exposure to ionising radiation, and repealing Directives 89/618/Euratom, 90/641/Euratom, 96/29/Euratom, 97/43/Euratom and 2003/122/Euratom. Off. J. Eur. Union L13/1.

Garnier-Laplace J., Copplestone D., Gilbin R., et al., 2008. Issues and practices in the use of effects data from FREDERICA in the ERICA Integrated Approach. J. Environ. Radioact 99, 1474–1483.

Garnier-Laplace J., Della-Vedova C., Andersson P., et al., 2010. A multi-criteria weight of evidence approach to derive ecological benchmarks for radioactive substances. J. Radiol. Prot 30, 215–233.

Garnier-Laplace J., Geras'kin S., Della-Vedova C., et al., 2013. Are radiosensitivity data derived from natural field conditions consistent with data from controlled exposures? A case study of Chernobyl wildlife chronically exposed to low dose rates. J. Environ. Radioact 121, 12–21.

Howard B.J., Beresford N.A., Andersson P., et al., 2010. Protection of the environment from ionising radiation in a regulatory context – an overview of the PROTECT coordinated action project. J. Radiol. Prot 30, 195–214.

IAEA, 2011. Radiation Protection and Safety of Radiation Sources: International Basic Safety Standards Sources. Interim Edition. General Safety Requirements Part 3 nGSR Part 3 (Interim). International Atomic Energy Agency,Vienna.

ICRP, 2008. Environmental protection – the concept and use of reference animals and plants. ICRP Publication 108. Ann. ICRP 38(4–6).

Lance E., Alonzo F., Garcia-Sanchez L., et al., 2012. Modeling population-level consequences of chronic external gamma irradiation in aquatic invertebrates under laboratory conditions. Sci. Tot. Environ 429, 206–214.

Larsson C.M., 2008. An overview of the ERICA Integrated Approach to the assessment and management of environmental risks from ionising contaminants. J. Environ. Radioact 99, 1364–1370.

R Development Core Team, 2010. R: a Language and Environment for Statistical Computing. R Foundation for Statistical Computing, Vienna. Available at: http://www.R-project.org/ (last accessed 12 February 2015).

Ritz C., Streibig J.C., 2005. Bioassay analysis using R. J. Stat. Software 12, 1–22.

UNSCEAR, 2008. Sources of Ionizing Radiation. Vol. II. Report to the General Assembly, Scientific Annex E. United Nations, New York.

Williams, C (Ed.), 2004. Framework for assessment of environmental impact (FASSET) of ionising radiation in European ecosystems. J. Radiol. Protect. 24.

SAGE

ICRP
Annals of the ICRP

Use of the ICRP system for the protection of marine ecosystems

D. Telleria[a], T. Cabianca[b], G. Proehl[a], V. Kliaus[c], J. Brown[d], C. Bossio[e], J. Van der Wolf[f], I. Bonchuk[g], M. Nilsen[d]

[a]*IAEA Assessment and Management of Environmental Releases Unit, Wagramer Str. 5, PO Box 100, 1400 Vienna, Austria; e-mail: D.Telleria@iaea.org*
[b]*Public Health England, Planned Exposure Group, UK*
[c]*Laboratory of Radiation Safety, Republican Scientific-Practical Centre of Hygiene, Republic of Belarus*
[d]*Norwegian Radiation Protection Authority, Norway*
[e]*Nuclear Regulatory Authority, Radiological Safety Assessments Division, Argentina*
[f]*Delft University of Technology, The Netherlands*
[g]*Ukrainian Radiation Protection Institute, Ukraine*

Abstract–The International Commission on Radiological Protection (ICRP) recently reinforced the international system of radiological protection, initially focused on humans, by identifying principles of environmental protection and proposing a framework for assessing impacts of ionising radiation on non-human species, based on a reference flora and fauna approach. For this purpose, ICRP developed dosimetric models for a set of Reference Animals and Plants, which are representative of flora and fauna in different environments (terrestrial, freshwater, marine), and produced criteria based on information on radiation effects, with the aim of evaluating the level of potential or actual radiological impacts, and as an input for decision making. The approach developed by ICRP for flora and fauna is consistent with the approach used to protect humans. The International Atomic Energy Agency (IAEA) includes considerations on the protection of the environment in its safety standards, and is currently developing guidelines to assess radiological impacts based on the aforementioned ICRP approach. This paper presents the method developed by IAEA, in a series of meetings with international experts, to enable assessment of the radiological impact to the marine environment in connection with the Convention on the Prevention of Marine Pollution by Dumping of Wastes and Other Matter 1972 (London Convention 1972). This method is based on IAEA's safety standards and ICRP's recommendations, and was presented in 2013 for consideration by representatives of the contracting parties of the London

This paper does not necessarily reflect the views of the International Commission on Radiological Protection.

Convention 1972; it was approved for inclusion in its procedures, and is in the process of being incorporated into guidelines.

Keywords: Radiological protection; Protection of the environment; Reference Animals and Plants; System of radiological protection

1. INTRODUCTION

Until publication of the 2007 Recommendations of the International Commission on Radiological Protection (ICRP, 2007), the system for radiological protection applied for the management of environmental releases of radionuclides was based on the evaluation of possible resulting radiation doses to humans (ICRP, 1991). It was considered that the limitation of exposures to humans would also provide an adequate level of protection to the environment (ICRP, 1991).

As a result of the growing societal concerns related to possible effects of ionising radiation on the environment, as well as pronounced awareness of environmental issues, considerations were initiated by the international radioprotection community with regard to possible radiological impacts on the environment. 'Protection of people and the environment' is identified as a key principle of safety (IAEA, 2006). ICRP now recommends broadening the radiological protection framework beyond the primary aim of contributing to an appropriate level of protection for human beings in an environmental context, to one that includes protection of the environment itself against the detrimental effects of radiation exposure, without unduly limiting the desirable human actions that may be associated with such exposure (ICRP, 2007).

2. OBJECTIVES OF PROTECTION OF THE ENVIRONMENT

IAEA defines the environment as the 'conditions under which people, animals and plants live or develop and which sustain all life and development; especially such conditions as affected by human activities' (IAEA, 2014). Usually, the environment includes biotic and abiotic components.

It is recognised that, similarly to humans, flora and fauna could be affected by ionising radiation resulting from planned or accidental releases.

The objectives of radiation protection of humans are intended to avoid deterministic health effects and to limit stochastic health effects (IAEA, 2014).

Otherwise, the objectives for the protection of the environment are more generic, and are not quantifiable in a straightforward manner. The aims of environmental protection are to prevent or reduce the frequency of deleterious radiation effects on biota to a level where they would have a negligible impact on the maintenance of biological diversity, the conservation of species, or the health and status of natural habitats, communities, and ecosystems (ICRP, 2007, 2008, 2014).

While the system of radiological protection of humans applies for individuals, consideration of individual animals or plants is not the aim for protection of the environment, where the target is higher organisational levels such as populations, community levels, and ecosystems.

3. ICRP'S APPROACH FOR PROTECTION OF THE ENVIRONMENT

ICRP has defined an approach to assess and control the effects of radiation on flora and fauna (ICRP, 2007, 2008, 2014) using the concepts of Reference Animals and Plants (RAPs) and Representative Organisms, consistent with the concepts of Reference Person and Representative Person for humans (ICRP, 1991, 2007), and corresponding dose criteria in the form of derived consideration reference levels (DCRLs).

The method to assess the level of protection of the environment where radioactivity releases may occur is to estimate dose rates resulting from radiation exposure of a set of RAPs representative of those more highly exposed, and to compare the resulting doses with a criteria. By this, it is possible to define a level of protection of the environment, or make decisions on managerial options.

The approach to environmental protection has to be commensurate with the overall level of risk, and compatible with other approaches being made to protect the environment (ICRP, 2008).

3.1. Reference Animals and Plants

In the system of radiological protection for humans, *Publication 103* (ICR, 2007) defined a model called 'Reference Person', and methods to calculate its doses that can be used in assessment of the radiological impact to members of the public. In a similar manner, for flora and fauna, ICRP has defined a small set of RAPs, including models for their dosimetry (ICRP, 2008). An RAP was defined (ICRP, 2008, 2014) as a hypothetical entity with the assumed basic biological characteristics of a particular type of animal or plant, as described to the generality of the taxonomic level of Family, with defined anatomical, physiological, and life-history properties that can be used for the purposes of relating exposures to dose, and dose to effects, for that type of living organism. The selection of the species, which may be used as indicators of the level of environmental protection, is discussed in ICRP (2014). The use of these particular species is somewhat subjective and is related to their connection with particular major ecosystems and, principally, to the existence of databases with information on radiation effects.

3.2. Derived consideration reference levels

For evaluating the level of radiological impact on flora and fauna, ICRP introduced DCRLs for the set of RAPs. The DCRLs do not represent limits; they should be considered as points of reference to inform on the appropriate level of effort that should be expended on environmental protection, dependent on the overall management objectives, the exposure situation, the actual fauna and flora present, and the numbers of individuals thus exposed (ICRP, 2007, 2008, 2014).

In a generic assessment, a level of protection established by setting reference criteria below or equal to the lower band of the DCRLs is considered to be an appropriate level of protection for flora and fauna (IAEA, in preparation).

DCRLs are presented in ICRP (2008, 2014).

3.3. Representative Organisms

Publication 124 (ICRP, 2014) indicates that a 'Representative Organism' is 'a particular species or group of organisms selected during a site-specific assessment, taking account of their assumed location with respect to the source'. The actual choice of Representative Organism will depend upon the purpose of the assessment (ICRP, 2008). In many cases, the actual Representative Organisms chosen for an assessment may be the same as, or very similar to, the RAPs, but in some cases, they may be very different. The selection of Representative Organisms for use in an assessment for protection could be based on assumptions of a generic character or site-specific characteristics (IAEA, in preparation).

4. IAEA'S METHOD FOR PROTECTION OF THE ENVIRONMENT

When considering environmental protection, IAEA (in preparation, 2015) adopted the concepts and use of RAPs, DCRLs, and Representative Organisms defined by ICRP (2007, 2008, 2014), and complemented the approach with the requisite to consider protection of flora and fauna and humans in an integrated manner (IAEA, 2011). This integration of human and flora and fauna protection can be considered, in practical terms, assuming a linkage in the exposure scenarios. Fig. 1 shows a scheme for estimating and evaluating exposures to humans and biota, and indicates the link between the exposure scenarios. All three exposure situations (i.e. planned, existing, and emergency) are considered. The assessment of exposures to humans and flora and fauna is based on measured or estimated radionuclide activity concentrations in the environment.

Using this approach, the managerial decisions related to the environment could be made based on considerations for the protection of humans and flora and fauna in an integral manner.

5. RADIOLOGICAL PROTECTION OF THE MARINE ENVIRONMENT IN THE FRAMEWORK OF THE LONDON CONVENTION 1972

IAEA is the international technical advisor in matters related to radioactive materials for the Convention on the Prevention of Marine Pollution by Dumping of Wastes and Other Matter 1972 (IMO, 1972).

The London Convention 1972 prohibits the disposal of radioactive wastes and other radioactive matter at sea. However, natural radionuclides are present in all

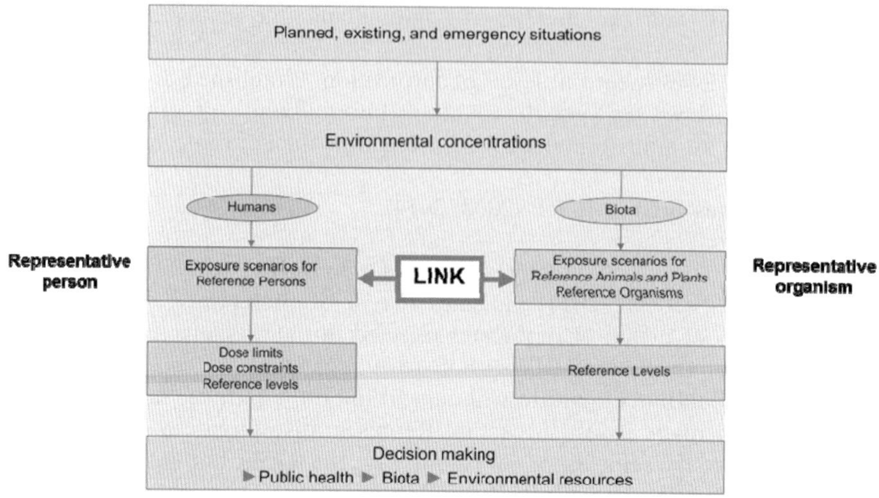

Fig. 1. Scheme illustrating the integrated approach proposed by the International Atomic Energy Agency to assess and manage the radiological protection of humans and the protection of the environment.

materials, including natural and inert materials, which can also contain artificial radionuclides from anthropogenic sources such as fallout due to past atmospheric nuclear testing. Therefore, there was a need to develop definitions and guidelines so that candidate materials containing less than de minimis levels of specific radioactivity (IAEA, 1999) can be regarded as 'non-radioactive', and may be disposed of at sea, subject to the other provisions of the Convention.

In 2003, IAEA was requested by the contracting parties of the London Convention 1972 to expand the method to determine if materials that are candidates for dumping have trivial radiological impact and, consequently, can be dumped in the framework of the Convention. Originally, this method considered human protection aspects (IAEA, 2003), but, in order to fill a conceptual gap and in accordance with the international trends on this topic, contracting parties requested that protection of the marine flora and fauna should be considered more explicitly.

For the purpose of the London Convention 1972, and thanks to the recent progress in this area by ICRP, IAEA could develop, with the cooperation of international experts, a method for assessing the impact on humans and the marine environment related to the dumping of materials containing radioactive residues.

This method is of a generic and conservative character, and is based on ICRP's approach for protection of flora and fauna combined with the approach to consider humans and the environment in an integrated manner as proposed by IAEA. The method was approved by the Contracting Parties to the London Convention in October 2013 (IAEA, 2015).

5.1. Calculation of radionuclide concentrations in water, sediment, and marine biota

A box model is used to simulate the dispersion and dilution of the radionuclides in the water column of the region surrounding a dumping site. Details are presented in IAEA (2003) and IAEA (2015). The radionuclide concentrations in seawater and sediments predicted by the model are average values over the entire volume of the box in the year of release. The model assumes instantaneous equilibrium between radionuclides in the soluble phase, and those adsorbed on particles suspended in the water column and on particles in the top sediment boundary layer.

5.2. Selection of Reference Animals and Plants

For the marine environment and under the exposure situations related to the framework of the London Convention 1972, IAEA indicates that the relevant ICRP RAPs are a marine fish, a marine crustacean, and a type of seaweed (IAEA, 2015). IAEA notes that this is a generic approach, and, although applicable to other similar exposure situations and regulatory frameworks in principle, it is important that careful consideration of the particularities of the exposure scenarios are taken into account before using the same methodology (IAEA, 2015).

5.3. Representative Persons and Representative Organisms

Representative Persons in the case of members of the public are those living near to the shoreline in the area of influence of the planned dumping activity, eating seafood coming from the same sea region where materials are dumped, and occupying the beach during some time in the year.

In the case of flora and fauna, the Representative Organisms are the RAPs located in the sea region of the dumping activity.

The exposure scenario to humans and flora and fauna is linked by these assumptions.

5.4. Exposure pathways to humans and marine flora and fauna

The following exposure pathways are considered for members of the public:

- external exposure to radionuclides deposited on the shore;
- ingestion of seafood caught in the area around the dumping site;
- inadvertent ingestion of beach sediments;
- inhalation of particles resuspended from beach sediments; and
- inhalation of sea spray.

In the case of marine flora and fauna, the following external and internal exposures are considered:

- external irradiation due to radioactive material diluted in the water and in the marine sediments; and
- internal exposure from incorporated radioactive material.

The simplifications in assessing internal and external exposures to flora and fauna are discussed in ICRP (2008).

5.5. Estimation of doses to members of the public and flora and fauna

Individual doses to members of the public are calculated starting from estimations of the annual average activity concentrations in water and sediments, resulting from the dumping of materials with radioactive residues.

For estimating the activity concentration, a generic box in the sea of 10 km × 10 km × 20 m located at the point of release is defined.

It is assumed that members of the public eat seafood from that box and, using consumption rates and the corresponding dosimetric factors, internal doses are estimated. For the internal and external doses related to the occupation of the beach by the public, the activity concentrations in beach sediments and sea spray is estimated using transfer factors.

Collective doses to the public are assessed using external and internal pathways, and generic factors determining the number of individuals affected (e.g. length of the coastline affected by the dumping activity, and annual amount of seafood caught in the area affected by a single dumping).

Doses to the defined marine RAPs are estimated for external and internal exposures using the activity concentration in water and marine sediments in the same generic box as for public and transfer parameters, and dosimetric factors provided in ICRP (2009) and ICRP (2008), respectively.

Based on this approach, a practical set of screening coefficients to estimate individual and collective doses to members of the public, and dose rates to marine RAPs (fish, crustacean, and seaweed), were calculated. For these screening coefficients, models and parameters which tend to overestimate the dose assessment were used in order to obtain conservative results.

IAEA (2015) provides these screening coefficients for the purpose of the London Convention 1972 and the details of the assessment described in this section, including models, assumptions, parameters, and sources of information.

5.6. Criteria for humans and marine flora and fauna

IAEA (2015) indicates that, under the framework of the London Convention 1972, candidate material may be considered for disposal, verifying other non-radiological requirements in the London Convention Guidelines (IMO, 2006), if:

- the estimated annual individual dose to members of public is equal to, or less than, 10 μSv;
- the estimated annual total collective dose is less than, or equal to, 1 man-Sv; and
- the estimated dose rate for each marine RAP is below, or equal to, the corresponding DRCL.

If the material satisfies all these criteria for members of the public and flora and fauna, it can be considered for dumping.

6. CONCLUSIONS

The 2007 Recommendations (ICRP, 2007) enhanced the system of radiological protection by providing an approach to assess and manage the protection of the environment, based on consideration of the radiological impact on flora and fauna. This approach includes the concepts of RAPs, DCRLs, and Representative Organisms (ICRP, 2008, 2014), consistent with the concepts of Reference Person, dose limits, constraints, reference levels, and Representative Person used to assess and manage the protection of the public against the effects of ionising radiation (ICRP, 2007).

IAEA considered protection of people and the environment in its Safety Standards (IAEA, 2006), and added the requisite of an approach that must consider humans and flora and fauna in an integrated manner (IAEA, 2014).

The London Convention 1972 asked IAEA to provide a method to assess the radiological impact to marine flora and fauna more explicitly, in order to authorise disposal of materials to the oceans from the radiation protection point of view.

IAEA has expanded previous work undertaken for the London Convention 1972 (IAEA, 2003), which considered human protection aspects alone, by incorporating explicit considerations for marine flora and fauna into the assessment (IAEA, 2015).

The work undertaken by IAEA, with the assistance of international experts, resulted in a practical methodology applicable for the London Convention 1972 (IAEA, 2015); this methodology was approved by the contracting parties in October 2013, and is in the process of being incorporated into guidelines (IMO, 2006).

The proposed assessment procedure is based on the state of the art in radiological protection of the environment, in line with ICRP recommendations and IAEA safety standards.

The application of these practical procedures would be appropriate to protection of the marine environment and use of the marine resources in a sustainable manner, without imposing unnecessary burden on the potential users.

REFERENCES

IAEA, 1999. Application of Radiological Exclusion and Exemption Principles to Sea Disposal, IAEA-TECDOC-1068. International Atomic Energy Agency, Vienna.

IAEA, 2003. Determining the Suitability of Materials for Disposal at Sea Under the London Convention 1972: a Radiological Assessment Procedure. IAEA-TECDOC-1375. International Atomic Energy Agency, Vienna.

IAEA, 2006. Fundamental Safety Principles, IAEA Safety Standards Series No. SF-1. International Atomic Energy Agency, Vienna.

IAEA, 2014. Radiation Protection and Safety of Radiation Sources: International Basic Safety Standards, General Safety Requirements Part 3, IAEA Safety Standards Series No. GSR Part 3. International Atomic Energy Agency, Vienna.

IAEA, in preparation. A General Framework for Prospective Radiological Environmental Impact Assessment and Protection of the Public. IAEA Draft Safety Guide DS427. IAEA Safety Standards Series. International Atomic Energy Agency, Vienna.
IAEA, 2015. Determining the Suitability of Materials for Disposal at Sea Under the London Convention 1972 and London Protocol 1996: a Radiological Assessment Procedure. IAEA-TECDOC-1759. International Atomic Energy Agency, Vienna.
ICRP, 1991. 1990 Recommendations of the International Commission on Radiological Protection. ICRP Publication 60. Ann. ICRP 21(1–3).
ICRP, 2007. The 2007 Recommendations of the International Commission on Radiological Protection. ICRP Publication 103. Ann. ICRP 37(2–4).
ICRP, 2008. Environmental protection: the concept and use of reference animals and plants. ICRP Publication 108. Ann. ICRP 38(4–6).
ICRP, 2009. Environmental protection: transfer parameters for reference animals and plants. ICRP Publication 114. Ann. ICRP 39(6).
ICRP, 2014. Protection of the environment under different exposure situations. ICRP Publication 124. Ann. ICRP 43(1).
IMO, 1972. Convention on the Prevention of Marine Pollution by Dumping of Wastes and Other Matter. Available at: http://www.imo.org/OurWork/Environment/LCLP/Pages/default.aspx.
IMO, 2006. Guidelines on the Convention of the Prevention of Marine Pollution by Dumping of Wastes. Report of the 21st Consultative Meeting of the Contracting Parties to the Convention on the Prevention of Marine Pollution by Dumping of Wastes and Other Matter, LC 21/13. Adopted in 1999, amended in 2003 (LC 25/16). International Maritime Organization, London.

SAGE

ICRP
Annals of the ICRP

Creation and application of voxelised dosimetric models, and a comparison with the current methodology as used for the International Commission on Radiological Protection's Reference Animals and Plants

K. Higley[a], E. Ruedig[b], M. Gomez-Fernandez[a], E. Caffrey[a], J. Jia[a], M. Comolli[a], C. Hess[a]

[a]*Department of Nuclear Engineering and Radiation Health Physics, Oregon State University, 100 Radiation Centre, Corvallis, OR 97330, USA; e-mail: Kathryn.Higley@oregonstate.edu*
[b]*Environmental and Radiological Health Sciences, Colorado State University, USA*

Abstract–Over the past decade, the International Commission on Radiological Protection (ICRP) has developed a comprehensive approach to environmental protection that includes the use of Reference Animals and Plants (RAPs) to assess radiological impacts on the environment. For the purposes of calculating radiation dose, the RAPs are approximated as simple shapes that contain homogeneous distributions of radionuclides. As uncertainties in environmental dose effects are larger than uncertainties in radiation dose calculation, some have argued against more realistic dose calculation methodologies. However, due to the complexity of organism morphology, internal structure, and density, dose rates calculated via a homogenous model may be too simplistic. The purpose of this study is to examine the benefits of a voxelised phantom compared with simple shapes for organism modelling. Both methods typically use Monte Carlo methods to calculate absorbed dose, but voxelised modelling uses an exact three-dimensional replica of an organism with accurate tissue composition and radionuclide source distribution. It is a multi-stage procedure that couples imaging modalities and processing software with Monte Carlo N-Particle. These features increase dosimetric accuracy, and may reduce uncertainty in non-human biota dose–effect studies by providing mechanistic answers regarding where and how population-level dose effects arise.

Keywords: Voxel phantom; Three-dimensional modelling; Reference Animals and Plants; Environmental protection

This paper does not necessarily reflect the views of the International Commission on Radiological Protection.

1. INTRODUCTION

Several approaches have been developed, tested, and compared for the limitation of radiation dose in non-human biota (NHB). The International Commission on Radiological Protection (ICRP) has developed a comprehensive approach that includes the use of Reference Animals and Plants (RAPs) for the assessment of dose to NHB.

ICRP's current approach to dosimetry calculations for NHB relies on simplified organism geometry. Organisms are modelled as ellipsoids made of homogeneous four-component soft tissue (ICRU, 1992). Some models also contain simple ellipsoidal organs that serve as targets for dosimetric calculations (ICRP, 2008). These rudimentary models are used to generate absorbed fractions that are, in turn, used to calculate dose conversion factors (DCFs) from radionuclides that are distributed uniformly throughout the organism. There are two scenarios that require consideration beyond that available with homogeneous models. The first is when comparing whole-body average dose rates for homogeneous and non-homogeneous distributions of incorporated radionuclides, and assessments of the uncertainties of whole-body doses due to those non-homogeneous sources. The second is when it is necessary or prudent to estimate organ dose rates from radionuclides that partition heterogeneously.

Voxel models are three-dimensional (3D) models created from radiological imaging modalities [e.g. computed tomography (CT) and magnetic resonance imaging (MRI)]. These models allow radioecologists to calculate organ-specific DCFs from heterogeneously distributed radionuclides. Voxel models are more robust and easier to defend than the previously used ellipsoidal models, but the additional effort required to make such models may not be justifiable. If simplified models are conservative, there is no need to create complicated models. However, if simplified models are not conservative, there is a need for the creation of voxel-derived dosimetric parameters.

2. DESCRIPTION OF HOMOGENEOUS MODELS

Since the 1970s, environmental dosimetry has relied upon the 'uniform isotropic model' as defined by Loevinger and Berman (1976). The uniform isotropic model consists of the following assumptions: the organism is in an infinite homogeneous medium; the activity is distributed uniformly throughout its body; and the densities of the medium and the organism's body are the same (ICRP, 2008). Fig. 1 shows an example of the homogeneous models in use by ICRP today, and Table 1 provides details on the dimensions of each RAP.

3. CONCEPT OF VOXEL PHANTOM

A voxel phantom is constructed with a series of image slices formatted in pixels with a given thickness, forming a vo͟lume pix͟el unit, or voxel unit, as shown in Fig. 2. The purpose of a voxel phantom model is to represent a body region of interest

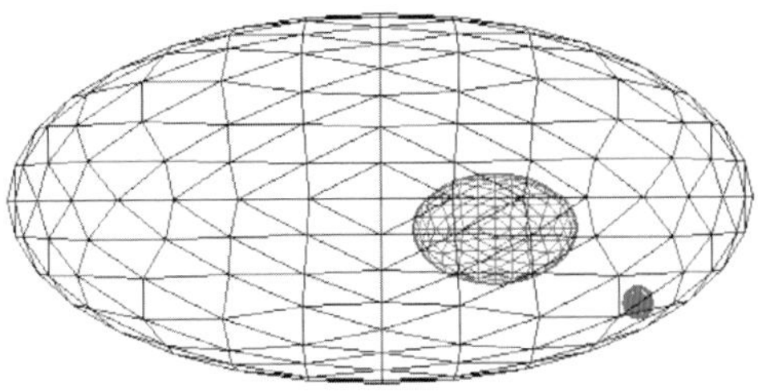

Fig. 1. Current International Commission on Radiological Protection geometric model of the Reference Animals and Plants with inner ellipsoid representing the internal structure (ICRP, 2008).

Table 1. Parameters of the International Commission on Radiological Protection's homogeneous model.

Animal/plant	Shape	Dimensions (cm)	Volume (cm^3)	Organ shape	Organ dimensions (cm^3)
Reference Deer	Ellipsoid	130×60×60	1960353.816	Ellipsoid	Variable
Reference Rat	Ellipsoid	20×6×5	2513.274123	Ellipsoid	Variable
Reference Duck	Ellipsoid	30×10×8	10053.09649	Ellipsoid	Variable
Reference Frog	Ellipsoid	8×3×2.5	251.3274123	Ellipsoid	Variable
Reference Trout	Ellipsoid	50×8×6	10053.09649	Ellipsoid	Variable
Reference Flatfish	Ellipsoid	40×25×2.5	10471.97551	Ellipsoid	Variable
Reference Bee	Ellipsoid	2×0.75×0.75	4.71238898	Ellipsoid	Variable
Reference Crab	Ellipsoid	20×12×6	6031.857895	Ellipsoid	Variable
Reference Earthworm	Ellipsoid	10×1×1	41.88790205	Ellipsoid	Variable
Reference Pine Tree	Ellipsoid	1000×30×30	3769911.184	Ellipsoid	Variable
Reference Wild Grass	Ellipsoid	5×1×1	20.94395102	Ellipsoid	Variable
Reference Brown Seaweed	Ellipsoid	50×50×0.5	5235.987756	Ellipsoid	Variable

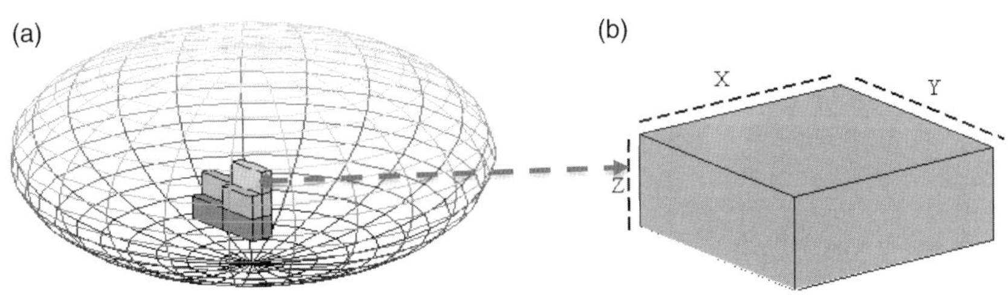

Fig. 2. (a) Example of a voxelised object inside a solid ellipsoid; (b) representation of a volume pixel or voxel unit.

(Zaidi and Xu, 2007). These image slices are obtained via medical imaging (e.g. CT or MRI), ensuring that the final phantom closely mimics the real internal anatomy of an organism. This method allows for detailed studies of the interaction of particles through different components of an organism (e.g. an individual organ, an organ system, or the whole body) to be conducted.

4. CONSTRUCTION OF A VOXEL PHANTOM

4.1. Selection of an imaging modality

CT is a non-invasive imaging technique that allows for visualisation of the internal structures of an organism without the superposition of over- and underlying structures (Hendee, 2002). CT scans are ideal for organisms where tissues have differing effective atomic numbers or density. In cases with these factors, contrast suffers and it may be difficult to resolve internal structures. MRI is well suited for these cases, as it is able to provide excellent contrast resolution in even those tissues with similar density and atomic number. In general, CT scans are acceptable for organisms with more bony structures and/or differentiation in tissue composition (e.g. Reference Crab). MRI scans are recommended when the organism consists primarily of soft tissue) (e.g. Reference Flatfish).

4.2. Required software suites

4.2.1. 3D-Doctor (or any other segmenting software)

3D-Doctor is a powerful and user-friendly image processing software for medical, scientific, and industrial imaging, developed by Able Software Corp. It specialises in processing, rendering, modelling, visualisation, and quantitative analysis of images from MRI, CT, positron emission tomography, and other industrial uses (Able Software Corp., 2012). Organism images are segmented on each image slice, outlining each identifiable organ. The software exports boundary (.bnd) files that are needed for input into a lattice generation tool (e.g. Voxeliser, discussed below).

4.2.2. Voxeliser

Voxeliser was developed in the python coding language by the Human Monitoring Laboratory of Canada (Kramer et al., 2010). This program reads boundary file geometry (.bnd) and creates repeated structures (e.g. universes) for Monte Carlo simulations. The limit on the number of voxels in MCNPX 2.5 or MCNP5 is 50 million. In order to stay within the limit, Voxeliser can adjust the array by using a compression factor or array reduction that will combine several voxels into one (Kramer et al., 2010). If a compression factor is applied, there will be a reduction of resolution of the final lattice.

4.2.3. Monte Carlo N-Particle (MCNPX 2.5, MCNP5)

MCNP is a radiation transport code that is capable of solving the transport equation in three dimensions for many particle types, including neutrons, photons, and electrons. It is also capable of coupled electron/photon/neutron transport calculations. The code generates a random number that describes the statistical life of each particle, including scattering (coherent or incoherent), fission, leakage, absorption interaction, and other particle–material interactions that may occur; thus, it can take quite some time for the code to run (X-5 Monte Carlo Team, 2008). MCNP has built-in cross-section libraries for different particles that allow the user to specify the elemental composition and density of each object.

4.2.4. Overall process

First, a CT or MRI image of the organism of interest is obtained. Those images are uploaded into the 3D-Doctor program and segmented. The boundary file is exported and subsequently uploaded into Voxeliser, which created the lattice geometry for input into MCNP. Absorbed fractions are obtained from the MCNP output.

5. MODEL COMPARISIONS

Thus far, voxel models have been created for seven of the 12 RAPs, all in adult life stages. Absorbed fraction (AF) data are publically available and easily obtainable for Reference Rat, Reference Frog, Reference Crab, and Reference Flatfish. AF data for Reference Trout, Reference Bee, and Reference Earthworm will be the subject of forthcoming publications. As mentioned previously, the geometry of the voxel phantom can vary depending on the object or organisms being analysed. One of the challenges in environmental radioprotection is the array of morphologies among organisms. From Table 2, a direct comparison can be made between the volumes of the homogeneous model shown in Table 1, and the voxelised models shown below. It should be noted that the volumes differ by an order of magnitude.

The ability to model realistic shapes and sizes with proper elemental compositions can improve the accuracy of calculations to determine dose rate in NHB as the range of particles is highly dependent on the physical properties of the tissue (e.g. effective atomic number and density). Fig. 3 shows a 3D rendering of the trout model, shown without the muscle tissue; this illustration underscores the level of detail present in a voxelised model.

It is evident from Fig. 3 that using a voxel-based model can be beneficial for many types of simulations, such as simulating a source distributed in multiple organs, or radionuclide transport through organ systems, to name a few. Table 2 shows the anatomical characteristics of four voxel phantoms (note that not every organ is listed, due to the large quantity of organs identified in each model).

5.1. Dose conversion factors

In order to determine if the efforts expended in creating voxel phantoms is justifiable, it is useful to assess the uncertainty in a simple model by calculating doses in a voxelised phantom. In order to conduct this comparison, two sets of absorbed fraction data were used: AFs from simplified, homogeneous models, and AFs from voxelised models for specific source:target pairs. In instances where the radionuclide of interest is distributed homogeneously throughout the organism (e.g. tritium), it is

Table 2. Parameters of various organs on voxel-based models.

Animal	Imaging modality	Whole-body volume (cm^3)	Organs identified	Volume (cm^3)	Mass (g)
Reference Crab*	CT	328.34	Gills	68.32	68.32
			Gonads	21.36	22.21
			Heart	2.17	2.26
			Hepatopancreas	29.79	29.79
Reference Flatfish*	MRI	961.113	Brain	0.628	0.695
			Liver	2.052	2.155
			Gills	2.298	2.298
			Stomach	1.447	1.447
Reference Trout†	CT	651.194	Brain	0.566	0.5886
			Oesophagus	13.75	14.3698
			Eyes	1.936	2.0715
			Heart	1.78	1.8354
Reference Frog‡	CT	235.544	Brain	0.103	0.107
			Heart	0.199	0.207
			Kidneys	0.21	0.219
			Intestine	0.275	0.286

CT, computed tomography; MRI, magnetic resonance imaging.
*Caffrey and Higley (2013).
†Hess (2014).
‡Kinase (2008).

expected that the simplified model and voxel model will show reasonably consistent results in dose rate calculations. For the comparisons detailed below, it was assumed that the radionuclide in question was perfectly partitioned into a single organ; this represents the most extreme case, and if the simplified model is conservative for this case, it can be assumed to be conservative for all cases. In order to assess the conservatism of the homogeneous model, dose rates were calculated for a theoretical 1-MBq source. The source was partitioned perfectly into each voxelised organ to obtain voxel-derived dose rates, and this was compared with dose rates arising from the same 1-MBq source distributed uniformly within the homogeneous model (Ruedig et al., 2015).

5.1.1. Dose conversion factor calculations

RAPs and radionuclides were selected based on the availability of dosimetric data. To this end, Reference Rat, Reference Flatfish, Reference Trout, and Reference Crab were included in this analysis (Mohammadi et al., 2011, 2012; Caffrey, 2012; Caffrey and Higley, 2013; Hess, MSc thesis in progress). Seven radionuclides were selected

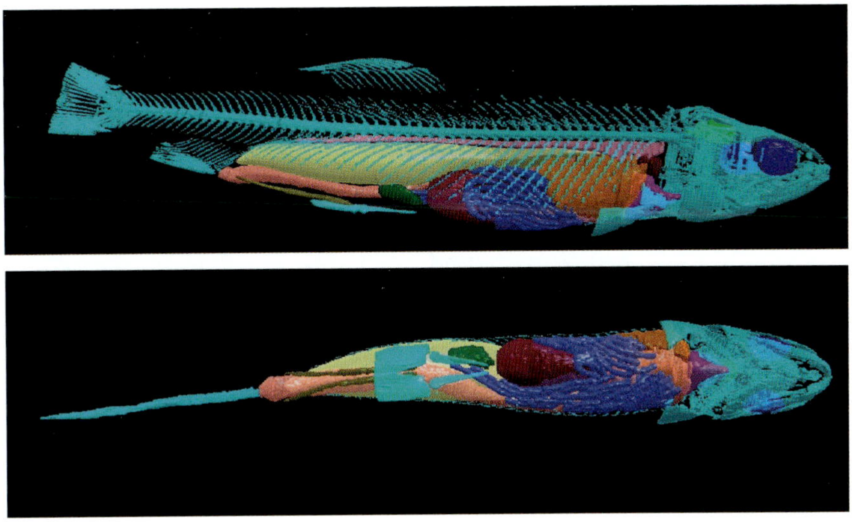

Fig. 3. Three-dimensional rendering of the trout model, shown without muscle tissue, from side (top) and above (bottom) (Hess, MSc thesis in progress).

for analysis as they represented a wide range of radiation types and energies, and are commonly encountered: ^{14}C, ^{36}Cl, ^{60}Co, ^{90}Sr, ^{131}I, ^{137}Cs, and ^{210}Po. Note that ^{137}Cs AF data were not available for Reference Rat. Biological behaviour is left for further study; as such, some organ:radionuclide pairs may not be indicative of reality.

Dose rates were calculated as follows:

$$DR\left[\frac{\mu\text{Gy}}{\text{days}}\right] = C\left[\frac{\text{Bq}}{\text{kg}}\right] * Y * BR * E[\text{keV}] * AF * e\left[\frac{\mu\text{J}}{\text{keV}}\right] * t\left[\frac{\text{s}}{\text{days}}\right]$$

where DR is the dose rate, C is radionuclide concentration, Y is the yield (unitless), BR is the branching ratio (unitless), E is the decay energy, AF is absorbed fraction (unitless), e is a conversion from keV to μJ, and t is a conversion from seconds to days.

The following assumptions were made (consistent with ICRP methodologies).

- Any radiation occurring in $<1\%$ of decays was omitted.
- As there are no AFs for positrons, the ε/β^+ decay (1.9%) was omitted in the ^{36}Cl calculations. The subsequent annihilation photons were included, and assumed to originate in the same organ as the parent nuclide.
- Daughter product emissions were included when the daughter half-life was >10 days (^{90}Sr/^{90}Y and ^{137}Cs/^{137m}Ba).
- Alpha emissions were assumed to be absorbed completely in the originating organs.
- Beta emitters with energies >100 keV were assumed to be absorbed completely in the originating organs.

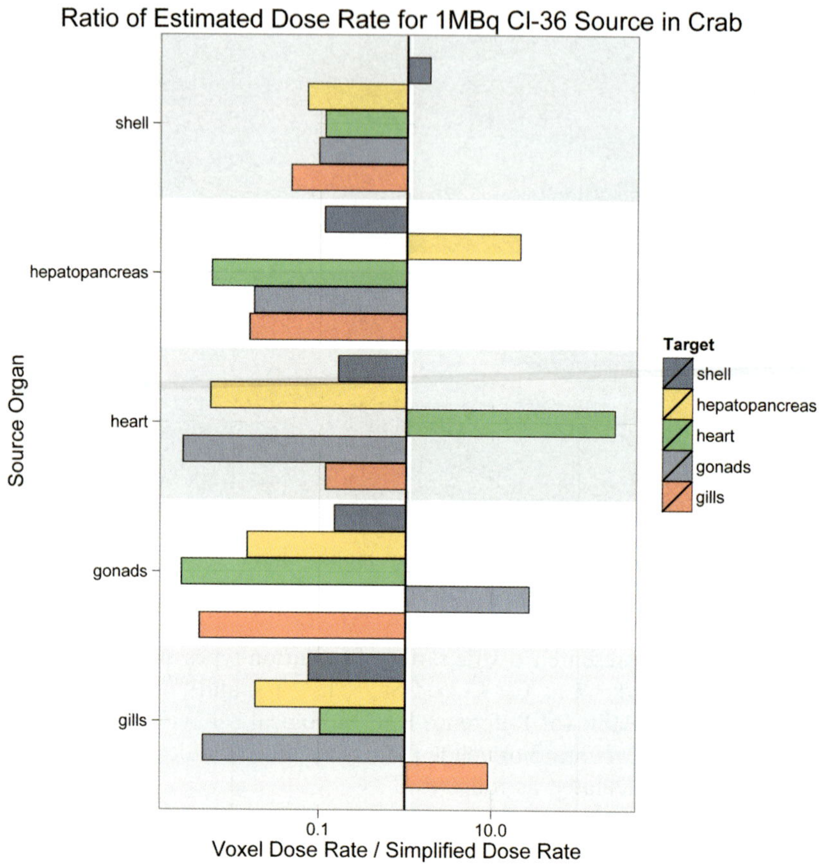

Fig. 4. Cl-36 source in Reference Crab.

- For beta emitters, average energy was taken to be one-third of the maximum.

AF data were obtained from published literature in the case of Reference Rat (Mohammadi et al., 2011, 2012) and Reference Crab (Caffrey and Higley, 2013), and from the Master's theses of Oregon State University students in the case of Reference Flatfish (Caffrey, 2012) and Reference Trout (Hess, MSc thesis in progress; Ruedig et al., to be published). DCFs for simplified models were taken from *Publication 108* (ICRP, 2008).

There is no consensus on the summation of individual organ doses (e.g. tissue weighting factors) to obtain a whole-body dose rate for NHB; as such, no attempt was made to sum the individual organ dose rates, and whole-body dose rates from simplified models were compared with individual organ dose rates from the voxelised model.

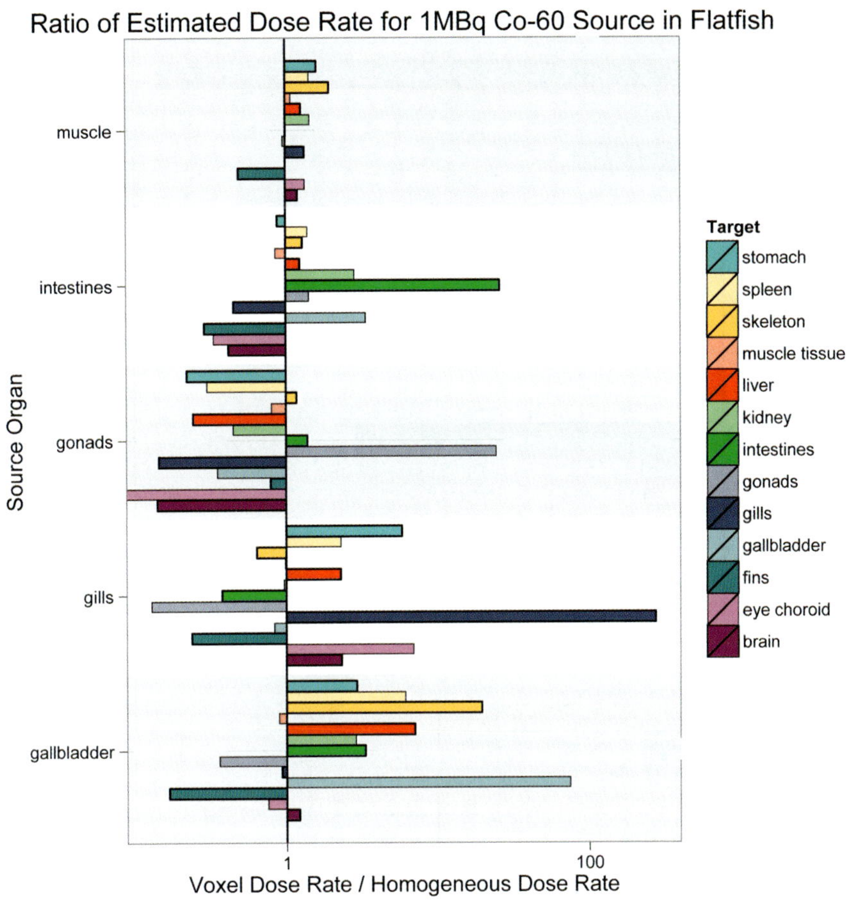

Fig. 5. Co-60 source in Reference Flatfish.

The results are offered as ratios of voxelised:simple dose rate. Thus, if the ratio is unity, the predicted dose rates are equivalent. For ratios greater than unity, voxelised dose rates are thus higher, and the simplified model is not conservative; for ratios less than unity, simplified dose rates are higher, and the simplified model is conservative. Each bar in Figs 4–7 is colour-coded for each source:target combination. The line at unity indicates where simplified and voxelised dose rates are equivalent. Bars to the right indicate source:target pairs where voxelised dose rates are higher; bars to the left indicate that the simplified model dose rate is higher. Bars to the left that appear to run off the chart are cases where the voxelised dose rates were essentially zero, and could not be plotted properly on a log scale.

As Figs 4–7 illustrate, there was no consistent trend seen across the organisms and radionuclides examined in this study. Fig. 4 shows estimated dose rates for a 1-MBq Cl-36 source in Reference Crab. Cl-36 is a beta emitter with a short range. In this

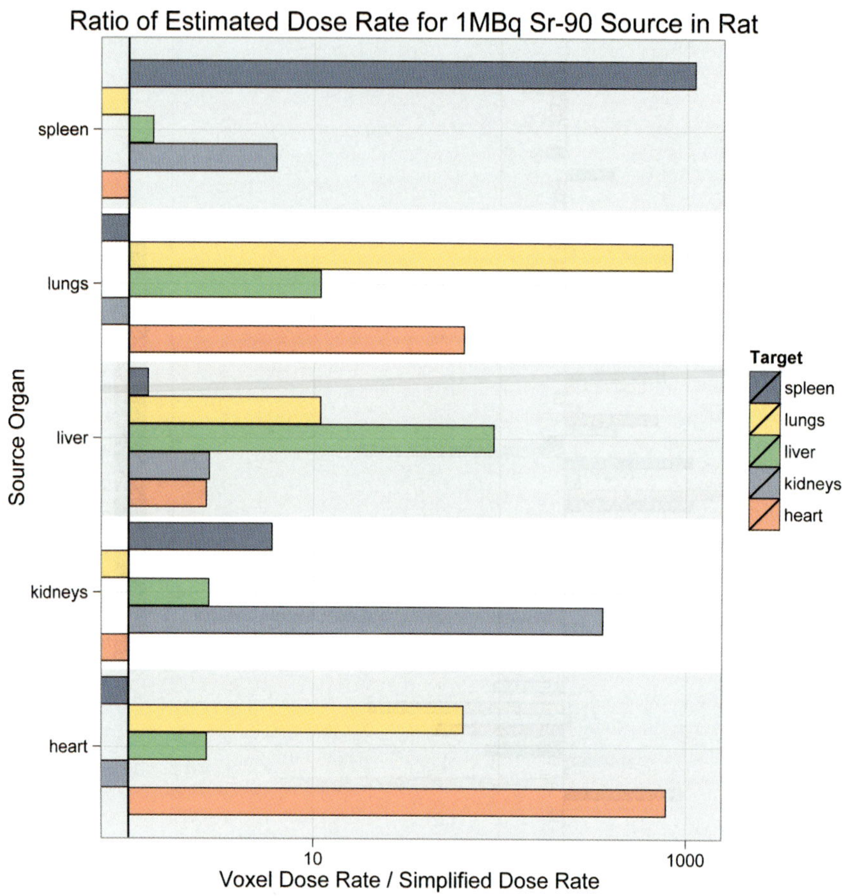

Fig. 6. Sr-90 source in Reference Rat.

case, the homogeneous model generally predicts higher dose rates, and thus is largely conservative. Fig. 5 depicts a 1-MBq Co-60 source in Reference Flatfish. Co-60 is a beta/gamma emitter, and has been known to concentrate in the kidneys. In this instance, there are no immediately discernible trends, likely due to the mass and positioning of the source organ in the fish. Fig. 6 shows a 1-MBq Sr-90 (including contributions from Y-90) source in Reference Rat. Sr-90 is another radionuclide that is likely to partition in specific organs. In this case, the simplified model would likely underestimate the organ dose rate, meaning that it is not conservative. Fig. 7 represents Cs-137 in Reference Trout. Caesium distributes in the soft tissues, and again, no discernible trends are immediately obvious.

Over 50 graphs such as those in Figs 4–7 were created. To show the data in a more concise manner, histograms describing the results were created. They are shown in Figs 8–11, and these log-scaled plots illustrate the distribution of deviations among

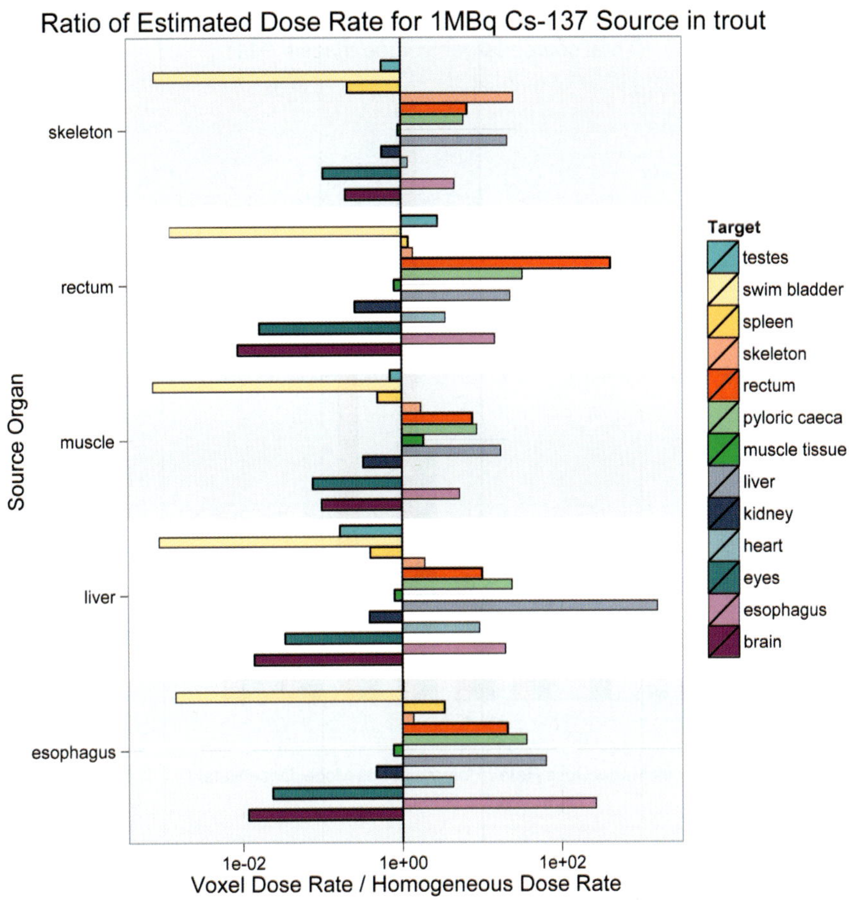

Fig. 7. Cs-137 source in Reference Trout.

all radionuclides for the four RAPs studied (Ruedig et al., to be published). Yellow bars indicate alpha emitters, green bars indicate pure beta emitters, and red bars indicate mixed beta/gamma emitters. Each of the histograms in Figs 8–11 contain data from all radionuclides included in the study: ^{14}C, ^{36}Cl, ^{60}Co, ^{90}Sr, ^{131}I, ^{137}Cs, and ^{210}Po. Fig. 8 compares 871 source:target values for Reference Flatfish, and found that the homogeneous model is primarily conservative for pure beta emitters, but less so for mixed beta/gamma emitters. Fig. 9 illustrates comparisons for 166 source:target values, and in this case, the homogeneous model is primarily conservative for mixed beta/gamma emitters, but not conservative for alpha and most pure beta emitters. Fig. 10 portrays the comparisons for 176 source:target values in Reference Crab, with the same results as seen in Reference Rat: the homogeneous model is conservative for mixed beta/gamma emitters, but not conservative for alpha emitters and the majority of beta emitters. Fig. 11 compares 871 source:target values

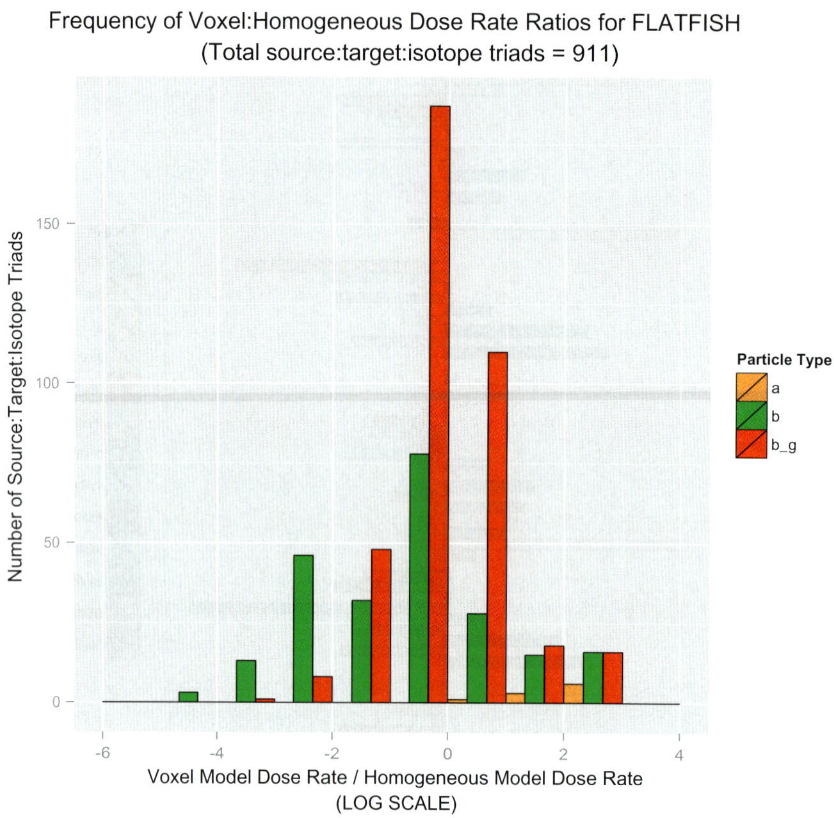

Fig. 8. Reference Flatfish: distribution of voxelised:simplified ratios for all radionuclides studied.

for Reference Trout, and has results akin to Reference Flatfish: the homogeneous model is conservative for pure beta emitters, but not conservative for mixed beta/gamma and alpha emitters.

5.2. Voxel model dose conversion factors compared with mass ratios

In 2008, Gómez-Ros et al. derived a relationship that allowed for the estimation of organ dose rates from ellipsoidal geometries, assuming both a central and eccentric point source (Gómez-Ros et al., 2008). In order to calculate an organ-specific dose rate from whole-body absorbed fractions, Gómez-Ros et al. defined the mass ratio (R) between an organism's whole-body dose rate ($\dot{D}_{wb}$) and its organ dose rate ($\dot{D}_s$) as follows:

$$R = \frac{\dot{D}_s}{\dot{D}_{wb}}$$

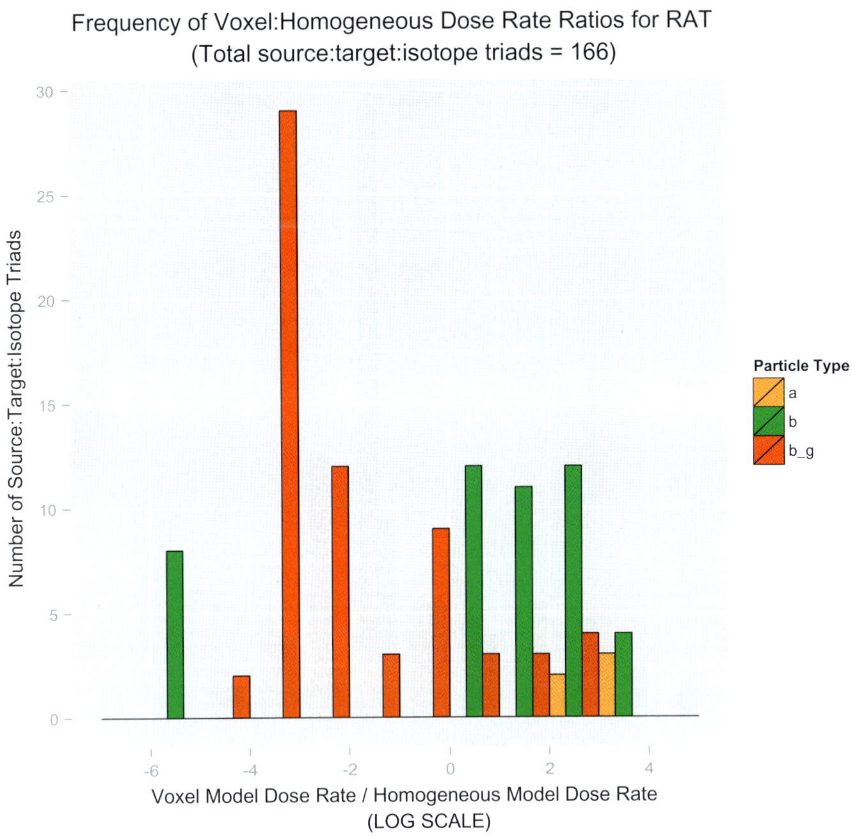

Fig. 9. Reference Rat: distribution of voxelised:simplified ratios for all radionuclides studied.

Gómez-Ros et al. (2008) showed that the maximum value of R is the ratio of body mass (M_{wb}) to organ mass (m_s), as follows:

$$\frac{\dot{D}_s}{\dot{D}_{wb}} \leq \frac{M_{wb}}{m_s}$$

This ratio suggests that the dose rate to a small organ from a heterogeneously partitioned source can be large compared with the hypothetical case of the same source distributed homogeneously throughout an organism's body. It also provides a rigorous methodology for estimating organ dose rates from existing dosimetric data for homogeneously distributed sources, by multiplying the whole-body dose rate by the ratio R for the specific organ of interest. Gómez-Ros et al. (2008) postulated that the ratio between organ and whole-body dose rates will be between 5 and 2000 for most organs, with the greatest values of R occurring for those organs that are lightest

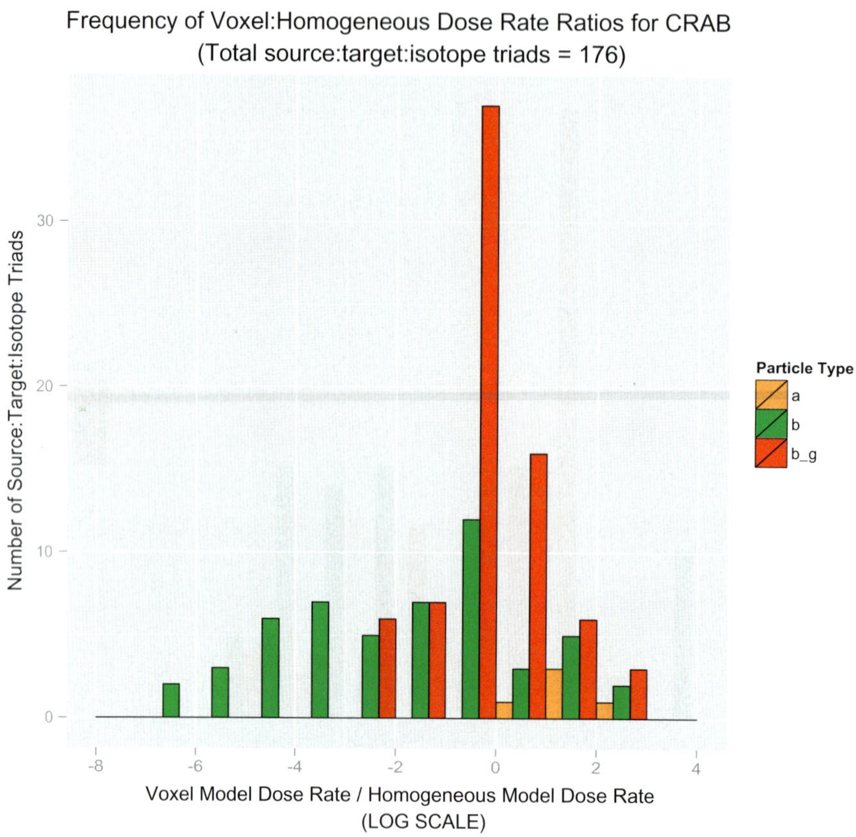

Fig. 10. Reference Crab: distribution of voxelised:simplified ratios for all radionuclides studied.

relative to body mass. These limits are reasonable, but it is worth noting that some organs known to partition radionuclides heterogeneously are much smaller (e.g. the thyroid of rainbow trout has an R of nearly 35,000) (Vanderploeg et al., 1975). Additionally, very light organs (e.g. the swim bladder of the rainbow trout) can substantially affect the value of R, resulting in an R value outside the expected range of 5–2000 estimated by Gómez-Ros et al. (2008).

Figs 12–14 show dose rates calculated via mass ratios compared with those calculated for each segmented portion of the organism for Reference Crab, Reference Flatfish, and Reference Trout, respectively. It can be observed in Figs 12 and 13 that the mass ratio value is remarkably consistent with the voxelised model. However, as shown in Fig. 14, the mass ratios for Reference Trout match the voxelised model poorly. This suggests that there are some instances in which mass ratios may be lacking.

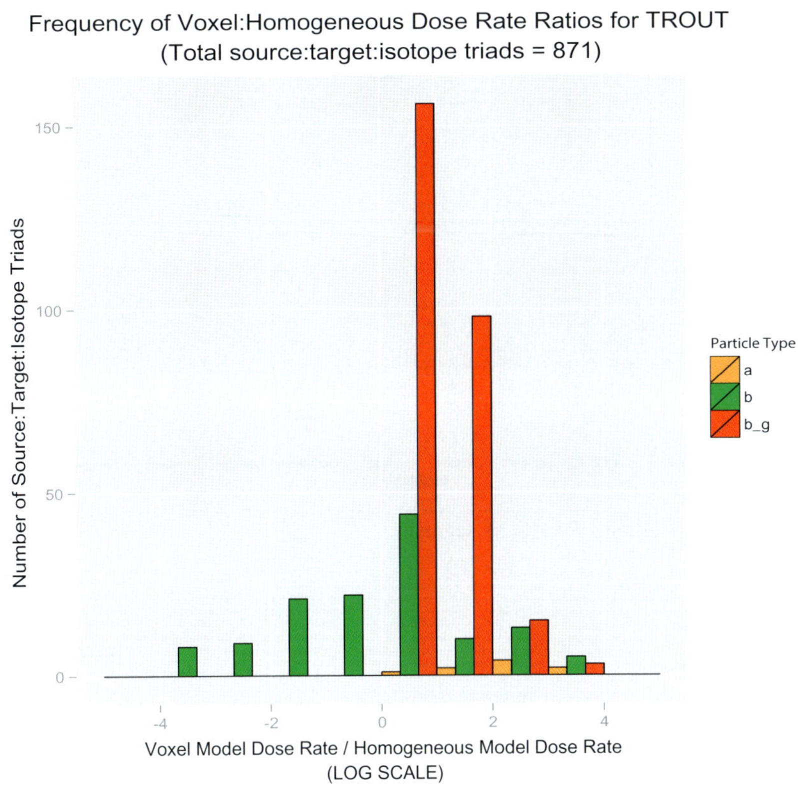

Fig. 11. Reference Trout: distribution of voxelised:simplified ratios for all radionuclides studied.

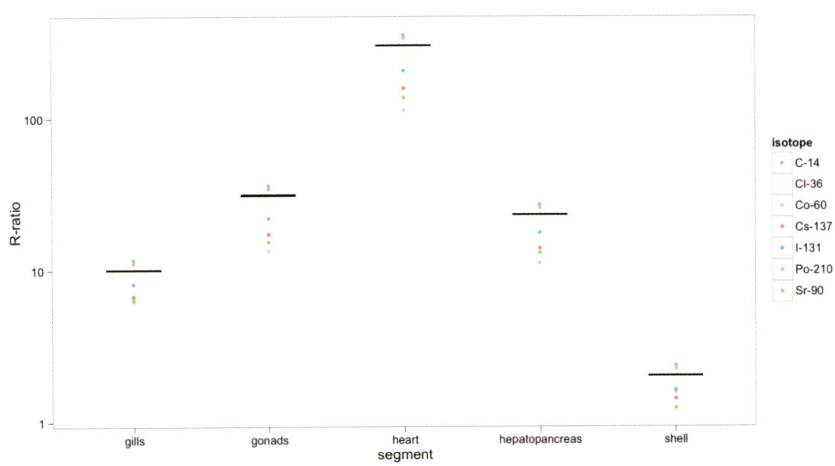

Fig. 12. Reference Crab: *R* ratio vs segment.

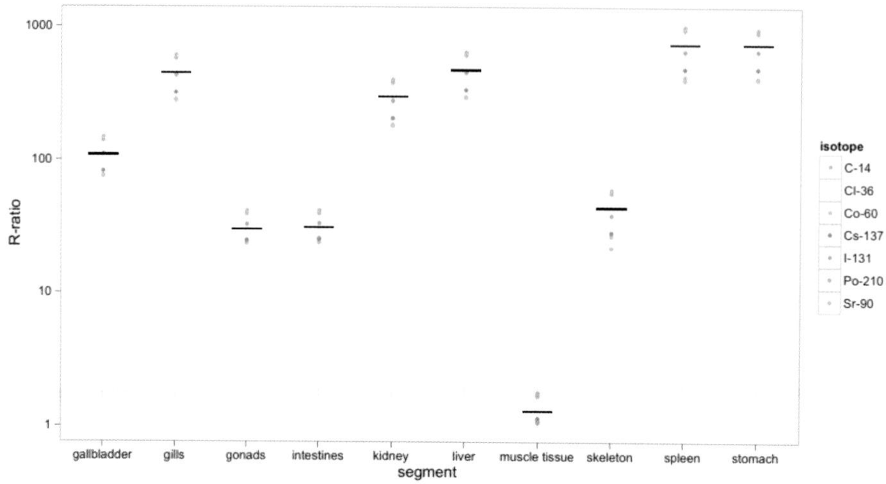

Fig. 13. Reference Flatfish: *R* ratio vs segment.

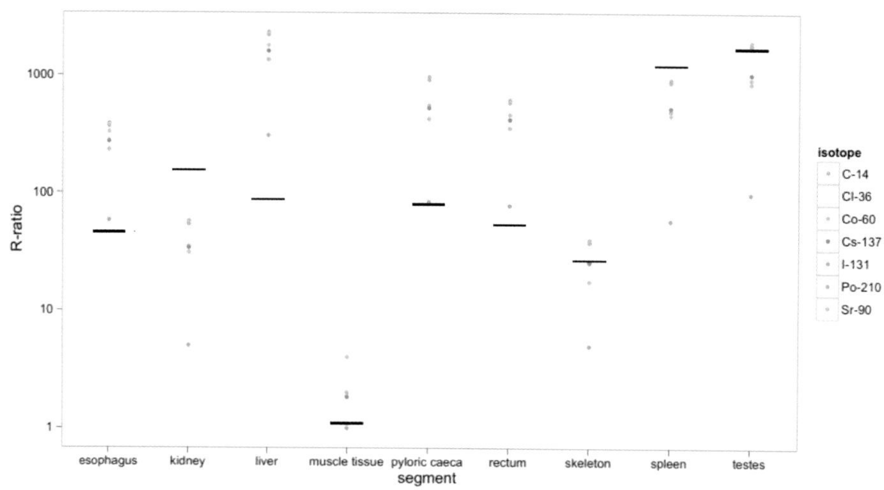

Fig. 14. Reference Trout: *R* ratio vs segment.

6. DISCUSSION

In a general sense, dose rates from voxelised models and simplified models only agree well for pure beta emitters. Beta-/gamma-emitting radionuclides show the greatest deviations in dose rates. These comparisons serve to illustrate the discrepancies seen in the comparison of homogeneous with voxelised models. In Reference

Rat, the simplified model was generally conservative, but in Reference Trout, voxelised dose rates were generally higher, regardless of radionuclide. The only generalisations that can be drawn from the data are that voxelised dose rates were consistently higher than homogeneous dose rates for alpha emitters. This is likely due to the short track length of alpha particles in tissue; alpha emitters are more likely to deposit all their energy in the source organ. In the homogeneous model, the energy is distributed over a much greater mass, and thus the overall dose rate is lower than that seen in the voxelised scenario. It is relevant to note that alpha emitters are among those radionuclides most likely to partition preferentially into specific organs, and thus the use of voxelised models for scenarios in which alpha emitters are present may be suggested.

7. CONCLUSIONS

The comparisons described above serve as a first-run examination of voxelised vs homogeneous models. The only clear conclusion that can be drawn with any level of certainty is that the models do not agree well, nor is one model consistently conservative.

This study should be taken for what it is: a first examination of voxelised vs homogeneous data. It neglects biology, and therefore organ:radionuclide pairs that are unlikely/impossible are considered in the analysis. In addition, most radionuclides rarely partition themselves perfectly, so the scenario above represents the most extreme case. More realistic comparisons that consider some of these nuances represent the next step in the process of determining the best method for expending resources on environmental protection.

REFERENCES

Able Software Corp., 2012. 3D-Doctor User's Manual. Lexington, MA: Able Software Corp.

Caffrey, E.A., 2012. Improvements in the Dosimetric Models of Selected Benthic Organisms. Master's thesis, Oregon State University, USA. Available at: http://ir.library.oregonstate.edu/xmlui/handle/1957/34305.

Caffrey, E.A., Higley, K.A., 2013. Creation of a voxel phantom of the ICRP reference crab. J. Environ. Radioact. 120, 14–18.

Gómez-Ros, J.M., Pröhl, G., Ulanovsky, A., et al., 2008. Uncertainties of internal dose assessment for animals and plants due to non-homogeneously distributed radionuclides. J. Environ. Radioact. 99, 1449–1455.

Hendee, W.R., 2002. Medical Imaging Physics, fourth ed. Wiley-Liss, Inc., New York.

Hess, C.A., 2014. Monte Carlo Derived Absorbed Fractions in a Voxelized Model of a Rainbow Trout. Master's thesis, Oregon State University, USA.

ICRP, 2008. Environmental protection: the concept and use of reference animals and plants. ICRP Publication 108. Ann. ICRP 38(4–6).

ICRU, 1992. Photon, Electron, Proton, and Neutron Interaction Data for Body Tissues. ICRU Report 46. International Commission on Radiation Units, Bethesda, MD.

Kinase, S., 2008. Voxel-based frog phantom for internal dose evaluation. J. Nucl. Sci. Technol. 45, 1049–1052.

Kramer, G.H., Capello, K., Chiang, A., et al., 2010. Tools for creating and manipulating voxel phantoms. Health Phys. 98, 542–548.

Loevinger, R., Berman, M., 1976. A Revised Schema for Calculating the Absorbed Dose from Biologically Distributed Radionuclides. Society of Nuclear Medicine, New York.

Mohammadi, A., Kinase, S., Saito, K., 2011. Comparison of photon and electron absorbed fractions in voxel-based and simplified phantoms for small animals. Progr. Nucl. Sci. Technol. 2, 365–368.

Mohammadi, A., Kinase, S., Saito, K., 2012. Evaluation of absorbed doses in voxel-based and simplified models for small animals. Radiat. Protect. Dosimetry 150, 283–291.

Ruedig, E., Beresford, N.A., Fernandez, M.E.G., Higley, K.A., 2015. A comparison of the ellipsoidal and voxelized dosimetric methodologies for internal, heterogeneous radionuclide sources. Journal of Environmental Radioactivity 140, 70–77.

Vanderploeg, H.A., Parzyck, D.C., Wilcox, W.H., et al., 1975. Bioaccumulation Factors for Radionuclides in Freshwater Biota. Environmental Sciences Division Publication No. 783. Oak Ridge National Laboratory, Oak Ridge, TN.

X-5 Monte Carlo Team, 2008. Volume I: Overview and Theory. Los Alamos, NM: X-5 Monte Carlo Team.

Zaidi, H., Xu, X.G., 2007. Computational anthropomorphic models of the human anatomy: the path to realistic monte carlo modeling in radiological sciences. Ann. Rev. Biomed. Eng. 9, 471–500.

SAGE

ICRP
Annals of the ICRP

Exposures and effects in the marine environment after the Fukushima accident

J. Vives i Batlle

Biosphere Impact Studies Unit, Belgian Nuclear Research Centre, Boeretang 200, 2400 Mol, Belgium; e-mail: jordi.vives.i.batlle@sckcen.be

Abstract–Radiation doses to marine biota near the Fukushima Daiichi nuclear power plant have been estimated for the immediate aftermath and subsequent period of the accident. Dose estimations using monitoring data have been complemented by means of dynamic transfer modelling, improving on the more traditional equilibrium transfer approach. Earlier assessments using equilibrium transfer models overestimated the exposures in the immediate aftermath of the accident, whereas dynamic transfer modelling brings them more in line with the doses calculated from monitored activity concentrations in the biota. On that basis, marine biota populations in the vicinity of Fukushima do not seem to be at significant risk. The situation in the late post-accident period shows a tendency for lower exposures, but radiocaesium in sediments and biota persists to this day, with some organisms inhabiting local hotspots. Little is known about how long radionuclides will continue to remain in the local environment, or the long-term effects on populations due to limited knowledge on the effects of chronic radiation exposures to marine organisms. Therefore, the marine environment at Fukushima needs further study. The Fukushima nuclear accident remains an ongoing problem for marine radioecology, requiring constant re-evaluation of the cumulative extent of contamination and effects on the environment for years to come.

Keywords: Radiological exposure; Environment; Fukushima; Non-human biota

1. INTRODUCTION

Assessing the impact of emergency releases on the environment requires detailed knowledge of the exposures and subsequent prediction of the likely effects, with both aspects posing challenges of their own. The first challenge in exposure assessment is

This paper does not necessarily reflect the views of the International Commission on Radiological Protection.

establishing radionuclide concentrations in the marine media – seawater, particulate matter, and seabed sediments – either by field monitoring or by means of dispersion modelling (Aoyama et al., 2012; Choi et al., 2013; Honda et al., 2013). In an area such as Fukushima, the problem is compounded by complex biogeochemical conditions (Buesseler et al., 2011; Vives i Batlle, 2011, 2012b). A mixture of radioisotopes with different decay half-lives is involved, and a good characterisation of the source term is required if dispersion modelling is intended (Bailly du Bois et al., 2012). Next, the uptake and turnover of radionuclides by the marine biota need to be determined. In the initial release phase, where monitoring data are not always available, the problem should be modelled dynamically because organisms are likely to be at disequilibrium with fluctuating activity concentrations in the environment (Vives i Batlle et al., 2008a; Kryshev and Sazykina, 2011; Psaltaki et al., 2013).

Once activity concentrations in biota are known, the problem becomes one of dose estimation. An international system for the radiological protection of the environment has been developed that provides the framework for such calculations (Beresford et al., 2007; Brown et al., 2008; ICRP, 2008a, 2009). There is some degree of consensus on dose rates that are unlikely to cause effects to flora and fauna. For example, the International Commission on Radiological Protection (ICRP) has proposed a 'derived consideration reference level' of 4–40 $\mu Gy\,h^{-1}$ for the most sensitive Reference Animals and Plants (RAPs) (ICRP, 2008b). The ERICA methodology proposes a screening dose rate at the ecosystem level of 10 $\mu Gy\,h^{-1}$ (Beresford et al., 2007; Brown et al., 2008; Andersson et al., 2009). The United Nations Scientific Committee on the Effects of Atomic Radiation (UNSCEAR) concluded that dose rates up to 400 $\mu Gy\,h^{-1}$ to a small proportion of individuals in aquatic populations would not have a detrimental effect at the population level (UNSCEAR, 1996).

The radiological protection system for the environment is put to the test when applied to emergency situations, due to four factors. The first factor is the aforementioned dynamic nature of the releases, preventing the use of equilibrium transfer parameters. Secondly, there is a need to assess exposures and effects for indigenous biological species which can be far removed from RAPs defined in the standard system (ICRP, 2008a). Thirdly, in an emergency situation, it is important to distinguish at least two phases differing in terms of the level and type of exposure to living organisms: an acute exposure or high dose phase over the few weeks after the accident, and a chronic exposure phase at lower dose where contamination levels of the environment change much more slowly, on a scale of months to years. Relevant acute and chronic benchmarks must be applied in each case. Lastly, radiation exposures in emergency situations may occur in combination with other contaminants and environmental stresses such as the Great Japan Tsunami itself, making the consequences of the radiation exposures themselves difficult to assess.

The present paper attempts to evaluate the immediate and ongoing impact of the accident at Fukushima Daiichi nuclear power plant following the earthquake and tsunami on 11 March 2011. The aspects influencing the ongoing analysis and re-assessment of the radiological impact of accidental releases of ^{131}I, ^{134}Cs, and ^{137}Cs

to marine biota are addressed, regarding both the immediate aftermath and subsequent period of the accident. Lessons and recommendations about how to assess and interpret exposures of, and effects on marine biota following an accidental discharge of radioactivity have been derived, and these are revisited for the benefit of investigators.

2. ASSESSMENT OF EXPOSURES TO NON-HUMAN BIOTA

2.1. Description of the accident

The reactor failures at Fukushima Daiichi nuclear power plant resulted in a major emergency release of radionuclides to the surrounding environment. A full description of the accident is given elsewhere (IAEA, 2011; PMJHC, 2011; IRSN, 2012; Povinec et al., 2013). Approximately 80% of the radioactive fallout occurred over the Pacific Ocean (Stohl et al., 2011). The magnitude of the marine release was realised shortly after the accident (Buesseler et al., 2011; Garnier-Laplace et al., 2011; IRSN, 2011; Vives i Batlle, 2011; Bailly du Bois et al., 2012; Tsumune et al., 2012). ^{131}I and radiocaesium were the main radionuclides released into the marine environment, although ^{129}Te, ^{129m}T, ^{132}Te, ^{136}Cs, and ^{133}I were also released along with trace amounts of plutonium isotopes (Zheng et al., 2013).

Due to contaminated water releases and the settling of airborne radioactive particles, some 10^{16} Bq of ^{137}Cs found their way into the Pacific Ocean, mainly between 11 March and 8 April 2011. Radionuclides spread widely over the coastal zone, decreasing by a factor of approximately 10^3 over the first 30 km from the release point and diminishing by factors of approximately 30 (^{137}Cs) and 200 (^{131}I) over the ensuing few weeks after the accident (Buesseler et al., 2011; Garnier-Laplace et al., 2011). By early June 2011, the short-lived radioisotopes had largely disappeared and ^{134}Cs and ^{137}Cs were the dominant radionuclides. Directly discharged ^{134}Cs and ^{137}Cs were transported predominantly southwards along the coastline (Aoyama et al., 2012), and quickly transported to the deep sea in the western North Pacific (Honda et al., 2013). Continued effluents from land resulted in sustained contamination levels in the area during July 2011, and activity concentrations in seawater fell sharply thereafter. Modelling simulations predicted enhanced surface ^{137}Cs activities offshore in the Kuroshio–Oyashio current zone up to 6 months after release and an order of magnitude less after 1 year (Dietze and Kriest, 2012), and this was confirmed by observation (Buesseler et al., 2012).

A fraction of the radionuclides released remains adsorbed to seabed sediments near the coast. Scavenging by in-falling biogenic particles and turbulent mixing are the likely causes (Shiomoto et al., 1998; Alekseev et al., 2006; Vives i Batlle, 2011). Most of this adsorption is estimated to have occurred during the first month, and radionuclides are expected to remain associated with sedimentary deposits for a long time (Choi et al., 2013). The association of radiocaesium with sedimentary deposits is of key importance, as it determines a pathway for the external exposure to benthic organisms.

2.2. Initial screening studies

An initial screening study suggested that maximum dose rates to marine biota in the immediate aftermath of the accident could have reached 0.2–5 Gy day^{-1}, with the highest exposures attributed to macroalgae (Garnier-Laplace et al., 2011). This study assumed equilibrium of the biota with the highest seawater activity concentrations measured. Another early assessment, also assuming equilibrium but carried out over the time period March–May 2011 (and consequently not assuming the maximum activity concentrations of the previous study), indicated that dose rates to fish and molluscs from the local coast did not exceed 420 $\mu Gy\ h^{-1}$ and were generally in the order of 80 $\mu Gy\ h^{-1}$ (Kryshev and Sazykina, 2011). At the level of exposure estimated in the former study, some radiation effects (mutagenic and reproduction) would be expected (Garnier-Laplace et al., 2011). However, as the dose rates reported were based on equilibrium with maximum concentrations, it was soon recognised that exposures may have been lower (Buesseler et al., 2011; Vives i Batlle, 2011) by at least one order of magnitude according to an early modelling study (Kryshev et al., 2012).

During the accident, radioactivity was released as a series of pulses and equilibrium between activity levels in the water, the sediments and the marine biota cannot be assumed. This is illustrated by a compilation of activity concentrations of ^{131}I, ^{134}Cs, and ^{137}Cs in seawater sampled in March–July 2011, reported by Tokyo Electric Power Company (TEPCO, 2011) and reprinted elsewhere (Buesseler et al., 2011). This shows that activity concentrations in seawater peaked after about 20

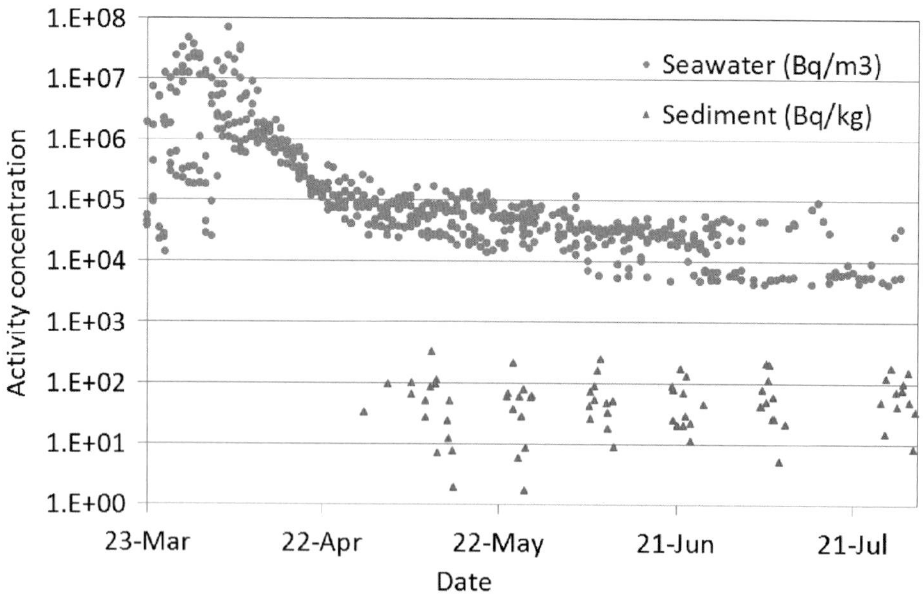

Fig. 1. Activity concentrations in seawater at four coastal stations near Fukushima (amalgamated) compared with activity concentrations in surface sediment (MEXT, 2011; TEPCO, 2011).

days, falling two orders of magnitude in 60 days (data for the first 12 days post-accident are not available). Additional activity concentrations in sediment between April and July 2011 (MEXT, 2011) indicate that activity concentrations in sediment do not have a similar decreasing trend post-peak, demonstrating the resilience of radionuclides in sediments (Fig. 1). The isotopic ratio $^{134}Cs/^{137}Cs$ in these samples is 0.97 ± 0.14 for seawater but somewhat lower at 0.81 ± 0.05 for sediment.

2.3. Assessment with monitoring data

Marine biota was not monitored immediately after the accident. Some of the earliest data actually originate from a Greenpeace survey of fish, algae, and molluscs sampled on 3–9 May and 23–24 June 2011 at various local ports as well as offshore from Fukushima Daiichi nuclear power plant (Greenpeace, 2012). The data show that the most exposed organisms were those receiving ^{131}I at locations closest to Fukushima Daiichi nuclear power plant (geometric mean approximately 10^3 Bq kg^{-1}, range 2–1.2 × 10^5 Bq kg^{-1}), particularly macroalgae. ^{134}Cs and ^{137}Cs were found to be in lower concentrations and in similar proportions (geometric mean approximately 10^2 Bq kg^{-1}, range 1–1.4 × 10^3 Bq kg^{-1}). This suggests that radioactivity levels in marine biota were below the maximum concentrations assumed by equilibrium models in some previous studies (Garnier-Laplace et al., 2011).

The earliest dose estimates using the above data were revised by assigning a marine reference organism to each of the species sampled, using the ERICA methodology (Beresford et al., 2007; Brown et al., 2008) as a basis. The activity concentrations were multiplied by a 'dose rate per unit concentration' (DPUC) value for internal exposure, derived from a template ERICA assessment simulation for the

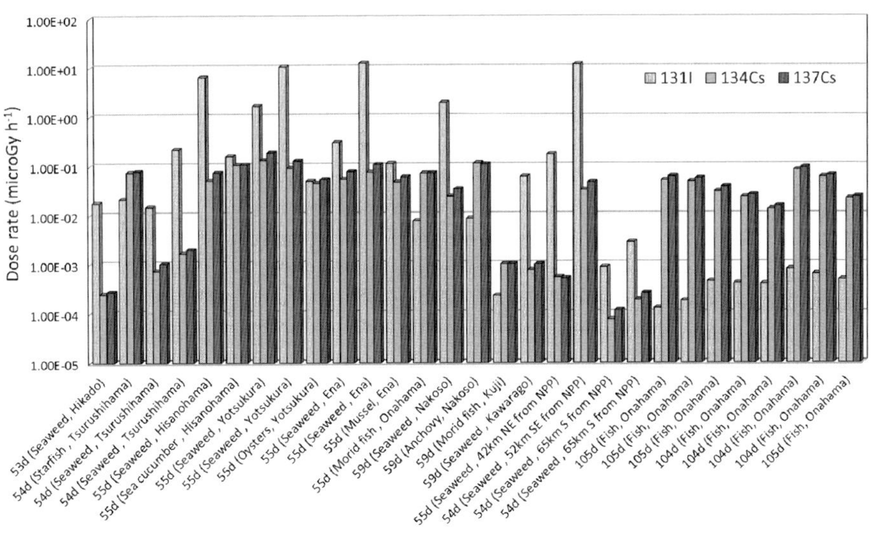

Fig. 2. Internal dose rates for marine biota sampled in early April and May 2011 (Vives i Batlle and Vandenhove, 2014). The time scale on the abscissa represents days since the accident.

relevant reference organisms. External dose rates were calculated similarly via external exposure DPUCs in combination with activity concentrations in the medium. This simple method captures the default parameters of the ERICA approach, such as dosimetry and occupancy factors. The highest internal dose rates calculated were 12.7 μGy h^{-1} for ^{131}I, 0.13 μGy h^{-1} for ^{134}Cs and 0.19 μGy h^{-1} for ^{137}Cs (Fig. 2). The highest external dose rates calculated were 0.006 μGy h^{-1} for ^{131}I, 0.075 μGy h^{-1} for ^{134}Cs, and 0.035 μGy h^{-1} for ^{137}Cs. Exposure in seaweed dominated, followed by molluscs and fish (Fig. 2).

2.4. Dynamic transfer modelling

Confirmation that the dose rates received by the biota would have been less than estimated in the earlier studies came from the modelling of radiation doses to marine biota using a dynamic transfer approach, which was applied to the first month after the accident where biological monitoring data are not available. Instead of assuming that the activity concentration A_O (Bq kg^{-1}) in an organism of mass M is proportional to the activity concentration A_W (Bq m^{-3}) in a surrounding volume of water V, the biota is modelled as reacting to ambient concentrations with a time delay. This is determined by the half-time of clearance following an intake of radioactivity. The key parameters are the biological half-life of elimination $T_{B1/2}$ and the equilibrium concentration ratio (CR). In its simplest form, the process can be represented by a simple model with two rate constants, k_W for uptake and k_O for elimination:

$$\frac{dA_W}{dt} = -(\mathrm{k}_W + \lambda)A_W + \mathrm{k}_O \frac{M}{V} A_O$$

$$\frac{dA_O}{dt} = \mathrm{k}_W \frac{V}{M} A_W - (\mathrm{k}_O + \lambda)A_O$$

where $k_O = \frac{\ln 2}{T_{B1/2}}$, $k_W = (k_O + \lambda)\frac{M}{V}$ CR, and λ is the radionuclide decay constant.

These equations have simple solutions and models designed to solve them, and more complex forms involving metabolism and/or the allometry of multicomponent release are described elsewhere (Sazykina, 2000; Vives i Batlle et al., 2007). The D-DAT assessment model and associated biokinetic parameter database (Vives i Batlle et al., 2008a; Watts et al., 2008) were used in this work, incorporating the ERICA dosimetric approach, with predefined activity concentrations in seawater as the starting point. An example of the modelling results is given in Fig. 3.

The model predicts ^{131}I internal dose rates of 20–25 mGy h^{-1} for macroalgae and 15–60 μGy h^{-1} for other species in the vicinity of the Daiichi stations, with a factor 30–40 times lower in the more distant Daiini and Iwasawa shore stations. ^{134}Cs and ^{137}Cs internal dose rates at Fukushima Daiichi nuclear power plant were calculated to range from 10 to 70 μGy h^{-1} for all species (20 times lower in the more distant stations). The most exposed organism was macroalgae, receiving ^{131}I near the Daiichi outlets, peaking 20–30 days post-accident and falling rapidly thereafter. Organisms outside Fukushima Daiichi nuclear power plant were predicted to have received <4 μGy h^{-1} of radiocaesium exposure and 750 μGy h^{-1} of radioiodine exposure (mostly in macroalgae). The internal dose rate dominates two to three

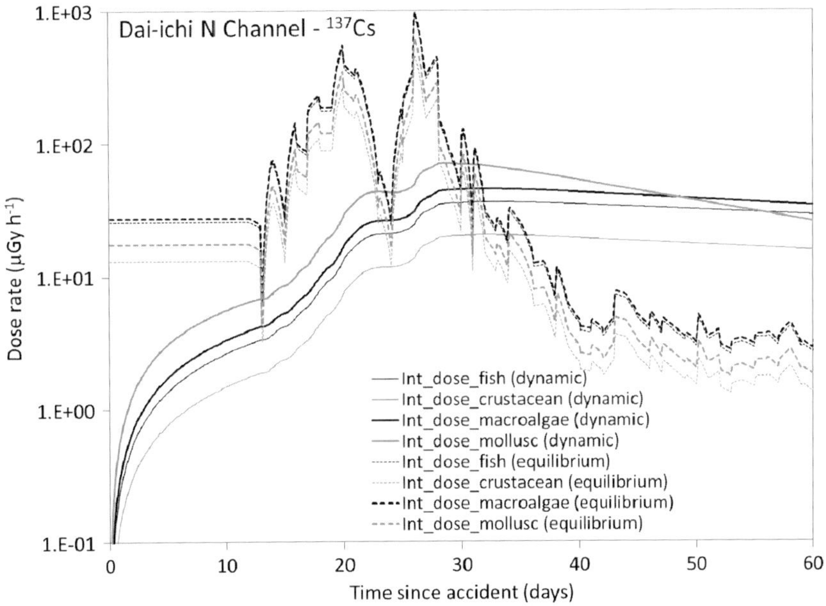

Fig. 3. Modelling predictions for the Fukushima Daiichi northernmost discharge point.

orders of magnitude over external dose rate. This calculation only considers the external exposure from seawater because no sediment monitoring data were available in the immediate aftermath of the accident. However, the build-up of radionuclides in sediment would not have equilibrated, and thus the error incurred would not have been high (this has been verified in later studies).

The ^{131}I cumulative doses over the initial 60-day period were modelled to be 6.5 Gy for ^{131}I for macroalgae and two to three orders of magnitude lower for fish, crustaceans, and molluscs for the stations nearer to Fukushima Daiichi nuclear power plant. The ^{131}I cumulative doses at the same station were 20–50 mGy, highest for molluscs. Outside the vicinity of Fukushima Daiichi nuclear power plant, cumulative doses were predicted to have been 460 mGy for ^{131}I and <3 mGy for ^{134}Cs and ^{137}Cs.

Where concentrations in the water increased sharply, dynamically modelled doses are lower than doses using the equilibrium CR approach due to the delayed build-up of activity in the biota. The opposite occurs over the subsequent late period, due to delayed retention. This is consistent with previous dynamic transfer modelling studies involving authorised ^{99}Tc pulses released by the Sellafield plant in Cumbria, UK (Vives i Batlle et al., 2008a). The differences between equilibrium and dynamic modelling are most pronounced for the biota with biological half-lives of >10 days (fish and molluscs), indicating a two-fold overestimation of magnitude by the equilibrium approach. The dynamic model brings the predicted dose rates more in line with dose rates calculated directly from monitoring data (within an order of magnitude for radiocaesium and a factor of two for radioiodine).

The above analysis is limited to a small monitoring dataset, exemplifying the issues faced when investigating emergency releases to the environment. Starting with screening studies and simple maximising assumptions, the assessment is refined as more data become available. A more comprehensive study of the situation is required, and this materialised with the UNSCEAR assessment in the case of Fukushima.

2.5. The UNSCEAR assessment

UNSCEAR decided early after the Fukushima accident to assemble a team of 60 international experts from 18 countries to assess radiation exposures and health effects due to the accident at Fukushima. Four expert groups (measurements, source term and dispersion, assessment of doses, and risk analysis) were established, with Japan providing an extensive dataset to the Committee.

The in-depth study undertaken by UNSCEAR resulted in updated estimates of radiation doses to humans and the environment. The assessment of exposures to marine biota covered exposures from the beginning of the accident to August 2012, based on an extensive dataset of over 500 sediment and 6000 seawater data points as well as 5000 samples of organisms from 210 species, supplemented by modelling of dynamic transfer to biota using the D-DAT model, among others. The doses and associated effects of radiation following the accident were evaluated against the Committee's previous evaluations of such effects (UNSCEAR, 1996, 2008).

At the time of writing this paper (November 2013), the full report with detailed scientific annexes had not yet been released, but it has since become available (UNSCEAR, 2014), followed by scientific publication of the marine biota assessment (Vives i Batlle et al., 2014). The study concluded that exposures of non-human biota following the accident were, in general, too low for acute effects to be observed. A possible exception of a transient nature is the relatively contaminated area in the vicinity of the discharge point, where effects on sensitive endpoints in individual plants and animals might have occurred in the weeks following the accident. However, impacts on population integrity would have been unlikely due to the short duration and the limited space area of the initially high exposures. Beyond that area, the potential for effects on biota was much lower.

2.6. Current situation

In recent times, it has been confirmed that radionuclide levels in fish off Fukushima are highly variable but remain elevated, potentially associated with delayed emissions. Radiocaesium levels up to several thousand Bq kg^{-1} in benthic fish have been reported, exceeding the Japanese radioactivity in food limit of 100 Bq kg^{-1} (Buesseler, 2012; Buesseler and Aoyama, 2012). A recent report from the Fukushima nuclear operator (TEPCO, 2013b) analysed local marine biota for evidence of changes in activity concentration in 2013. This report shows that seawater contamination near Fukushima Daiichi nuclear power plant has decreased from approximately 10^5 Bq l^{-1} in March 2011 to a few Bq l^{-1} in late 2012. Around 600 Bq kg^{-1} total Cs were found in sediment 1 km offshore from Ukedo, Namie Town, and sediment Cs levels have not changed significantly, remaining at a

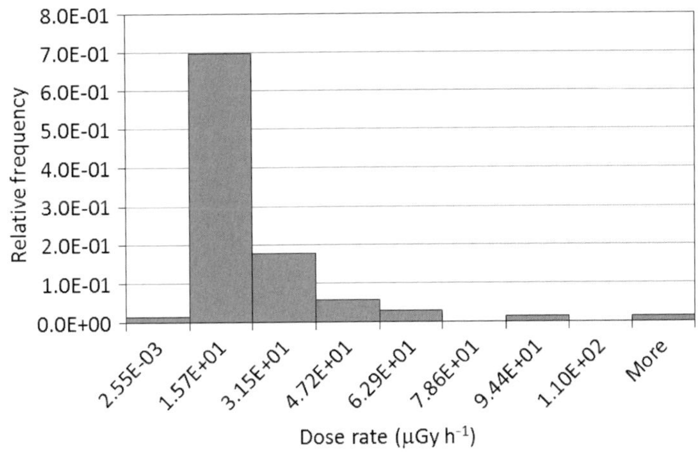

Fig. 4. Histogram of dose rates to marine biota sampled in 2013 (TEPCO, 2013b).

few hundred Bq kg^{-1}. In late February–early March 2013, levels of radiocaesium as high as 5×10^5 Bq kg^{-1} were found occasionally in greenling from fishing baskets and in gill nets at the port entrance to Fukushima Daiichi nuclear power plant.

Using the ERICA methodology, the highest internal dose rates (greenling samples of 17/02/13, 21/02/13, and 04/03/13) can be calculated to be 17–44 μGy h^{-1} for ^{134}Cs and 32–82 μGy h^{-1} for ^{137}Cs. Most dose rates are clearly below this maximum: the data are log-normally distributed ($\sigma = 0.59$, $\mu = 0.32$), and the calculated median is $e\mu = 1.38$ μGy h^{-1} with 95% of results under 50 μGy h^{-1} (Fig. 4). The external dose rates estimated with limited seawater and sediment data are significantly lower at 0.12 μGy h^{-1} for ^{134}Cs and 0.05 μGy h^{-1} for ^{137}Cs.

The highest dose rates in marine biota from 2013 were still below the 400 μGy h^{-1} UNSCEAR benchmark for the most exposed individuals of an aquatic population, below which population detriment is not expected. From this preliminary analysis, it seems unlikely that sporadic high concentrations in greenling close to Fukushima Daiichi nuclear power plant signal significant exposures to whole fish populations. However, the highest estimates are of the same order of magnitude as radiation dose rates giving a 10% effect in comparison with a control (EDR_{10}) in some species and endpoints (Knowles and Greenwood, 1994, 1997; Garnier-Laplace et al., 2010), and they exceed the ERICA screening dose rate of 10 μGy h^{-1}. Although the biota analysed at these locations are not under threat at the population level, a follow-up study is clearly advisable in order to evaluate exposures and effects in organisms inhabiting local hot spots more systematically.

3. ASSESSMENT INTERPRETATION AND DISCUSSION

In the case of Fukushima, it seems reasonable to assign a 2-month period for the acute phase of the accident, based on the presence of short-lived ^{131}I at that time, inducing high dose rates in some biota. During this period, there is the possibility that so-called 'acute effects' (rapid biological alteration leading to irreversible damages) would have occurred to the most exposed marine organisms, but there is no firm scientific evidence that they did indeed occur. During the late (or chronic) phase, radiocaesium isotopes remained. The potential effects in marine species during this phase are difficult to predict because they are scarcely documented. For example, limited data available for fish indicate that dose rates lower than 4000 $\mu Gy\,h^{-1}$ are unlikely to affect survival. The lowest value of chronic dose rate with a 10% effect on reproduction is 47 $\mu Gy\,h^{-1}$ in plaice (Garnier-Laplace et al., 2010). For marine invertebrates, the lowest value of EDR_{10} is found at 36 $\mu Gy\,h^{-1}$ for annelids (Knowles and Greenwood, 1994). To the author's knowledge, there are no chronic effects data for marine plants in the literature.

With the information available, it is concluded that the doses likely to have been incurred in the chronic phase are generally below the levels necessary to cause a long-term measurable effect on biota populations, as suggested by other researchers (Buesseler et al., 2012). In the acute phase, one would only expect transient effects for macroalgae exposed to ^{131}I close to the discharge point, but these effects have not been confirmed, to date, by actual observation. At locations further away from Fukushima Daiichi nuclear power plant, exposures fell below the thresholds for which population effects are deemed likely. This includes the elevated concentrations in some individual greenling specimens, as stated previously.

3.1. Uncertainties and limitations

Exposure assessments to non-human biota are affected by uncertainties in both dose estimates and effects benchmarks. A previous study (Vives i Batlle and Vandenhove, 2014) showed that the uncertainties in dosimetry calculations are approximately $\pm 25\%$ for internal exposures and $\pm$ 150% for external exposures. The model-based assessment has an additional uncertainty estimated to be within a factor of two (unpublished data), mainly due to uncertainties in the $T_{B1/2}$ and CR parameters.

Initial assessment efforts have several limitations. A number of radionuclides known to have been released are absent from the biota monitoring data (i.e. ^{89}Sr, ^{90}Sr, ^{129}Te, ^{129m}Te, ^{136}Cs, and the actinides). Exposures from sediment are not included in external dose calculations for the acute phase of the accident. The $T_{B1/2}$ and CR parameters used in the studies described so far are generic rather than site-specific, but a recent publication containing $T_{B1/2}$ values for radiocaesium in 16 species of Fukushima marine biota (Iwata et al., 2013) offers the prospect to improve this. In general, the monitoring data available for assessment in the aftermath of an emergency are often limited, and a publicly available, quality-assured, comprehensive database would be a major benefit. Efforts such as the Fukushima Monitoring Database (IAEA, 2012) or the Historical Artificial Radionuclides in the Pacific Ocean and its Marginal Seas database (Aoyama and Hirose, 2004), which include global and local radionuclide data associated with the Fukushima accident, are highly welcome.

The scarcity of radiation dose effect data with which to establish a clear benchmark for assessment adds a new dimension to the problem. Applying radiation dose benchmarks to an accidental situation covering an acute and a chronic phase is still novel and fraught with difficulties, especially when comparing and contrasting average exposures with doses to the most exposed organisms. There is also a need to develop studies to observe any potential effects in the field.

The radiological protection system for non-human biota is designed to protect at the population level. However, the possibility of more subtle effects at the individual level cannot be ruled out completely. Long-term effects over several generations, for instance on reproduction, are also possible in theory (Vives i Batlle, 2012a). In addition, non-human biota have a wide range of interspecies radiosensitivity, and they may react according to a complex interplay between absorbed doses (or dose rates) and radiotoxic responses, expressed at different levels of biological and ecological organisation. Although not yet observed, future studies should keep watch for these possible effects.

3.2. Ongoing and future issues for consideration

In October 2013, radioactivity leaked from storage tanks at Fukushima Daiichi nuclear power plant, emphasising the difficulties encountered by the operators in the management of the crisis (TEPCO, 2013a). There is a possibility that some effluent could enter the sea through the groundwater system. A range of abatement measures are being proposed to address this problem (Coughlan, 2013), but the possibility of enhanced radioactivity concentration at some coastal points cannot be excluded.

Meanwhile, dispersion of radionuclides through water will continue, and some radionuclides will continue to migrate to the benthic environment where bottom-dwelling species will continue to become exposed. There is a need to conduct long-term and long-range dispersion modelling studies to determine the residence times and deposition rates of Fukushima radionuclides in the North Pacific Ocean. These studies should cover radionuclides such as plutonium, which has received little attention to date (Zheng et al., 2013).

Detailed fieldwork is also needed in order to validate modelling studies and to calculate the radionuclide inventories in sediments. Investigations in estuaries are important because they are highly dynamic environments with particular biogeochemical characteristics affecting potentially sensitive, brackish species.

Lastly, there is a need to conduct long-term effect studies in tandem with revisiting current assessments in order to assess the potential effects of environmental radiation at the population level.

3.3. Ethical considerations

Fukushima is, and will likely remain, the main problem for marine radioecology over the next 20 years. In the meantime, the challenge of communication during nuclear emergencies remains a concern (Perko, 2011), and the scientific community has a role to play in informing the public in a trustworthy, fair way about its meaning and its consequences. Amid the journalistic stories and political positioning, there is a need to remain objective, producing scientific papers aimed at a scientific audience with the objective of helping Japan to deal with the consequences of the Fukushima accident.

The impact to the overall environment, including the abiotic component, is far beyond the scope of current studies. However, radiation dose as a magnitude cannot capture impacts to the actual value that mankind assigns to the environment. For example, 'small' dose rates of $2\,\mu Gy\,h^{-1}$ imply a numerically large ^{137}Cs activity concentration of $10^4\,Bq\,kg^{-1}$ in fish which, using the ERICA CRs and sediment/water distribution coefficients (FREDERICA, 2006; Beresford et al., 2007; Brown et al., 2008) would be in equilibrium with $10^5\,Bq\,m^{-3}$ in water and $5 \times 10^5\,Bq\,kg^{-1}$ in sediment. Such levels are comparable to levels offshore Sellafield at their peak period during the late 1970s (Vives i Batlle et al., 2008b), and cannot be dismissed. The applicability of the radiation protection framework to the abiotic environment is a matter of legitimate scientific debate (IAEA, 2002) which ought to continue with Fukushima.

4. CONCLUSIONS

This paper has described ongoing efforts to investigate the impact of the Fukushima nuclear accident on marine biota, including retrospective assessments of the first year and perspectives. It has been shown that the radiation doses received by marine biota in Fukushima are generally below the amounts necessary to cause a measurable effect on populations. It would seem that the possibility of effects on non-human biota is, at most, constrained in space to the vicinity of Fukushima Daiichi nuclear power plant, and in time to the initial phase of the accident when radioiodine exposures to marine biota (especially to macroalgae) were at their peak.

If this paper has reduced some of the initial estimates of exposures from previous studies, it has also indicated that enhanced levels of radioactivity in biota persist in the vicinity of Fukushima Daiichi nuclear power plant. There is a need to characterise local hotspots and assess the exposures to the most exposed biota. There are other areas where assessment is not complete. For example, our current understanding of the biological impacts of radiation on chronically exposed marine biota is based on limited effects data for the range of exposures considered.

The assessment of environmental impacts of Fukushima tests the international system for environmental protection from radiation. There are several reasons for this. Firstly, the transfer of radioactivity from environmental media to biota was initially a highly dynamic process, rendering the equilibrium transfer approach inapplicable during the acute phase. Secondly, the wide range of Japanese indigenous species that need to be covered is a challenge to the RAPs predefined in the assessment system (ICRP, 2008a). Lastly, application of radiation dose benchmarks to an accidental situation covering an acute and a chronic phase is novel and, as such, requires further development.

Ultimately, the environmental impact of Fukushima to the marine environment remains an open subject and a challenge for marine radioecology in the decades to come, signalling the direction for future investigations.

REFERENCES

Alekseev, A.V., Khrapchenkov, F.F., Baklanov, P.J., et al., 2006. Oyashio Current–GIWA Regional Assessment 31: Regional Definition. Global International Waters Assessments (GIWA) Regional Report, Kirjastusaktsiaselts MATS, Tallinn, pp. 13–21. Available at: http://www.unep.org/dewa/giwa/areas/reports/r31/giwa_regional_assessment_31.pdf (last accessed 16 March 2014).

Andersson, P., Garnier-Laplace, J., Beresford, N., et al., 2009. Protection of the environment from ionising radiation in a regulatory context (PROTECT): proposed numerical benchmark values. J. Environ. Radioact. 100, 1100–1108.

Aoyama, M., Hirose, K., 2004. Artificial radionuclides database in the Pacific Ocean: HAM database. Sci. World J. 4, 200–215.

Aoyama, M., Tsumune, D., Uematsu, M., et al., 2012. Temporal variation of ^{134}Cs and ^{137}Cs activities in surface water at stations along the coastline near the Fukushima Dai-ichi Nuclear Power Plant accident site, Japan. Geochem. J. 46, 321–325.

Bailly du Bois, P., Laguionie, P., Boust, D., et al., 2012. Estimation of marine source-term following Fukushima Dai-ichi accident. J. Environ. Radioact. 114, 2–9.

Beresford, N.A., Brown, J., Copplestone, D., et al., 2007. D-ERICA: an Integrated Approach to the Assessment and Management of Environmental Risks from Ionising Radiation. A Deliverable of the ERICA Project FI6R-CT-2004-508847. Swedish Radiation Protection Authority, Stockholm. Available at: https://wiki.ceh.ac.uk/download/attachments/115017395/D-Erica.pdf?version=1 (last accessed 16 March 2015).

Brown, J.E., Alfonso, B., Avila, R., et al., 2008. The ERICA tool. J. Environ. Radioact. 99, 1371–1383.

Buesseler, K., 2012. Fishing for answers off Fukushima. Science 338, 480–482.

Buesseler, K., Aoyama, M., Fukasawa, M., 2011. Impacts of the Fukushima nuclear power plants on marine radioactivity. Environ. Sci. Technol. 45, 9931–9935.

Buesseler, K., Aoyama, M., 2012. Fukushima Results. Presented at the Fukushima Ocean Impacts Symposium, Exploring the Impacts of the Fukushima Dai-ichi Nuclear Power Plants on the Ocean, 12–13 November 2012, Tokyo, Japan.

Buesseler, K.O., Jayne, S.R., Fisher, N.S., et al., 2012. Fukushima-derived radionuclides in the ocean and biota off Japan. Proc. Natl. Acad. Sci. USA 109, 5984–5988.

Choi, Y., Kida, S., Takahashi, K., 2013. The impact of oceanic circulation and phase transfer on the dispersion of radionuclides released from the Fukushima Dai-ichi Nuclear Power Plant. Biogeosciences 10, 4911–4925.

Coughlan, A., 2013. Japan will build wall of ice to stem Fukushima leak. New Sci. 219, 7.

Dietze, H., Kriest, I., 2012. ^{137}Cs off Fukushima Dai-ichi, Japan – model based estimates of dilution and fate. Ocean Sci. 8, 319–332.

FREDERICA, 2006. FREDERICA Radiation Effects Database. Available at: http://www.frederica-online.org/mainpage.asp (last accessed 16 March 2015).

Garnier-Laplace, J., Della-Vedova, C., Andersson, P., et al., 2010. A multi-criteria weight of evidence approach for deriving ecological benchmarks for radioactive substances. J. Radiol. Protect. 30, 215–233.

Garnier-Laplace, J., Beaugelin-Seiller, K., Hinton, T.G., 2011. Fukushima wildlife dose reconstruction signals ecological consequences. Environ. Sci. Technol. 45, 5077–5078.

Greenpeace, 2012. Marine testing lab results 1, 25 May 2011 (analysis report). Available at: http://www.greenpeace.org/international/Global/international/publications/nuclear/2011/Report%20SCK%20CEN.pdf (last accessed 16 March 2015).

Honda, M.C., Kawakami, H., Watanabe, S., et al., 2013. Concentration and vertical flux of Fukushima-derived radiocesium in sinking particles from two sites in the Northwestern Pacific Ocean. Biogeosciences 10, 3525–3534.

IAEA, 2002. Ethical Considerations in Protecting the Environment from the Effects of Ionizing Radiation. Technical Document 1270. International Atomic Energy Agency, Vienna.

IAEA, 2011. IAEA International Fact Finding Expert Mission of the Fukushima Dai-ichi NPP Accident Following the Great East Japan Earthquake and Tsunami. Tokyo, Fukushima Dai-ichi NPP, Fukushima Dai-ni NPP and Tokai Dai-ni NPP, Japan, 24 May–2 June 2011. International Atomic Energy Agency, Vienna.

IAEA, 2012. IAEA Fukushima Monitoring Database. International Atomic Energy Agency, Vienna. Available at: https://iec.iaea.org/fmd/ (last accessed 16 March 2015).

ICRP, 2008a. The concept and use of reference animals and plants for the purposes of environmental protection. ICRP Publication 108. Ann. ICRP 38(4–6).

ICRP, 2008b. Environmental protection: the concept and use for reference animals and plants. ICRP Publication 108. Ann. ICRP 38(4–6).

ICRP, 2009. Transfer parameters for reference animals and plants. ICRP Publication 114. Ann. ICRP 39(6).

IRSN, 2011. Synthèse actualisée des connaissances relatives à l'impact sur le milieu marin des rejets radioactifs du site nucléaire accidenté de Fukushima Dai-ichi. Institut de radioprotection et de sûreté nucléaire, Fontenay-aux-Roses. Available at: http://www.irsn.fr/FR/Actualites_presse/Actualites/Documents/IRSN-NI-Impact_accident_Fukushima_sur_milieu_marin_26102011.pdf (last accessed 16 March 2015).

IRSN, 2012. Fukushima, One Year Later – Initial Analyses of the Accident and its Consequences. Report IRSN/DG/2012-003. Institut de radioprotection et de sûreté nucléaire, Fontenay-aux-Roses.

Iwata, K., Tagami, K., Uchida, S., 2013. Ecological half-lives of radiocesium in 16 species in marine biota after the TEPCO's Fukushima Daiichi nuclear power plant accident. Environ. Sci. Technol. 47, 7696–7703.

Knowles, J.F., Greenwood, L.N., 1994. The effects of chronic irradiation on the reproductive performance of *Ophryotrocha diadema* (Polychaeta, Dorvilleidae). Mar. Environ. Res. 38, 207–224.

Knowles, J.F., Greenwood, L.N., 1997. A comparison of the effects of long-term β- and γ irradiation on the reproductive performance of a marine invertebrate *Ophryotrocha diadema* (Polychaeta, Dorvilleidae). J. Environ. Radioact. 34, 1–7.

Kryshaev, A., Sazykina, T., 2011. Evaluation of the irradiation dose rate for marine biota in the region of the destroyed Fukushima reactor (Japan) in March–May 2011 Atom. Énerg. 111, 41–45.

Kryshev, I.I., Kryshev, A.I., Sazykina, T.G., 2012. Dynamics of radiation exposure to marine biota in the area of the Fukushima NPP in March–May 2011. J. Environ. Radioact. 114, 157–161.

MEXT, 2011. Readings of Environmental Radioactivity Level (English Version). Ministry of Education, Culture, Sports, Science and Technology of Japan, Tokyo. Available at: http://www.mext.go.jp/english/incident/1303962.htm (last accessed 16 March 2015).

Perko, T., 2011. Importance of risk communication during and after a nuclear accident. Integr. Environ. Assess. Manag. 7, 388–392.

PMJHC, 2011. Report of Japanese Government to the IAEA Ministerial Conference on Nuclear Safety. The Accident at TEPCO's Fukushima Nuclear Power Stations. Prime Minister of Japan and His Cabinet, Tokyo.

Povinec, P.P., Hiroshe, K., Aoyama, M., 2013. Fukushima Accident: Radioactivity Impact on the Environment. Elsevier, New York.

Psaltaki, M., Brown, J.E., Howard, B.J., 2013. TRS Cs CRwo-water values for the marine environment: analysis, applications and comparisons. J. Environ. Radioact. 126, 367–375.

Sazykina, T.G., 2000. ECOMOD – an ecological approach to radioecological modelling. J. Environ. Radioact. 50, 207–220.

Shiomoto, A., Hashimoto, S., Murakami, T., 1998. Primary productivity and solar radiation off Sanriku in May 1997. J. Oceanogr. 54, 539–544.

Stohl, A., Seibert, P., Wotawa, G., et al., 2011. Xenon-133 and caesium-137 releases into the atmosphere from the Fukushima Daiichi nuclear power plant: determination of the source term, atmospheric dispersion, and deposition. Atmos. Chem. Phys. Discuss. 11, 28319–28394.

TEPCO, 2011. Detection of Radioactive Materials from Seawater near Fukushima Dai-ichi Nuclear Power Station. Press Releases N 1–47, 21 March–9 May 2011. Tokyo Electric Power Company, Tokyo. Available at: www.tepco.co.jp/en/press/corp-com/release/.

TEPCO, 2013a. Leakage from B Area South Tank. Tokyo Electric Power Company Online Report, 3 October 2013. Tokyo Electric Power Company, Tokyo. Available at: http://www.tepco.co.jp/en/nu/fukushima-np/handouts/2013/images/handouts_131003_01-e.pdf (last accessed 16 March 2015).

TEPCO, 2013b. Result of Radioactive Nuclide Analysis around Fukushima Daiichi Nuclear Power Station. Tokyo Electric Power Company, Tokyo. Available at: http://www.tepco.co.jp/en/nu/fukushima-np/f1/smp/2013/index03-e.html (last accessed 12 March 2015).

Tsumune, D., Tsubono, T., Aoyama, M., et al., 2012. Distribution of oceanic 137Cs from the Fukushima Dai-ichi nuclear power plant simulated numerically by a regional ocean model. J. Environ. Radioact. 111, 100–108.

UNSCEAR, 1996. Sources and Effects of Ionizing Radiation. United Nations Scientific Committee on the Effects of Atomic Radiation Report to the General Assembly, with Scientific Annex: Effects of Radiation on the Environment. Publication E.96.IX.3. United Nations, New York.

UNSCEAR, 2008. Sources and Effects of Ionizing Radiation. Report to the General Assembly, with Scientific Annexes; Annex E: Effects of Ionizing Radiation on Non-human Biota. United Nations, New York.

UNSCEAR, 2014. Sources, Effects and Risks of Ionizing Radiation. Report to the General Assembly with Scientific Annexes Volume 1, Scientific Annex A: Levels and effects of radiation exposure to the nuclear accident after the 2011 great east-Japan earthquake and tsunami. United Nations Scientific Committee on the Effects of Atomic Radiation, Vienna.

Vives i Batlle, J., Wilson, R.C., McDonald, P., 2007. Allometric methodology for the calculation of biokinetic parameters for marine biota. Sci. Tot. Environ. 388, 256–269.

Vives i Batlle, J., 2011. Impact of nuclear accidents on marine biota. Integr. Environ. Assess. Manag. 7, 365–367.

Vives i Batlle, J., Wilson, R.C., Watts, S.J., et al., 2008a. Dynamic model for the assessment of radiological exposure to marine biota. J. Environ. Radioact. 99, 1711–1730.

Vives i Batlle, J., Bryan, S., McDonald, P., 2008b. A process-based model for the partitioning of soluble, particulate and sediment fractions of plutonium and radiocaesium in the Eastern Irish Sea near Sellafield. J. Environ. Radioact. 99, 1464–1473.

Vives i Batlle, J.', 2012a. Dual age class population model to assess radiation dose effects to non-human biota populations. Radiat. Environ. Biophys. 51, 225–243.

Vives i Batlle, J., 2012b. Radioactivity in the marine environment. In: Meyers R.A. (Ed.) Encyclopaedia of Sustainability Science and Technology. Springer Verlag, New York, pp. 83878425.

Vives i Batlle, J., Vandenhove, H., 2014. Dynamic modelling of the radiological impact of the Fukushima accident on marine biota. In: Proceedings of the International Symposium on the Occasion of the 50th Anniversary of the Belgian Association for Radiological Protection BVS-ABR, 8–10 April 2013, Brussels, Belgium. Annalen van de Belgische Vereniging voor Stralingsbescherming 38, 299–312.

Vives i Batlle, J., Aono, T., Brown, J.E., et al., 2014. The impact of the Fukushima nuclear accident on marine biota: retrospective assessment of the first year and perspectives. Sci. Total Environ. 487, 143–153.

Watts, S.J., Vives i Batlle, J., Wilson, R.C., et al., 2008. Dynamic modelling for the assessment of radiation impacts on biota and man using Excel VBA, and related extensions of the ERICA assessment methodology. In: Strand, P., Brown, J., Jolle, T. (Eds.), Proceedings of the International Conference on Radioecology and Environmental Radioactivity, 15–20 June 2008, Bergen, Norway, pp. 289–293.

Zheng, J., Tagami, K., Uchida, S., 2013. Release of plutonium isotopes from the Fukushima Daiichi nuclear power plant accident: what is known and what needs to be known. Environ. Sci. Technol. 47, 9584–9595.

SAGE

ICRP
Annals of the ICRP

Radiation protection issues on preparedness and response for a severe nuclear accident: experiences of the Fukushima accident

T. Homma, S. Takahara, M. Kimura, S. Kinase

Japan Atomic Energy Agency, Nuclear Safety Research Centre, 2–4 Shirakata-shirane, Tokaimura, Nakagun, 319-1195 Ibarakiken, Japan; PH +81 (29) 282 6862; FAX +81 (29) 282 5408; e-mail: homma.toshimitsu@jaea.go.jp

Abstract–Radiation protection issues on preparedness and response for a severe nuclear accident are discussed in this paper based on the experiences following the accident at Fukushima Daiichi nuclear power plant. The criteria for use in nuclear emergencies in the Japanese emergency preparedness guide were based on the recommendations of International Commission of Radiological Protection (ICRP) *Publications 60* and *63*. Although the decision-making process for implementing protective actions relied heavily on computer-based predictive models prior to the accident, urgent protective actions, such as evacuation and sheltering, were implemented effectively based on the plant conditions. As there were no recommendations and criteria for long-term protective actions in the emergency preparedness guide, the recommendations of ICRP *Publications 103*, *109*, and *111* were taken into consideration in determining the temporary relocation of inhabitants of heavily contaminated areas. These recommendations were very useful in deciding the emergency protective actions to take in the early stages of the Fukushima accident. However, some suggestions have been made for improving emergency preparedness and response in the early stages of a severe nuclear accident.

Keywords: Fukushima accident; Radiological situation; Protective actions; Emergency response; Potential exposure

This paper does not necessarily reflect the views of the International Commission on Radiological Protection.

1. INTRODUCTION

The arrangements for nuclear emergencies in Japan were developed incrementally following the accident at Three Mile Island nuclear power plant in the USA in 1979. In 1980, the Japanese Nuclear Safety Commission (NSC) issued an emergency preparedness guide as one of the measures to maintain preparedness for severe accidents, such as the accident at Three Mile Island. The Chernobyl accident in 1986 did not have a particularly significant impact on the emergency management system due to differences between the reactor types. Several weaknesses in the emergency response system, such as prompt initial actions, collaboration of national and local governments, strengthening of the emergency response system, and clarification of the licensee's responsibilities, were identified following the Tokaimura criticality accident in 1999. After the criticality accident, the Act on Special Measures Concerning Nuclear Emergency Preparedness was enacted in December 1999 and enforced in June 2000. While the legal framework of emergency preparedness and response in Japan had been established before the Fukushima accident, several weaknesses of the emergency response system were also found in the light of radiation protection. The decision-making process for implementing protective measures relied heavily on computer-based predictive models. Criteria for terminating urgent protective actions and for long-term protective actions, such as temporary relocation, had not been prepared in the NSC emergency preparedness guide.

The NSC emergency preparedness guide was based on recommendations in *Publications 60* (ICRP, 1991) and *63* (ICRP, 1993). The International Basic Safety Standards for Protection against Ionizing Radiation and the Safety of Radiation Sources by the International Atomic Energy Agency (IAEA) considered these recommendations and guidance issued by ICRP, and provided guidelines, such as generic optimised intervention levels for urgent protective actions and generic action levels for foodstuffs (IAEA, 1996). NSC adopted these guidelines for generic intervention and action levels in the emergency preparedness guide. The 2007 Recommendations of ICRP (ICRP, 2007) evolved from the previous process-based approach of practices and interventions to an approach based on the characteristics of radiation exposure situations. Applications of ICRP recommendations for both emergency and existing exposure situations were issued as *Publications 109* (ICRP, 2009a) and *111* (ICRP, 2009b), respectively. At the time of the Fukushima accident, the changes in the approach by ICRP had not been incorporated into international or national standards, and therefore could have been part of the problem for the response to the Fukushima accident.

This paper briefly describes the initial lessons learned from the Fukushima accident. In particular, the paper discusses some radiation protection issues on emergency preparedness and response in the early stages of a large-scale severe nuclear accident.

2. RADIOLOGICAL SITUATION IN THE EARLY STAGES OF THE FUKUSHIMA ACCIDENT

At the Fukushima Daiichi nuclear power plant, the circulation of sea and land breezes is a predominant factor for dispersion of released radioactivity. At night, a west wind blows towards the Pacific Ocean. In the early morning, it shifts to the south, and subsequently to the south-west and west. From afternoon to evening, it blows to the north-west. Fig. 1 shows the ambient dose rate measurements at seven locations relatively distant from the site, and describes the radiological situation in the early stages of the Fukushima accident. The first peak in ambient dose rate was observed in Minamisoma City, approximately 20 km north of the plant, at approximately 20:00 h on 12 March 2011. This may have been due to transportation of the radioactive plume by the strong southerly winds when Unit 1 was vented, and the subsequent hydrogen explosion in the reactor building. The releases from Unit 3 on 13 and 14 March 2011 were mainly transported towards the Pacific Ocean.

At midnight on 14 March 2011, Unit 2 underwent a core melt, and the onsite radiation levels began to rise. Further, at approximately 06:00 h on 15 March 2011, a sound of explosion was suspicious near Unit 2, and the reactor building of Unit 4 was damaged. At 04:00 h, outside the reactor site, radiation levels as high as 24 μSv h^{-1} were observed in Iwaki City, approximately 50 km south of the site. During this time, the prevailing winds, which were from the north, transported the radioactive plume to

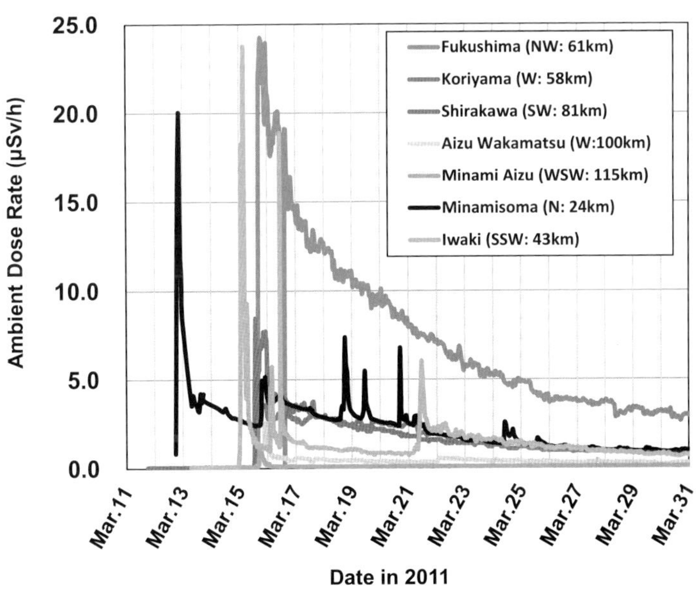

Fig. 1. Ambient dose rates observed at seven cities around the Fukushima plant in the early stages of the accident.

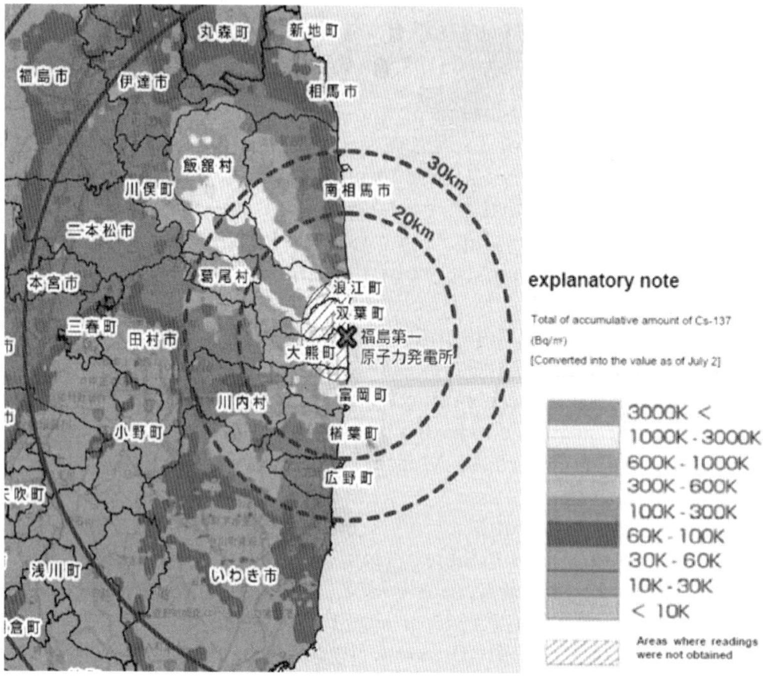

Fig. 2. Deposition distributions of caesium-137 from airborne monitoring by the Japanese Ministry of Education, Culture, Sports, Science, and Technology and the US Department of Energy.

Ibaraki Prefecture, Southern Kanto areas, and even as far as Shizuoka Prefecture. In the afternoon, the radioactive plume reached the cities of Shirakawa, Koriyama, and Fukushima, and airborne radiation levels rose. In the evening of 15 March 2011, the entire prefecture of Fukushima experienced rain and snow, depositing the radioactivity on the ground during passage of the plume. Areas to the north-west of the nuclear power plant were thus heavily contaminated. The Ministry of Education, Culture, Sports, Science, and Technology (MEXT) of Japan, in cooperation with the US Department of Energy, conducted airborne monitoring within a 100-km radius of the power plant. According to the radiation map produced by this monitoring, the total depositions of caesium-134 and caesium-137 on the ground were strongly affected by the precipitation that fell when the plume passed over, and showed a non-uniform distribution (Fig. 2).

3. URGENT PROTECTIVE ACTIONS

In order to avert potential radiation exposure to the public, the Japanese authorities implemented the precautionary urgent protective actions to advise individuals within 3 km of the plant to evacuate, and those within 10 km to remain indoors on 11

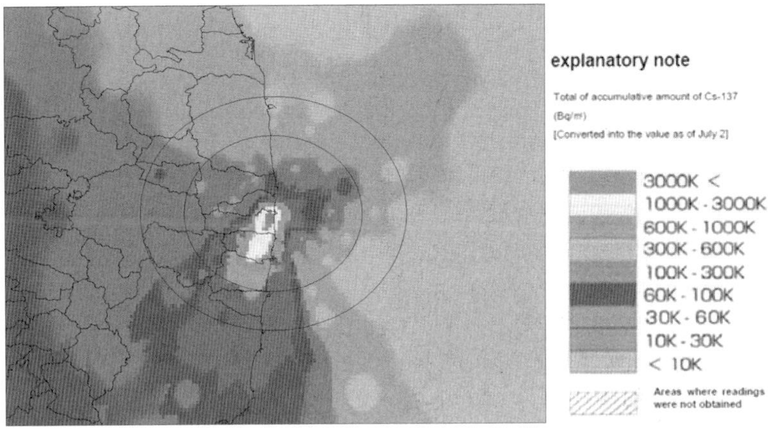

Fig. 3. Deposition distributions of caesium-137 calculated by OSCAAR using the source terms provided by MELCOR.

March 2011. This was determined based on the plant conditions, as the reactor of Unit 1 had not been cooled. The Japanese Government officially reported that the evacuation had been completed at midnight on 12 March 2011. The evacuation zone was extended to a 10-km radius early on 12 March 2011, and then later to a 20-km radius after the hydrogen explosion in Unit 1. This was because the pressure in the primary containment vessel may have been increasing, and also to prepare for any possible risks that would occur simultaneously at multiple reactors, including the other units at the nuclear power plant. The 'in house' sheltering zone between 20 and 30 km of the plant was established on 15 March 2011. The Nuclear Industrial and Safety Agency (NISA) announced that the evacuation of the 20-km zone had been completed at 19:00 h on 15 March 2011. These precautionary urgent protective actions were taken based on the plant conditions.

In emergency exercises to date, recommendations on the implementation of urgent protective actions have been made based on projected dose ('model predictions') with intervention levels. The NSC emergency preparedness guide indicates that the projected dose will be evaluated by models that predict source terms of the accident and also the dose to the public. The disaster management basic plan assigns organisations with the role and responsibility to prepare and implement the computer prediction system. The emergency response support system, which predicts the progression of the accident at the plant, could not be used in the Fukushima accident because necessary information could not be obtained from the plant. In addition, the system for prediction of environmental emergency dose information (SPEEDI), which conducts a numerical forecast of atmospheric conditions, air concentrations, and depositions of released radioactivity and dose distributions, could not be used because the source terms could not be obtained via the emergency response support system.

After the Fukushima accident, the issue relating to disclosure of the results generated by SPEEDI to the public was raised in the media, and also discussed in terms of the national diet. The criticism claimed that if the information generated by SPEEDI had been provided more quickly, it could have helped local governments and the public to choose a more appropriate route and direction of evacuation. However, neither quantitative nor qualitative analyses have been presented to support such discussions. Fig. 3 shows the results of caesium-137 deposition patterns calculated by the Level 3 PSA code, OSCAAR developed at the Japan Atomic Energy Agency (Homma et al., 2005) using the source terms presented in the Government report on the Fukushima accident, submitted to IAEA in June 2011. The source terms were calculated using a severe accident analysis computer code MELCOR by NISA. The differences between Fig. 2 and 3 are mainly due to the release timing of caesium-137 calculated by MELCOR, and also the uncertainty of the atmospheric dispersion model that was used in the calculations. This highlights the difficulty of developing protective action recommendations based on computer model predictions.

When implementing urgent protective actions, there is no time to undertake detailed exposure assessments in real time. There are also extremely large uncertainties associated with predictive models. It is therefore necessary to determine, in advance, a set of internally consistent criteria for taking such actions, and, based on these criteria, to derive appropriate 'triggers' for initiating them in the event of an emergency, as described in *Publication 109* (ICRP, 2009a). IAEA also recommends that precautionary urgent protective actions should be taken before or shortly after a major release based on the plant conditions; implementation should not be delayed until after a release. The IAEA Safety Guide (IAEA, 2011) provides the basis for the emergency classification schemes, and an example of emergency action levels that are the criteria for classification.

4. RESTRICTIONS ON FOOD AND WATER

After heavy contamination of the ground in areas to the north-west of the Fukushima Daiichi nuclear power plant on 15 March 2011, high levels of radioiodine and radiocaesium were detected in foodstuffs and water. Radioiodine in tap water was first detected on 16 March 2011 in Fukushima City. On 20 March 2011, water samples taken in Iitate Village were found to have iodine-131 levels of 965 Bq kg^{-1}, exceeding the provisional regulatory value of the Japanese Ministry of Health, Labour, and Welfare (MHLW) (300 Bq kg^{-1}). For this reason, MHLW restricted consumption of water from private water supply systems. Kinase et al. (2011) reported that the estimated averted thyroid equivalent dose to infants in Iitate Village was approximately 8 mSv, although the effective half-life of iodine-131 in tap water was approximately 2.8 days. In addition, on 21 March 2011, MHLW announced that people should avoid using tap water for infants, including use in baby formula. This measure was enacted throughout March 2011 in several prefectures around the Fukushima plant, including Tokyo.

On 16 March 2011 in Kawamata Town, raw milk was found to contain 1190 Bq kg^{-1} iodine-131, which exceeded the provisional regulatory value of 300 Bq kg^{-1}. Furthermore, on 18 March 2011, in Hitachi of Ibaraki Prefecture, 54,100 Bq kg^{-1} of iodine-131 and 1931 Bq kg^{-1} of radiocaesium were detected in spinach. These also exceeded the provisional regulatory values. Since 21 March 2011, when Fukushima Prefecture enacted restrictions on the distribution of raw milk, it has enacted similar distribution and consumption restrictions on vegetables. At a later stage of the accident, radiocaesium was found in shiitake mushrooms in April 2011 and in tea plants in May 2011. Moreover, in July 2011, beef cattle that had been fed with contaminated rice stalks had unacceptably high levels of caesium-137, and distribution of beef was restricted in four prefectures.

Radioactivity in food and drinking water caused significant anxiety amongst the public, and resulted in rumours being spread. Conceptually, protective actions on food consumption can be implemented for each different type of exposure situation because the transfer of radionuclides into the food chain depends upon the time and characteristics of the contamination. In the early stages after an accident, a quick response is needed to avert the ingestion of dose due to elevated levels of radioactivity, such as radioiodine that has been deposited directly on agricultural products and drinking water. Restrictions can be triggered using an operational intervention level, such as the gamma dose rate from a contaminated surface suggested by IAEA (2011). During intermediate and longer term stages after an accident, criteria for foodstuff restrictions should be considered in the process of optimisation for the whole protective strategy, considering factors such as radiological and nutritional impact, reference level and contribution of ingestion dose, realistic estimates based on dietary habits, market dilution, and harmonisation with internationally agreed standards.

5. EARLY PROTECTIVE ACTIONS

On 17 March 2011, the highest ambient dose rate (170 μSv h^{-1}) was observed at a point outside the evacuation zone, approximately 30 km from the site. NSC collected monitoring data around the heavily contaminated areas, and advised the Nuclear Emergency Response Headquarters that the residents should be asked to voluntarily evacuate from those areas where a relatively high dose was expected. After reviewing the situation, NSC considered modifying the urgent protective actions implemented, and the process of implementing temporary relocation of residents in those areas. As there were no criteria for long-term protective actions in the NSC guide, the revised recommendations of *Publications 103*, *109*, and *111* (ICRP, 2007, 2009a,b) were taken into consideration in determining the temporary relocation of the inhabitants in the heavily contaminated areas.

On 22 April 2011, the areas beyond 20 km from the site, where annual cumulative dose for the inhabitants would potentially reach 20 mSv, were established as 'deliberate evacuation areas'. The criterion of 20 mSv was used based on the bottom end of the reference level of 20–100 mSv in emergency exposure situations. Inhabitants in the deliberate evacuation areas were asked to leave by the end of May 2011. At the same time, the remaining sheltering areas, between 20 and 30 km from the site, were established as 'evacuation-prepared areas in case of emergency', with the inhabitants prepared to shelter or evacuate in the event of a further emergency. In addition, on 16 June 2011, hot spot areas with a dose above 20 mSv year^{-1} were identified and established as 'specific spots recommended for evacuation'. Inhabitants in these hot spots did not have to evacuate immediately, but had to take precautions in their daily activities in order to prevent their dose exceeding 20 mSv.

The new criterion of annual dose of 20 mSv year^{-1} was used as an intervention level for justifying temporary relocation instead of a reference level during implementation of the optimisation process for exposures in emergency exposure situations. In addition, during the discussion on the implementation of temporary relocation, a further issue was raised related to the criteria that would be applied to the use of contaminated playgrounds by children in April 2011. MEXT selected 20 mSv year^{-1} from the dose band for an existing exposure situation recommended by ICRP. This was because ICRP described how the transition from an emergency exposure situation to an existing exposure situation may take place at different geographical locations at different times, such that whilst some areas are managed as an emergency exposure situation, others are managed as an existing exposure situation. This level was selected as a starting point for optimisation. MEXT subsequently advised the prefecture to reduce the dose to 1 mSv year^{-1}, but selection of the same value as that used for relocation raised concern amongst the public.

6. POTENTIAL EXPOSURE

One of the lessons learned from the Fukushima accident is that loss of land use resulting in social disruption in communities cannot be prevented, but appropriate emergency preparedness and response can prevent radiation health effects following a severe reactor accident. This may be an issue for nuclear safety, such as severe accident management, but from the viewpoint of radiation protection, risk attributes from potential exposure should be discussed thoroughly to protect individuals, society, and the environment.

In December 2003, the Special Committee on Nuclear Safety Goals established by NSC proposed that qualitative and quantitative safety goals should be applied consistently to all types of nuclear activities, and future efforts should be made to investigate performance goals for each field of utilisation of nuclear energy in the Interim Report on the Discussion of Safety Goals (NSC, 2003). The proposed quantitative goal should be objective and common to various nuclear facilities and activities that give rise to potential adverse health effects from radiation exposure to the public. Among various types of adverse effects of nuclear accidents, individual risks of both early fatality and cancer fatality to members of the public were established as risk indicators for the quantitative safety goals.

The Special Committee on Nuclear Safety Goals had considerable discussions about various risks such as collective risk, societal risk, and risk to individuals

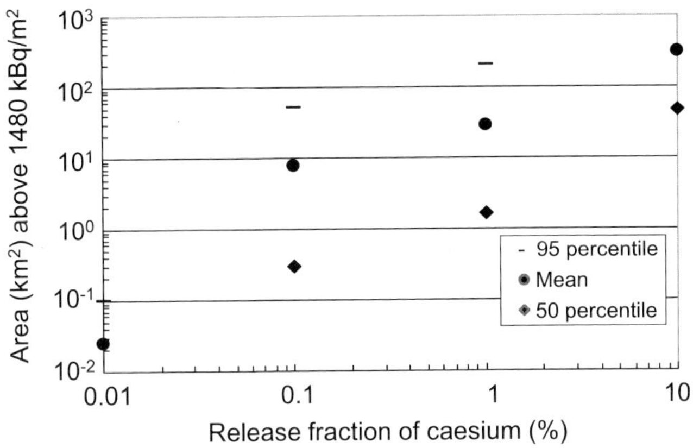

Fig. 4. Area of land contamination that corresponded to permanent relocation vs release fraction of caesium to the inventory.

covered by safety goals. Fig. 4 shows the exploratory calculations using the Level 3 PSA code, OSCAAR to determine the extent of land contamination as a function of different releases and weather conditions. It shows that the release limit at approximately 0.1% fraction of caesium to the core inventory (3200 MWth) ensures that a larger societal impact which would result in long-term ground contamination does not occur beyond site boundaries. However, the Special Committee on Nuclear Safety Goals concluded that no criteria were selected in terms of societal risk because it was difficult to quantify the overall societal impact rather than human health effects, and there was no benchmark of societal risk levels to be restricted.

The risk of a severe accident in a nuclear power plant can be managed as a potential exposure in the context of planned exposure situations. In the safety goals for nuclear installations in Japan, only individual health risks were used as metrics. The Fukushima accident, however, indicated the importance of societal risk, such as countermeasures needed to protect individuals and land contamination, rather than individual health risks. ICRP should consider whether it should further elaborate the recommendations for protection from potential exposure.

7. CONCLUSIONS

A general lesson learned from the Fukushima accident, as well as the accidents at Three Mile Island and Chernobyl, is that there was an implicit assumption that such severe accidents could not happen, and thus sufficient attention had not been paid to preparedness for the accidents by the operators and the regulatory authorities. Although arrangements for preparedness and response for severe nuclear accidents in Japan had not been well established at the onset of the Fukushima accident, the ICRP recommendations described in *Publications 60*, *63*, *103*, *109* and *111* (ICRP,

1991, 1993, 2007, 2009a,b) were very useful in deciding the emergency protective actions to take in the early stages of the Fukushima accident.

Among several initial lessons learned from the Fukushima accident, it is recommended that arrangements should be established for taking precautionary urgent protective actions before a release based on the plant conditions.

The following suggestions have been identified to improve emergency preparedness and response in the early stages of an accident:

- to make practical recommendations for control of contaminated foodstuffs and water;
- to clarify the intended use of the concept of dose limits, dose constraints, and reference levels for protection of the public; and
- to consider whether the recommendations for protection from potential exposure should be further elaborated.

REFERENCES

Homma, T., Tomita, K., Hato, S., 2005. Uncertainty and sensitivity studies with the probabilistic accident consequence assessment code OSCAAR. Nucl. Eng. Technol. 37, 245–258.

IAEA, 1996. International basic safety standards for protection against ionizing radiation and for the safety of radiation sources. Safety Series No. 115, International Atomic Energy Agency, Vienna.

IAEA, 2011. Criteria for use in preparedness and response for a nuclear or radiological emergency: General Safety Guide, GSG-2. International Atomic Energy Agency, Vienna.

ICRP, 1991. 1990 Recommendations of the International Commission on Radiological Protection. ICRP Publication 60. Ann. ICRP 21(1–3).

ICRP, 1993. Principles for intervention for protection of the public in a radiological emergency. ICRP Publication 63. Ann. ICRP 22(4).

ICRP, 2007. The 2007 Recommendations of the International Commission on Radiological Protection. ICRP Publication 103. Ann. ICRP 37(2–4).

ICRP, 2009a. Application of the Commission's recommendations for the protection of people in emergency exposure situations. ICRP Publication 109. Ann. ICRP 39(1).

ICRP, 2009b. Application of the Commission's recommendations to the protection of people living in long-term contaminated areas after a nuclear accident or a radiation emergency. ICRP Publication 111. Ann. ICRP 39(3).

Kinase, S., Kimura, M., Takahara, S., Homma, T., 2011. Evaluation of averted doses to infants by tap water restrictions after the Fukushima Daiichi nuclear power plant accident. Trans. Atom. Energy Soc. Jpn. 10, 149–151 (in Japanese).

NSC, 2003. Interim Report on the Discussion of Safety Goals Special Committee on Nuclear Safety Goals of NSC, December 2003. Available at: http://www.nsc.go.jp/mokuhyo/h1512.pdf (in Japanese).